McGraw-Hill Series on Computer Communications (Selected Titles)

Remote
Access
Networks
PSTN, ISDN, ADSL, Internet, and Wireless

Chander Dhawan

McGraw-Hill
New York • San Francisco • Washington, D.C. • Auckland • Bogotá
Caracas • Lisbon • London • Madrid • Mexico City • Milan
Montreal • New Delhi • San Juan • Singapore
Sydney • Tokyo • Toronto

Library of Congress Cataloging-in-Publication Data

Dhawan, Chander.
    Remote access networks : PSTN, ISDN, ADSL, Internet and wireless / Chander Dhawan.
        p.   cm. — (The McGraw-Hill series on computer communications)
    Includes index.
    ISBN 0-07-016774-5
    1. Computer networks. 2. Internetworking (Telecommunication)
    3. Computer terminals—Remote terminals. I. Title. II. Series.
    TK5105.5.D515 1998
    384.3′2—dc21
                                        97-41018
                                            CIP

McGraw-Hill

*A Division of The **McGraw·Hill** Companies*

The sponsoring editor for this book was Steven Elliot and the production supervisor was Pamela Pelton. It was set in Vendome by North Market Street Graphics.

Printed and bound by R. R. Donnelley & Sons Company.

McGraw-Hill books are available at special quantity discounts to use as premiums and sales promotions, or for use in corporate training programs. For more information, please write to Director of Special Sales, McGraw-Hill, 11 West 19th Street, New York, NY 10011. Or contact your local bookstore.

 This book is printed on recycled, acid-free paper containing a minimum of 50% recycled, de-inked fiber.

This book is dedicated to my family.

To my wife Bina, who amazes me with her confidence and positive outlook. With an uncanny balance that only she can figure out, she can switch from one role to another with ease and aplomb—a perfect mother, a perfect wife, and an outstanding professional in her business life of personal financial services. Although she did not enjoy the deprivation of my attention toward her while I spent endless days writing this book, she stood by me, nonetheless.

To my elder daughter Sonia, who will complete her Ph.D. in Psychology as my book goes to print. Armed with intelligence, determination, and a strong sense of service to the community, she starts her mission of solving cross-cultural problems and helping people cope with intergenerational caregiving stresses.

To my younger daughter Priya, who became my partner and in-house editor. She has started out her professional career in business after I handed in my manuscript. With strong personal values, she finds eastern philosophy of life applicable to the modern complex world of today, as it did thousands of years ago.

And finally, this book is dedicated to my late father-in-law, Ram Parkash Sachdev, who was an outstanding example of high morals, ethical standards, and decency. He would have been happy to see this book. However, my mother-in-law, Shakuntala Sachdev, continues to carry forward his values and traditions.

CONTENTS

Contents

Contents

Contents

Contents

Contents

FOREWORD

Just a few years ago the remote access phenomenon was barely perceived. In the early 1990s, when the Shiva LanRover and the Ascend MAX were first released, virtually all industry analysts saw remote access and Internet access as potentially interesting but peripheral applications that would surely be dwarfed by backbone phenomena such as ATM.

Yet today we realize that ad hoc remote networking—connecting to a variety of global resources both private and public at any time, from any location—is much more than the next great killer application. In fact, remote access is one of the primary drivers of the next great wave of global technology evolution, with impact not only on the computer and networking industries, but on the telephony and media industries, health and medicine, education—even on the fundamental way that people will live and think about living.

Over the next decade, systems and applications software, development tools, and resource management systems will all develop with extreme rapidity to support the remote networking model. Worldwide service providers will increase available bandwidth massively and pervasively. The result will be far different from today's remote access systems and Internet services: rather, we will see a proper platform for a global, multimedia, data-based model of communications with significant structural implications for many, if not all, of the world's major industries. Person-to-person, business-to-consumer, and business-to-business relationships will all be deeply affected.

For business leaders and IT professionals, mastering remote access is becoming a competitive and financial necessity. In *Remote Access Networks: PSTN, ISDN, ADSL, Internet, and Wireless,* Chander Dhawan provides a comprehensive basis for achieving this mastery. At all times, his text remains attentive to three key subjects: the economic drivers and realities of the remote access applications; the key technologies, whether proven or merely promised; and the actual implementation necessities and issues that will affect real-world deployment.

To create a source reference that is both thorough in its scope and technically sound in its detail requires tremendous effort. Readers, users, and vendors alike will benefit from Mr. Dhawan's diligence.

Frank A. Ingari
Chairman of the Board
Shiva Corporation

PREFACE

The idea of writing my second book came to me while I was still finishing my first book on mobile computing entitled *Mobile Computing—A Systems Integrator's Handbook*. During the early part of 1996, I was beginning to feel that while mobile computing was moving much slower than projected, it was the remote access industry which was galloping forward at a much faster pace. The Internet revolution suddenly and overwhelmingly created a strong demand for connecting remote workers, telecommuters, and small offices to both internal and external information resources. There were many articles being published in trade magazines, and the topic was being discussed regularly in technology seminars. Consequently, I suggested the topic to Jay Ranade, the Series Editor for McGraw-Hill.

Having felt the need for a professional book and developing a table of contents on the subject was one thing; writing it, however, was a daunting task. Before I really started writing the book, I was awarded two consulting contracts. Besides, there were personal and family preoccupations that left very little time for creative writing. Paid consulting work seemed to get higher priority over the future royalties of a book. The personal satisfaction of writing and the dissemination of knowledge to others seemed to fall behind the demands of customers who had deadlines of their own.

Network and Internet technology have been moving very fast for the past few years. Although I had done a lot of research earlier for my first book—there is a strong relationship between the topic of my first book and this book—I had to conduct additional and extensive research into the remote access industry. I had only one chapter on the topic of remote access in my first book; this book had to have seventeen. More important, I had to acquire a depth of understanding of many new technologies in order to write about them.

As a result of these requirements and conflicting priorities, I missed a couple of deadlines for the book. However, my McGraw-Hill staff stood by me and allowed me to schedule priorities in such a way that I have finally finished the book.

For a period of 12 months, off and on, I collected background information, researching the subject, surfing the Internet, discussing the issues with industry experts, putting ideas down on paper, and illustrating them with graphics. Then during the last three months, I limited my consulting work and spent a majority of my time on the book. Although it was a long struggle (I am glad it is over), looking back, I realize that I have learned a lot and have enjoyed the end result—a finished manuscript that has turned into a book.

The Organization of the Book

The book follows the life cycle of a typical remote access project. It is organized into five parts.

The first part deals with the market, applications, and economics of remote access. Here, I discuss the remote access industry, remote access technical concepts, market size, business applications, and a methodology for the development of a remote access business case.

The second part deals in detail with the technology and architecture behind remote access—remote control and remote node, in particular. BBS (terminal emulation) and mail gateways are also discussed from a historical perspective, even though they have gone into the background of our industry. I also introduce the need for developing a technology architecture in the context of broad mobile computing requirements that include wireless implementations. Chapter 11 discusses integration with enterprise network architecture. I have delayed discussion of this subject until part three has been covered.

In the third part, I review network options—PSTN, ISDN, ADSL/Cable Modems, Internet, and Wireless. Evaluating network options and designing solutions based on the selected option constitute an important part of the book.

In the fourth part, I look at ATM-type switching and enterprise network considerations in developing a remote access solution. Then, I compare various network options in Chap. 12.

In the fifth and final part of the book, I discuss systems design, application design, security design, bandwidth management, and network management issues. This is the systems engineering glue that binds everything together from implementation and operational points of view.

Evolution of Simple Concepts to Advanced Technologies

In Chap. 4, the book introduces remote access from a chronological perspective: from bulletin board and terminal emulation implementations to the more current design of advanced RAS switches. However, many design concepts and products in this industry are still undergoing evolutionary change and have not fully matured into durable components of enterprise network infrastructure.

So far, the industry has built RAS products in isolation with an objective of making LAN applications available to remote users on public shared networks without considering the inherent differences in the speeds of the two media. This must change because there is no convergence in sight in terms of the bandwidth of LANs and switched networks. Therefore, I have proposed and given significant substance to the idea of mobile-aware application design. This will encourage readers to use these ideas to crystallize their own, and thus to formulate them into functional applications systems. I also encourage readers to investigate the client/agent/server concept in remote access application design. Second, I strongly recommend that network professionals integrate the remote access technology into their enterprise network infrastructure.

Intentional Repetition

The reader will encounter a certain amount of repetition in different chapters. This is intentional for two reasons: first, it keeps the subject matter together and avoids repeated references to different parts of the book. Second, information has been repeated, where appropriate, in order to emphasize its importance.

Feedback on the Book

Obviously, as remote access technology evolves, the information in this book will have to be updated. As well, in spite of the efforts of the editorial staff, errors may creep into the final copy, or perhaps certain statements will be challenged by experts and specialists. Whatever the reason, I would like to receive your comments, feedback, and corrections. My e-mail address is: cdhawan@mobileinfo.com or cdhawan@netsurf.net; my telephone number is (905) 881-8537; and my fax number is (905) 881-3589.

I hope this book meets its objective of being comprehensive and that it meets your information needs.

—CHANDER DHAWAN

ACKNOWLEDGMENTS

Writing this book was a long and arduous exercise. I could not have accomplished it without the assistance and cooperation of many professional colleagues, vendors, friends, and family members. Therefore, it is my distinct pleasure to acknowledge the following individuals and groups who helped me in this task:

I owe a special thanks to Steve Elliot, Senior Editor at McGraw-Hill and Jay Ranade, Series Editor, McGraw-Hill Series in Communications, who encouraged me to write this book soon after finishing my first book on mobile computing. Both of them have shown tremendous understanding and patience while I was trying to juggle my schedule between my professional consulting work (which pays the bills) and creative writing, which is time-consuming but satisfying and rewarding in a different way.

I also wish to note the willing cooperation of many vendors, industry forums, and colleagues who provided information on their technologies, white papers, and product-information brochures. In particular, I want to thank Angelo Santinelli, vice president of Shiva, who gave me permission to use Shiva's white papers as the source material for chapters on security design and bandwidth management. ADSL Forum allowed me to use their Web tutorial for Chap. 8. I also want to acknowledge Greg Ma, a well-known networking expert and a Principal at IBM Consulting Group in Toronto, who contributed Chap. 11 on "Remote Access and Switching—In Enterprise Networking Context." Additional credit goes to Ascend, Citrix, RIM, Metricom, Puma, and Symantec.

While I have acknowledged sources in different places throughout the text, I would be taking undue credit if I did not also mention here the authors of the numerous books and articles that served as sources of information. While many of the ideas and opinions expressed in the book are my own, I have undoubtedly learned a great deal from books by Fritz, Wong, and Kessler. In certain instances I have used the ideas and concepts from their books and illustrated them with minor modifications.

Special thanks go to my lovely daughter Priya who did most of the graphics work for the book. She is an expert in Microsoft Powerpoint and graphics. It is her work that stands out more than my writing. She also did proofreading and in-house editing. She made valuable suggestions and enhancements, and corrected the grammar to ensure that sentences and paragraphs made the same sense to others as they did to me when I was writing.

Acknowledgments

To the editorial and production staff at McGraw-Hill, who converted my manuscript into a finished product, I want to say, "Thank you for your patience, understanding, and efforts in producing this book." Wayne Coleson, my editing supervisor at North Market Street Graphics, was meticulous in his editing. He took care to ensure that the finished product that you have in your hands met the high standards that you expect from McGraw-Hill. Thank you, Wayne.

Finally, special thanks go to my family. To my lovely wife Bina, my very bright elder daughter Sonia, and younger daughter Priya who is bright and beautiful, I say: "Thank you for your understanding and lack of attention all these months. This book belongs to you as much as it does to me. You sacrificed the time and attention that this book usurped."

REMOTE
ACCESS
NETWORKS

Remote Network Access— Market, Applications, and Economics

1

Remote Network Access—An Overview

Industry pundits and marketing visionaries have been promising ubiquitous access to corporate information through wireless networks for many years. It did not materialize as they predicted. Meanwhile, the remote network access industry just took off. It was simply because industry responded to customer demand with easy-to-install solutions using affordable technology.

—Chander Dhawan

About This Chapter

Remote access is one of the fastest growing segments of the internetworking marketplace. In this chapter, we define remote access and review various factors driving this industry. Foremost among these drivers is the change in the way business is being conducted today. Increased competitiveness, organization downsizing, and increased emphasis on customer service are bringing businesses closer to the customer. Information is generally centralized—hence the need for remote access. Social and environmental pressures are also driving this industry.

Needs of businesses for remote access of central information vary with the size of the organization. We comment on the size and growth of the market, assess the current environment, and look at the future trends in this industry.

1.1 Defining Remote Network in Mobile Computing Context

Mobile computing is a fascinating concept that is still trying to find a legitimate place in the network computing world. Through the use of telecommunications and wide area networks of various kinds, wireline and wireless, mobile computing connects computers and people on the move to information servers that reside in a different, albeit central location. *Mobile computing* is a broad and comprehensive term: It includes end-user devices, networks, computing applications, communications software, and the discipline associated with the design and implementation of such solutions. That is the topic we discussed in my previous book, *Mobile Computing—A Systems Integrator's Handbook.*

The term *remote network access,* on the other hand, is a subset of the mobile computing concept. It signifies the technology that is associated with providing network connectivity between two computers, typically a remote computer and a central information server, on a temporary basis. The information exchange can be between two peers, as well, as is the case with videoconferencing. The distinction between mobile computing and remote access is more in the use of these terms by industry and vendors, rather than in their dictionary meanings. Therefore, we shall try to explain what we mean by remote network access (RNA as an acronym) or simply, remote access. In many trade circles, the term RAS (remote access server) is often used. We prefer the term RNA for the technology and RAS for the server itself. Essential attributes of RNA are as follows:

1. RNA refers to connectivity solutions requiring temporary and intermittent network connections, rather than a continuous connection between an end user and an information server that is typically, but not necessarily, on a LAN. As such, it does not include networking solutions based on private dedicated networks, or corporate LAN/WAN internetworking infrastructure based on routers, switches, and hubs. This does not imply that routers or switches cannot, or should not, provide RNA capabilities. In fact, they do; we discuss this issue in chapter 6.

2. RNA is more than a remote LAN connectivity solution. Remote LAN access (RLA) would be a more appropriate term if the access was limited to LAN-based information services. In this book, we include access to information stored on legacy information servers, such as mini or mainframe computer-based superservers, whether these servers are accessed through a LAN gateway or directly through conventional communications servers/switches.

RNA solutions may be based on either wireless or switched wireline connection. Most common remote network connections utilize public switched networks (dial-up), ISDN, ADSL, cable modems, and, of course, the Internet. In this book, we concentrate on nonwireless connections and provide a brief overview of wireless network connection issues. We recommend that the reader refer to other books on the subject. One such book is the author's previous book, entitled *Mobile Computing—A Systems Integrator's Handbook* (McGraw-Hill, published in November, 1996) for a detailed discussion.

3. Business applications may be the same for RNA or mobile computing. In fact, many applications such as e-mail and database access are common. On the other hand, many mobile computing applications, such as public safety or law enforcement, have on-line transaction processing (OLTP) characteristics, and require virtually continuous wireless network connection. These applications are not suitable for RNA technology.

4. Although there is greater emphasis on the network connection in any discussion of RNA, this does not imply that application development and software issues are not important. The majority of RNA vendors focus on providing remote and mobile workers with access to LAN applications in a transparent manner, just as if they were desktops on the office LAN. To a large extent, this transparency eliminates the need for application integration. Nonetheless, we feel that the mobile-aware design concept, introduced by the author in his mobile computing book mentioned earlier, is relevant and necessary in RNA solutions, especially with slow-speed PSTN and wireless connections.

Therefore, RNA can be defined as *the physical (hardware) and logical (software) connectivity of both fixed and mobile remote client nodes to a central information server using temporary network connection.* Physically, it denotes the hardware (modems, communications servers, or concentrators) and software (client software in the remote end-user device and communications control software, including firmware, in the communications server.)

In the present state of the industry, RNA products do not provide the ultimate mobility that wireless networks provide; however, it does give mobile users access to mobile computing applications, and corporate information wherever there is a telephone jack—whether in a customer's office, at home, or in a hotel. This limitation is expected to change, and we comment on this in Chaps. 5 and 10.

The same business imperatives that are galvanizing mobile computing are spurring the growth of RNA (distributed operations, telecommuting business process reengineering, downsizing, etc.). Nevertheless, there is one major difference between the two: RNA solutions are cheaper to install and easier to use with an existing suite of LAN applications. Also, the proliferation of conventional, wireline public switched telephone systems, in the past, and cellular networks that can emulate the wireline switched connection during the last ten years, has been so extensive that you have access to either a real or a virtual telephone jack in many locations.

1.2 The Business Requirements Driving Remote Network Access

There are a number of business requirements that are driving the demand for remote network access solutions. The following requirements are significant:

Information Becomes a Key Competitive Tool.   As product creation becomes automated by using mass manufacturing techniques, or is relegated to developing world countries, research and development, marketing, distribution, and services components become the predominant values in the developed world. Availability of information—anywhere and any time to permanently connected or temporarily connected workers—is a key tool in this new world. RNA facilitates distribution of this information to the latter category of mobile workers and business partner organizations.

Increased Competitiveness. Increased competition in the marketplace is forcing organizations to pay greater attention to customer service. This means the field and sales force workers have to spend more time with their customers. They must be able to respond to their customers' queries instantly and confirm sales orders on the spot. This is possible only when they can access information from customers' offices and other remote locations. Field and sales force automation is perhaps the most popular mobile application for RNA.

Organizational Downsizing. The present-day downsized organization depends on fewer permanent staff at the office and more temporary and contract staff operating out of homes, at customer locations, or project sites. Since they do not have permanent offices, they must access information remotely.

Replacement of Old, Static Monolithic Organizational Model by a More Dynamic Multi-layered Workforce Model. Increasingly, the workforce in a modern organization consists of a core of full-time staff at the head office supported by telecommuters operating out of their homes and temporary contract staff. All these groups must have access to the same set of information and services—such as e-mail, order entry, and database queries—wherever they are.

The barriers between organizations are also breaking down. Larger organizations are allowing customers, suppliers, and business partners to access less-sensitive operational business information, such as inventory information, order shipment status, and so forth.

Remote Offices Demanding Access to the Same Information as the Head Office Workforce. To provide superior service to their customers, satellite offices, smaller remote offices, or even regional offices that are permanently connected to the corporate networking infrastructure are demanding that they be able to exchange operational information, or upload and download database information with the central site. Replication and synchronization of data in the disconnected computer networks are becoming increasingly important requirements.

RNA Helping Business Process Reengineering (BPR) to Eliminate Intermediate Business Processes. Many redesigned business processes depend on remote workers entering information at the source without the help of administrative staff and order entry personnel. Many mobile applications are being considered for eliminating redundant processes and streamlining others.

External Business Partners Demanding Temporary Connection. Customers, suppliers, and business partners are demanding electronic exchange of information through bidirectional access to each other's data.

Figure 1.1 shows the mobile workers inside the organization, telecommuters, and other extraorganizational workers, who are demanding temporary connection for e-mail, operational business applications, and document sharing.

1.3 Social and Environmental Pressures Driving Remote Network Access

Besides the business requirements discussed earlier, there are a number of social forces that are driving the movement toward remote network access. We discuss the following reasons:

Pollution and Smog. Many cities are reaching their limits in terms of transportation capacity. Therefore, many municipalities and state legislators are

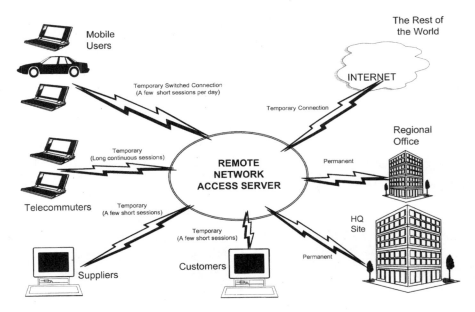

Figure 1.1
Remote network access—business user community.

realizing that moving everybody to and from the downtown core in the morning and evening leads to air pollution, smog, and other ecological problems. Moreover, travel time between the office and home is unproductive and economically costly to the region or the nation. Therefore, these organizations are implementing regulations that promote telecommuting of workers who can work out of their home. These telecommuters need remote network connection for picking up their e-mail from the office, transferring documents, and exchanging files with their colleagues in the office.

Increasing Mobility and the Changing Structure of Workforce.   Ubiquitous remote access to central information resources is changing the way we think of work and pleasure, home and office, and intra- and interorganizational communication. Switching from one mode to another and from one state to another is no longer as discrete as it used to be. The concept of mobile computing and remote access gives new meaning to the phrase *instant communication, anytime and anywhere.* It caters to an inherent desire to stay in touch, at the touch of a button, so to speak, while at the same time satisfying a basic human need to be mobile and independent. What the cellular telephone did for people on the move—allowing them to be able to reach out and talk to each other whenever they want—remote network access promises to do for mobile workers.

Physical Proximity Is No Longer Important for Shared Workgroups.   As a result of the mobile computing evolution in the workforce, organizations are not insisting on the physical proximity of workgroups or face-to-face meetings among them. As a result of using modern technology and the Internet, modern organizations can pick the best talent from different countries or even continents where the mobile employees prefer to live. The virtual workgroup formed across physically disparate locations can share information, documents, and even have electronic face-to-face contact through inexpensive videoconferencing, possible through remote access technology.

Cradle-to-Grave Employee-Employer Relationship Changing.   The relationship between employers and employees is changing radically. The notion of long-term and permanent employment, with an employer who provides an office that you can call your own, is giving way to temporary and shared Plexiglas offices. However, supervised electronic contact, information access, and information sharing are being enhanced as a substitute in this new mobile world of remote access.

1.4 Technological Factors Driving Remote Network Access

There are a number of reasons that remote network industry has moved past the wireless network-based mobile computing industry. We list the following major reasons:

Affordable RNA Solutions.   RNA solutions started out by extending LAN applications to remote workers through inexpensive hardware/software solutions. Per port connection costs were only in hundreds (or low thousands) of dollars; users could easily absorb that without doing a detailed analysis.

Commoditizing the RNA Solution by Embedding It in the OS.   Microsoft and IBM have started embedding the client portion of the RNA software in the operating software. This gives ready access to the RNA software to the user who tries it, likes it, and pressures the organization to provide it on a structured and planned basis.

Easy-to-Install and -Use Solutions.   Entry-level RNA solutions are relatively easy to install. It takes hours or days to install a small RNA solution. It can be done by an average network administrator. Once installed, the user interface is quite transparent, as on the desktop. On the surface, this is very attractive, certainly from the end user's perspective.

Bottom-Up Marketing by Vendors to End Users.   The marketing strategies of the RNA vendors are aimed directly at the end users. Selling to end users and supporting these RNA installations without getting IT planners involved have worked very effectively from their perspective, although it might have caused some headaches for IS operational staff. The end users have become vendors' allies in their attempt to extend the solution to more people and across the enterprise.

Internet Revolution.   During the last two years, the Internet revolution has set standards of ease and universality with respect to providing information to the public. The Intranet follow-on to the Internet was precipitated by users asking why it is that their internal IS organization is unable to provide operational information that they need to conduct their business. It may seem ironic that end users are able to exchange e-mail more easily with outside organizations than they are able to do within the organization. This is so because the high start-up costs of research, development, testing, and implementation must be borne by an organization to implement effective Intranet communication; the public-shared Internet infrastructure, on the other hand, makes it relatively easy to connect to the corporate LAN, without spending months or years to plan and install a private remote access infrastructure.

First PSTN, Then ISDN, and Now ADSL. Remote network access started with dial-up connections, using public switched telephone networks (PSTNs). However, 28,800 bps speed on PSTN did not match the 10 Mbps LAN speed, even though the former link was dedicated to a single user. Waiting in the wings was ISDN, a technology invented more than a decade ago. This provided a faster and more affordable link, when it involved faster data transfer. Emerging soon will be another interesting telecommunications link technology—known as digital subscriber line (DSL), high data rate DSL (HDSL) or asymmetric digital subscriber line (ADSL), depending on its implementation. This will allow the ordinary twisted copper wire medium strung throughout our homes and businesses to be used for 1 to 6 Mbps speed. With this speed, many more bandwidth-intensive applications—such as videoconferencing and multimedia—will become available to the remote mobile workers.

Plethora of Horizontal Applications for Mobile Workers. Several horizontal applications, notably e-mail, file or data replication, sales force automation, and service representative dispatching have been developed by third-party software houses to spur the remote access industry. In fact, the sales-force automation (SFA) industry has taken off. The first generation SFA applications that relied on disconnected sales workers are expected to be replaced by second-generation versions that will have bidirectional exchange of data.

Specialized Vendors Providing Vertical Applications. Vendors, such as Telxon, Symbol, and Psion have developed complete vertical industry solutions for mobile workers. Some of these are based on wireless networks, others on PSTN links. (See Fig. 1.2.)

1.5 User Organization Scenarios— Small, Medium, and Large Enterprise

Different organizations look at remote access requirements in different ways. The size of the organization and the complexity of their network

Figure 1.2
External factors
affecting remote
access industry.

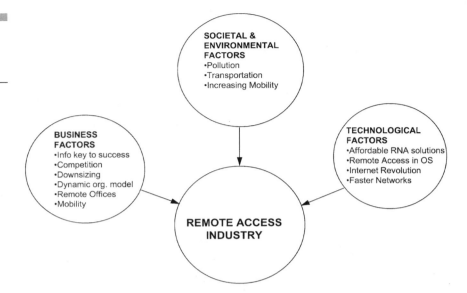

infrastructure affect the technical solutions. Accordingly, organizations make different demands on the type of solutions that fit their needs. Here we describe these differences based primarily on the size of an organization. As the size increases, the type of remote access solution becomes more complex.

The small organizations or remote offices of larger organizations have LAN-based application systems with typically less than 100 users, some of whom may be traveling on sales or service calls. Some of the workers are part-time employees needing telecomputing facilities. This organization would need a simple remote access solution with which they can access their LAN applications. At any time, no more than 10 active users may be logged on remotely. They cannot afford to modify applications for mobile workers. Low-end RNA solutions will suffice for their needs.

The second category is RNA for medium-size organizations or departmental applications of a large enterprise. The organization size is usually between 100 and 500 users, with remote users accounting for approximately 25 percent of the user population. Besides LAN-based desktop applications and a few SQL DBMS applications, they have some operational applications such as accounting, order entry, inventory control, and the like, on a minicomputer that runs on UNIX, DEC's VAX/VMS, or IBM's AS/400 platforms. They exchange information. Some of the more sophisticated organizations use EDI for purchasing, invoicing, shipping (and so forth) information exchange with larger companies or their suppliers. Their customers may demand access into their operational systems; this must be provided on a controlled and secured basis. The mobile sales force and telecommuters are important to this organization to the extent that some of them can justify wireless access. Application integration of the mobile client applications with operational applications or legacy minicomputer applications becomes significant. (See Fig. 1.3.)

The third category is for the larger organizations that want an enterprise-wide type of solution. Their user base is in the thousands, with mobile population itself exceeding 500, quite often exceeding 1000 users. They have desktop and departmental applications on multiple client-server platforms. Operational and decision-assist applications reside on legacy platforms such as IBM mainframes, AS/400, DEC VAX/VMS, Tandem fault-tolerant clusters, or UNISYS A series. The organization has in place a private LAN-WAN internetwork based on Router technology. They have a desire to exploit Internet or Intranet infrastructure for remote access. Many corporate accounting, customer service, and order-entry applications have been implemented on large centralized mainframes using IBM IMS/VS or DB2 SQL databases. These applications run on different host computers, sometimes on different platforms. Ideally, a mobile sales worker in the field would like to access information from these distributed databases with a single query. Individual departments or regional sales offices have their own local LAN where the e-mail server resides. The organization has invested in a large private wide area telecommunications network that may utilize routers, telecommunications switches, and LAN switches. Protocol

Figure 1.3
Remote network access solution for (a) small and (b) medium organizations.

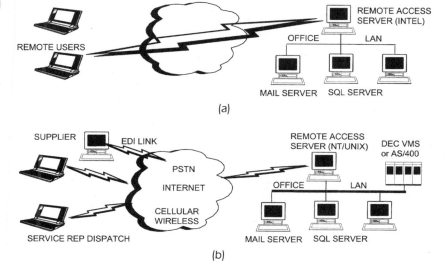

conversion is a major issue. They have a need to implement mobile computing solutions based on both wireline and wireless networks. Application integration is extremely complex in such environments. (See Fig. 1.4.)

1.6 RNA Market—A Growth Industry

RNA is one of the fastest growing markets in information technology. There are many reasons for this growth. The single most important factor is user demand for connectivity to home offices and corporate information resources—a demand that the RNA vendor community has met with relatively inexpensive and simple entry-point solutions. Thus, with only minimal budgets, corporations can easily provide remote users dial-up access. Once a few users in any particular setting experience the convenience and efficiency of remote access, others are quick to join the bandwagon—in the process, demanding access not just to e-mail but to all the corporate applications and data. Fueling this demand is the availability of inexpensive, fast ISDN links that give remote connections LAN-like speeds. After a slow start in the 80s and no growth at all for years, the number of ISDN links in the United States grew by 24 percent

Figure 1.4
Remote access solution for a large organization.

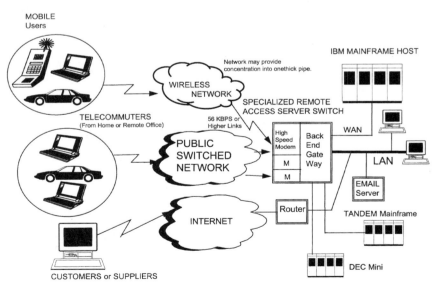

Note: Vendor names are for illustration purposes only. It does not imply an exclusive role of named vendors in architecture.

in 1996. The Internet revolution has, of course, accelerated this growth considerably during the 1994 to 1996 period. According to a Dataquest study, the total remote access—related market (remote access routers, remote access servers, remote access concentrators, and modems) will grow from $2.6 billion in 1995 to $12 billion in year 2000. Of this, remote access servers, modems, and client software represent about one-third share of the total market.

Besides systems integration and network usage services, RNA market can be divided into at least three categories:

- Remote access server (RAS) market
- Remote access modem market for PCs, notebooks, and new network computers (NCs)
- Remote client software market

We provide summarized projections of the three submarkets that highlight the growth of this industry. These projections are not necessarily aimed at the vendor community, who can obtain more detailed estimates and greater insight from market research companies such as IDC, LINK, BIS, and Infonetics. Our book is aimed at professionals in end-user organizations and our intent in giving these numbers is to encourage these organizations to implement remote access solutions after proper planning.

1.6.1 Remote Access Server Market

IDC expects the remote access server market to grow from approximately $1.0 billion in 1995 to $4.9 billion by the year 2000, as shown in Fig. 1.5*a* and *b*. The server submarket consists of two components: fixed port remote access servers and RAS concentrators. Fixed port remote access servers comprise asynchronous technology platforms that are sold in fixed configurations of a certain number of ports, such as 4, 8, 16, and so on. Typically these ports correspond to the active number of simultaneous switched connections from the remote users to the LAN. The second component of this market, the RAS concentrator, is a relatively new incarnation of server that offers ISDN PRI and/or channelized T1 ports. A RAS concentrator takes several incoming analog or ISDN-based switched connections and multiplexes them into a single thick pipe, generally T1 or ISDN-based primary rate interface (PRI). This is among the fastest growing submarkets.

Figure 1.5
(a) RAS server market
in revenue dollars. (b)
Worldwide RAS ship-
ment volume.
(Source: International
Data Corporation,
1996)

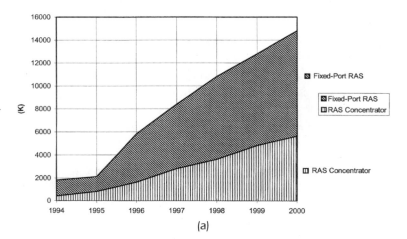

Worldwide Remote Access Server Port Shipments, 1994-2000

(a)

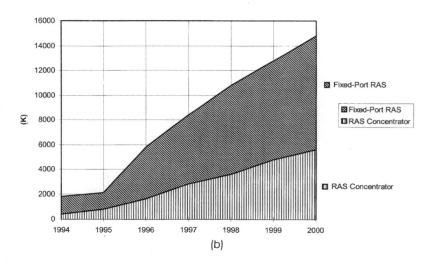

Worldwide Remote Access Server Port Shipments, 1994-2000

(b)

1.6.2 Modem Submarket

Internet use and the remote access market have driven the modem submarket beyond many industry projections. Demand for increasing speed
from end users has caused many to upgrade either to 28,800 bps/34,400
bps modems or to ISDN adapters. The adoption of ADSL in the high-end

modem market will perhaps lead to another quantum jump in speed and a significant upgrade market. The modem demand consists of client-side demand at the remote side (in ISA or PC card form factors) and server-side demand. On the server side, card-based modem rack systems are being replaced by either multiple modems on a single card or a concentrator. We illustrate the growth of the mobile remote access market by Fig. 1.6*a* and *b* that shows growth of PC Card (PCMCIA) modems.

1.6.3 Client Remote Access Software

While the dollar value of this software is much smaller than the previous two components of the RNA market, the actual numbers are much better indicators of the demand for remote access.

Figure 1.6

(a) Worldwide modem market—indicator of PSTN-based remote access.
(b) Worldwide PC Card modem market—indicator of mobile remote access.

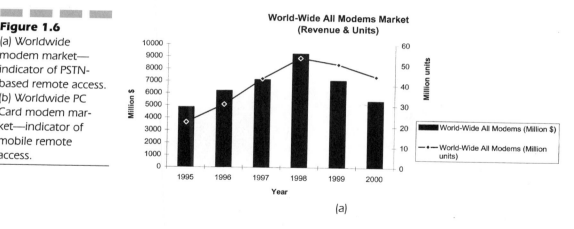

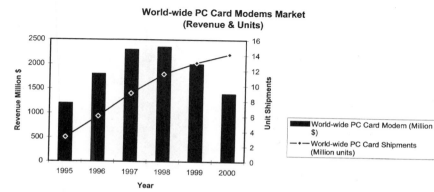

TABLE 1.1

Worldwide Remote
Access Client
Software Market,
1995–2000

	1995	1996	1997	1998	1999	2000	1995—2000 CAGR
Revenue (million $)	163	226	277	322	356	387	18.9%
Installed base (000)	5,482	8,618	12,933	18,637	25,712	33,857	43.9%

SOURCE: IDC Report.

1.7 Assessment of the Current Situation

The RNA market has grown rapidly. The initial group of remote access vendors in the marketplace was small, with a simple objective and a narrow focus to meet end users' needs. They simply wanted to provide a solution so that technical support personnel could solve problems from home or from a remote site without being physically there. They did not do any market research or analyze customer requirements and then architect the remote access products. Little did they realize that their solutions would find a much bigger home. Then came the burgeoning demand from mobile workers and telecommuters. A few more vendors got into the act and started offering easily installable solutions. Their target customers were end-user departments and LAN administrators. They kept themselves away from the information system (IS) groups of these organizations. It was bottom-up growth. Until 1997, the big internetworking

Figure 1.7
IDC supporting the
remote user.

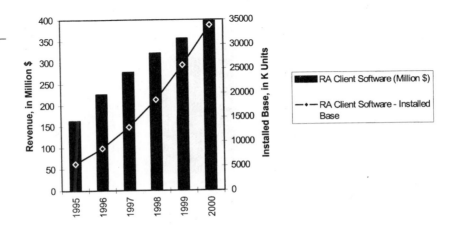

World-wide Remote Access Client Software 1995-2000

players stayed on the sidelines and found this market somewhat elusive, though the size of the market has been attracting their attention. As a result, we have started seeing some collaborations, acquisitions, and mergers between companies—the biggest merger being between 3-COM and US Robotics. These consolidations will no doubt continue.

There are several different characteristics we can associate with the evolution of the RNA industry:

- Transparency to existing LAN applications
- Ease of installation
- Keeping it simple
- Reactive upgrading of technology
- Isolation of application integration from communications solution
- Haphazard growth
- Lack of enterprise-wide architecture compatible with other technologies

1.7.1 Transparency to Existing LAN Applications

One of the key objectives of RNA solutions was to ensure transparency to LAN applications. Whether it was the Symantec's pcAnywhere client or the Shiva's hybrid communications server, the vendors had to ensure that applications did not know the difference between an on-LAN or off-LAN (i.e., remote) connection.

While this approach was sound initially, there is a need to make the applications mobile-aware so that we can deal with differences in LAN speeds and remote link speeds, especially for wireless networks. As we see later in the book, this can be achieved through several new techniques, such as remote caching and intelligent agent software.

1.7.2 Ease of Installation

Ease of installation of remote access solution is an important reason for RNA's quick adoption by the users. The vendors provide a completely bundled solution consisting of communications hardware and server software. The configuration definition is quite simple. For a small-size organization, any trained LAN administrator can implement such a solution in a few hours. This means that you can start getting benefits right away. For medium-size organizations, Value Added Resellers (VARs) specializing in remote access can help.

1.7.3 Many Options to Suit Each Customer

As discussed in Sec. 1.5, customer needs vary with the size of the organization and mobile workforce. Accordingly, there is a variety of RNA solutions available that customers can implement. There are software-only solutions that you can implement on existing hardware. Novell provides an NLM (Netware Load Module) version that you can implement on an existing file server. Microsoft provides an RAS software on NT platform, with client remote access software embedded in Windows 95. One can start with these solutions. Entry-level multiport hardware solutions from Shiva can be installed for less than $10,000 to service up to 100 users for simple e-mail applications. More sophisticated mobile applications require more extensive hardware with ISDN and T1 communications links, security firewalls, and Internet connectivity. There is tremendous granularity in the choice of solutions.

1.7.4 Reactive Upgrading of Technology

Instead of building the most sophisticated product with all the bells and whistles that will be required in the future, the industry has opted very wisely to develop additional functionality and features in a gradual manner, as their customers' requirements become more sophisticated. This has helped customers to start off with simpler solutions for a controlled group. As the user population grew, and information systems departments insisted on requirements such as planning, security, and tariff management, the vendor community has responded relatively quickly.

1.7.5 Isolation of Application Integration from Communications Solution

One of the reasons for slow growth in mobile computing based on wireless networks is that applications have to be integrated with communications software. Each of the networks has its own communications drivers and APIs; this necessitates changes to the applications. The big players continue to promote proprietary network interface applications. Perhaps this is justifiable for efficiency, but support for open TCP/IP transport protocol exposes them somewhat to future competition. The RNA industry kept the two (communications software and applications) separate, a strategy which made life simple for both LAN administrators, who could meet user demands by acquiring a completely bundled solution, and application developers, who kept on developing their applications on LAN without worrying about remote access requirements.

1.7.6 Lack of Enterprisewide Architecture Compatible with Other Technologies

In spite of distinct advantages enjoyed by tactical solutions provided by RNA vendors, it is being recognized that solutions for the enterprise require a more architected approach, as well as use of the existing network infrastructure, integration with legacy information servers on mainframes, capacity issues, mobile-aware design, and transparent coexistence with wireless network-based solutions. IDC studies have shown that all the vendors are gunning for the corporate RNA market. This will require deeper research and development expertise, understanding of other pieces in enterprise connectivity solutions, and compatibility. In the next section, we venture to speculate on attributes of this future solution in the enterprise.

1.7.7 Somewhat Haphazard Growth So Far Needs Professional Direction

It is our observation that remote access technology has grown in a haphazard fashion, and that network managers were caught largely unaware of what was going on in their user departments. Initially, users were being exploited by remote access vendors pushing products that may or may not have been appropriate for the enterprise. Often, their low cost and ease of use resulted in wholesale adoption of remote access solutions without either proper research for the optimal solution for the organization or going through a systems planning exercise.

This is not to suggest that there is anything wrong with users experimenting and trying new applications during the formative stages of a new discipline. Far from it. It is just such bold experimentation that leads to breakthroughs and the emergence of new technology.

However, in this particular case, we believe that we have now reached a critical mass for this technology in small-, medium-, and large-size organizations. To progress beyond this point in an orderly and planned manner, we need to thoroughly grasp the business, technical-design, and management issues involved in providing remote access to the users. It is time for IT and network management staff to take charge and start advising those who are implementing remote network access and integrated network management.

1.8 The Future Outlook

RNA solutions are still in the first or second generation of their evolution. We expect further evolution of the technology and are listing the following attributes of a future RNA solution. The industry has started addressing some of these requirements in their second-generation products, but we will see substantial focus on these features in the future.

Enterprise Solutions.   In the first three years, RNA vendors addressed the needs of small organizations and departmental solutions within larger organizations. Lack of control by IS organizations within the enterprise has led to many different incompatible solutions using different network security schemes and, quite often, very little network management. Now that the size of the problem has increased, end-user organizations are demanding integrated solutions that both meet the needs of the enterprise in a consistent fashion and are part of the overall network architecture. We expect more emphasis by the industry on solutions that address these requirements.

Concentration and Multiplexing over Fixed Port Design.   We expect to see more emphasis on communications servers that rely on concentration and multiplexing techniques in their design as compared to fixed port design. This will allow organizations to utilize T1 and ISDN PRI connections. This will result in smaller footprints, lower costs of hardware, and better utilization of network link capacity, leading to a higher user-to-port ratio for same traffic levels. Of course, the usage itself will increase, and new bandwidth-intensive applications will come on stream; this will lead to lower user-to-port ratios. The second trend will be stronger, resulting in an overall decrease in user-to-port ratios.

More Bandwidth-Intensive Applications.   Multimedia applications, videoconferencing, network-centric JAVA applications, and data replication will increase the amount of traffic on RNA links by an order of magnitude. Second- and third-generation RNA solutions will have to deal with this increase of traffic by higher-capacity servers with superior compression techniques built in.

Security.   Security continues to be among the top three concerns of network managers, alongside functionality and central management. Proprietary schemes of the first-generation design will give way to the standards-based approach. Until then, pragmatic security design, based on currently available products, is essential.

Integrated Solutions.   Stand-alone RNA solutions will be replaced by more integrated solutions for the enterprise. This implies that a technical solution that integrates the functions of a router, LAN switch, and RNA communications server would emerge as a serious option for those organizations that want fewer boxes to manage and maintain. This trend will change the complexion of internetworks of today. We can already see germination of this trend from certain acquisitions and mergers going on in the industry. Vendors such as CISCO with strong router technology will benefit.

From PSTN to ISDN and ISDN on to ADSL.   More and more RNA servers, except the very low-end ones, have incorporated ISDN support. During 1997 to 1998, ADSL will become a viable link option. We expect high-end solutions to incorporate this capability in their products in the future. In due time, perhaps in 1999, ADSL will become mainstream. (You should refer to Chap. 8 for further analysis.)

Wireless Network Support.   Mobile computing solutions based on wireless networks employ specialized servers, called mobile communications server switches (MCSSs), because of proprietary radio interfaces and APIs. Since some of the end users need wireless connectivity where there is no wireline jack, it makes sense to us that RNA servers should support wireless networks. We have seen vendors, such as Shiva, announce their intent to support wireless networks such as CDPD. We can expect more wireless network support embedded into RNA servers.

Standards-Based Mobile Management.   Mobile management is an extremely important issue as the RNA industry moves toward the enterprise. The vendors have deliberately shied away from tackling the problem because of its complexity and lack of standards. The Mobile Management Task Force (MMTF) within the Internet Engineering Task Force (IETF) have now created MIB standards which can be incorporated into their products. Companies such as Epilogue are providing development tools to ease this effort.

Controlling Uncontrolled Telecommunications and Support Costs.   Network managers are unable to get a good handle on network usage costs with current RNA solutions. We expect more and more vendors to provide tariff management capabilities in their future products as Shiva has already done.

LAN a Conduit of Choice for Accessing Legacy Information.   According to industry estimates, despite significant advances in client-server applications, more than 50 percent of corporate operational information continues to reside on superservers such as IBM mainframes, minicomputers, and UNIX servers from a variety of vendors. In the past, dial-up access has been provided through terminal emulation in many cases. This approach is being phased out. We expect that LANs will become the conduit of choice for accessing legacy information. As an example, an SNA server on a host-attached LAN will become the vehicle for accessing legacy information from IBM mainframes.

Internet.   Internet will play an increasing role in RNA solutions. First, Internet will be used as a front-end connectivity mechanism for remote users. Second, RNA solutions implemented in the organization will have a thicker telecommunications pipe into the Internet.

Industry Saturation.   Since the RNA industry has grown very fast, everybody who needs to have access will have a remote connection. This will lead to saturation of demand in the next two or three years.

Industry Consolidation.   The previously mentioned trends will lead to consolidation of RNA vendors, causing the demise of the weaker players and buyouts or mergers of others. Just like any other industry, only the strong vendors will survive.

Summary

In this chapter, we have given an overview of the remote access technology and industry. We started off with a definition of RNA. Then we looked at the driving forces behind sudden growth in remote access. In this context, we analyzed these factors from business, social, and technological viewpoints. We presented industry forecasts on the growth in the market size for various remote access markets. We assessed the current situation and finally we speculated on future trends. The objective was to give an introduction to the subject.

2

Remote Access Applications

Give the mobile users the same applications as they have on the desktop and they will use them instantly.

—*Chander Dhawan*

About This Chapter

Describing remote access applications is difficult, especially when the principal objective of the industry is to provide the same applications to the remote worker as provided on the LAN. This could include thousands of applications; yet there are some applications that are demanded more often by the mobile workers, and there are still others that are more suitable for remote access because of the limitations of technology. In this chapter, we discuss some of these applications that fit the profile of popular remote access applications. First, we start with horizontal industry applications, such as e-mail and paging. Then we move on to applications in specific vertical markets.

We also describe the differences between desktop and remote applications. In this context, we discuss the concept of mobile-aware applications. We discuss application integration issues in Chap. 13.

2.1 Characteristics of Remote-Worthy Applications

Certain business applications are more suitable for remote access implementation than others. We use the term *remote-worthy* to describe any application that has characteristics that make it particularly suitable for use with remote personal computers connected to central information resources. These characteristics are as follows:

- A significant percentage of users either are telecommuters or spend a lot of time away from the home office. Accordingly, they need access to the same information in their virtual offices at home as they would in their real offices.

- Remote users do not usually access the information residing on a central information server throughout the day, minute by minute, but occasionally, several times during the day. Therefore, they do not require a continuous and permanent connection; that is, they are not on a dedicated private network.

- The application requires only a small, portable and lightweight notebook or carry-on computer device (sometimes mounted in a user's vehicle, van, or truck). The users transport this device. This represents the user's permanent virtual office.

- There is a significant economic value, public service, or public safety benefit resulting from information provided while the user is away from the office or a fixed place of work. The benefits may be of different types; for example, extra travel is eliminated, productivity is enhanced, business process cycle time is reduced, patients' lives are saved, crime is prevented, and so forth.

- Only a small amount of data from a central information server is needed at the remote site on a regular basis. This characteristic has as much to do with network economics as with the need to do the task.

- Applications requiring large digital data transfer (e.g., videoconferencing) have a significant payback by way of reduced travel costs in busy cities and across the nation.

- Remote field offices need to update their local copy of databases by refreshing from a central data repository on a regular basis. Data replication is an important remote access application.

These characteristics help us determine whether a specific LAN application is a candidate for being offered economically to remote users.

2.1.1 First-Generation Remote Access Applications—Simply LAN Applications

A key objective of remote access technology was to extend the LAN applications to remote workers through a switched network connection. These applications were not modified in any way to suit the specific needs of mobile workers; nor did the LAN application developers allow for the slower speed of the communications link. They expected the network providers to solve the problem of network speed. However, the technical realities of network capacity are that switched network speeds will continue to be slower than LAN speeds for the foreseeable future. ISDN and DSL will improve the situation in the future, and more LAN applications will become affordable in remote locations than what is feasible today. However, when ADSL becomes mainstream for remote access, 100 Mbps Ethernet will be mainstream on the desktop. This gap will continue. Moreover, a remote worker has different work patterns and priorities than somebody working inside the office. Therefore, not all LAN applications are suitable for remote access without modifications. We expect this to change over a period of time when both the needs of mobile workers and remote applications become as important as those who are permanently in the office. This will happen with the next generation of mobile-aware remote applications.

2.1.2 Defining a Mobile-Aware Application Design

We deal with this subject more fully in Chap. 12. However, we introduce the concept here, because application designers who are either building custom applications or buying vendor-developed applications should keep this design requirement in mind. Applications may simply not fly, and may never be rolled out across the enterprise if they are not mobile-aware. Briefly, applications are worthy of a mobile-aware label only if they have been designed initially (or modified subsequently) to recognize the following important attributes of mobile users:

1. The application recognizes that mobile users are often in a hurry, and accordingly provides a fast path through the application dialogue.

2. The application provides sufficient but not excessive amounts of data to enable users to complete tasks without time-consuming and expensive interactions with remote servers.

3. The application recognizes that remote access networks are relatively slow and significantly more expensive as compared to LANs, because they incur a usage charge for every minute of network use. Therefore, it limits the user to exchanging only essential data that he/she needs to do the task at hand. This is especially true with wireless networks or even slower PSTN networks at 28,800 bps.

We are suggesting that the mobile-aware concept (introduced in my previous book, *Mobile Computing—A Systems Integrator's Handbook*) will become better understood as mobile computing and remote access get adopted by more organizations. In this effort, an intelligent-agent-based application development model (described in Chap. 5 and emphasized in Chap. 12) will allow modification of these LAN applications without complete redevelopment.

2.2 Application Categories

We describe remote access technology applications in the following three categories:

- Shrink-wrapped horizontal remote access technology applications
- Generic horizontal applications requiring extensive customization
- Vertical industry-specific applications

2.2.1 Horizontal Remote Access Technology Applications

In this category we discuss application suites that can be used in broad segments of various industries. Since these applications are based on common business processes, they are available in shrink-wrapped packages with only minor customization required—or allowed.

2.2.1.1 ELECTRONIC MAIL. There is no doubt that electronic mail (also called electronic messaging) is the single most popular remote access application. The Internet revolution has almost instantly moved this application to a state where it has become indispensable. No longer are we at the mercy of an internal IT organization's priority schedule because Internet-based e-mail has become easy to install and inexpensive to provide. In fact, pressure is on for integrating internal e-mail with Internet e-mail.

In order to provide a high level of customer service, mobile workers and sales professionals must stay in touch with home offices and customers. In many circumstances, e-mail is the most efficient mode of human communication, where intimate personal interaction is not required and human rapport need not be established. E-mail is precise and leaves an electronic trail of messages and responses. It is an excellent complement to voice communication. Even in public safety and law enforcement applications, it can replace routine and nonurgent communication that otherwise would take place on the radio. According to an 1996 InfoWorld study, quoted by IDC in one of its reports, 88 percent of all remote workers use e-mail (see Table 2.1). (Additionally, 81 percent use remote computing to transfer files and 62.5 percent access databases.)

TABLE 2.1

Type of Applications Used by the Remote User

	% of Total Respondents
E-mail	88.0
File transfer	81.0
Database access	62.5
Network management/remote administration	50.0
Terminal emulation	38.0
Corporate intranet access	38.0
Workgroup applications	37.5
Order entry	28.0
Other	3.0
No answer	0.5

SOURCE: Infoworld's Remote Access in the Enterprise Study, 1996.

E-mail can be received on or sent from a variety of end-user devices, including PC notebooks, PDAs, WIN-CE compatible handheld computers (such as those from Casio, Compaq, and Sharp), two-way pagers (such as Research In Motion's Inter@ctive Pager or Motorola's PageWriter), and PCS digital telephones (such as Nokia's PCS1900). A number of e-mail packages are available, but Lotus's cc:Mail (or current version of it), Microsoft's MS-Mail, and QUALCOMM's Eudora (popular on the Internet) are among the most popular among remote users. Figure 2.1 shows an example of mobile users accessing e-mail servers through various switched networks. In the case of wireless packet networks, wireless modems such as Ericsson's Mobidem emulate the AT command set to open connections between client and server modems.

ISSUES FOR REMOTE ACCESS E-MAIL IMPLEMENTATIONS.   The following issues should be considered when implementing a remote-access-based e-mail application:

Internal versus External E-mail Service.   If the organization uses an internal e-mail standard that the mobile user community is familiar with, you should consider it as a preferred e-mail platform, especially if it pervades the organization. You should investigate if the internal e-mail standard can be extended to mobile workers as well. If it is too difficult and expensive to do so, Internet or an external e-mail service should be considered.

Figure 2.1

Remote access for accessing LAN-based e-mail.

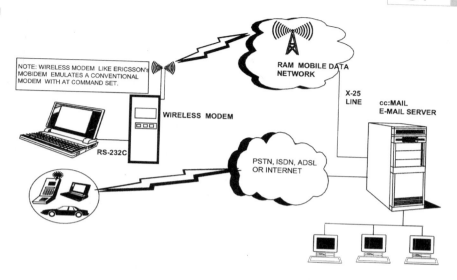

Fortunately, interoperability between mail packages, and the translation of e-mail contents from one platform to another popular platform has made significant progress during the last few years. Also, the Internet mail standard is becoming more popular every day, and may become the de facto standard.

Mobile Versions of E-mail Client Software.   The availability of mobile-aware versions of e-mail software should be investigated. Both Lotus and Microsoft have mobile-aware versions of their products available. These products offer distinct advantages for mobile workers who are in a hurry and want to get urgent messages, especially if they are using a wireless network. This capability is achieved by providing customizable filters that allow users to set up the priority schemes of messages, and whether they simply want headers to be downloaded.

E-mail as a "Starter" Mobile Application.   E-mail is likely only one—albeit the first—application that professional and sales users may eventually implement. Decisions as to what remote access network option to use for e-mail should be made in the context of other applications that will also be offered to mobile users.

Integration of E-mail with Other Operational Applications.   In many cases, application developers find it very convenient and easy to use e-mail as the core application that can be extended to link to other applications. In other cases, by embedding electronic forms in e-mail, mobile workers can transfer operational data—such as expenses, sales orders, purchase requisitions, and so forth—from the field to central DBMS applications. While this application development model may be less efficient from a computer processing point of view, it could cut down the lead time by eliminating any intermediate data entry process. Groupware packages, such as Lotus NOTES and Microsoft Message Exchange, use mail as a transport for many simple applications.

Table 2.2 highlights information about e-mail in a remote access context:

TABLE 2.2 Remote Access Applications—Electronic Mail	What does an e-mail application do?	▪ Allows remote users to send and receive e-mail while they are away from home offices. Some vendors provide connectivity to in-house desktop e-mail packages; others, especially network providers, such as RadioMail, offer proprietary e-mail packages. Most vendors provide access to the Internet mail for exchanging mail with rest of the world.
	Desirable features of a mobile e-mail application	▪ Extends workgroup e-mail to mobile users through PSTN or wireless networks.
		▪ Pages to remote users when urgent messages are waiting to be received.
		▪ Connects with Lotus cc:Mail, MS-Mail, and provides Internet gateway.
		▪ Transfers encoded files as attachments.
		▪ Voice mail interface à la telephone with text-to-speech conversion.
	Products available	▪ Lotus cc:Mobile, Lotus Notes, MS Mobile, NS Exchange, RadioMail, QualComm Eudora, and Xcellenet Mail.
	Cost—S/W License	▪ $50—$75 per user for client software only; server software extra. Some vendors bundle client and server licenses.
	Usage fees	▪ Included in Internet access but with wireless networks, cost varies over a wide range. "Rule of thumb" cost = $0.15 on RAM to $0.60 on cellular for a 600-character message in the United States.
	Benefits and payback	▪ Most popular remote access applications.
		▪ More efficient than voice mail and more selection features for high priority messages.
		▪ 5 to 10% productivity improvement reported in customer studies.

TABLE 2.2 CONTINUED Remote Access Applications— Electronic Mail	Network support	▪ PSTN, ISDN, and Internet for wireline networks.
		▪ Wireless network support for ARDIS, RAM, and cellular network support (e.g., cc:Mobile, Lotus Notes, MS-Mail).
	Typical platforms	▪ DOS, Windows 3.1/95, Windows NT, Unix, NetWare.
	Compatibility with internal mail systems	▪ Many e-mail service providers include interfaces with popular industry packages, such as cc:Mail, MS-Mail.
	Special considerations	▪ Intelligent-agent-based client/server versions appearing soon.
		▪ Users like to use the same e-mail software in the field (ideally a mobile-aware version with fast pass-through) as they do in the office.

2.2.1.2 VIDEOCONFERENCING—ELECTRONIC MEETING WITHOUT WALLS. We have said that remote access brings information from the central information server to the mobile worker. That relatively narrow definition needs to be expanded because remote access can really bring information and people together in a virtual fashion. With advanced compression technology, and high-speed switched access into the Internet, through ISDN now and ADSL links/cable modems in the future, videoconferencing is no longer an expensive application that requires T1 links. In fact, it is becoming a viable application through remote access today, though with some limitations that will go away soon.

WHAT IS VIDEOCONFERENCING? *Videoconferencing* is the use of interactive multimedia and live telecommunications for the purpose of exchanging ideas and delivering information with instant feedback. Videoconferencing combines:

▪ Video images (the people participating in the electronic meeting)

▪ Audiocommunication (voice, sounds, or music)

▪ Scanned or computer-generated images (photographs, graphs, charts)

▪ Computer-to-computer exchanges of information

With inexpensive videoconferencing available through remote access, distant places become closer and you can meet with your colleagues instantly.

AN INTERESTING VIDEOCONFERENCING APPLICATION.   There are virtually hundreds of videoconferencing applications today. We describe here an interesting application in the public sector used for prisoner arraignments. Videoconferencing technology helps many courts and local governments cut staff time spent in transporting prisoners and reduces security risks in courthouses. The courts use videoconferencing to conduct arraignments, pretrial releases, interviews, mental health hearings, pretrial conferences, and other events without requiring all parties to be at the same location. In the past, most videoconferencing systems in courts used cameras, monitors, and T1 links; new technology now allows video to be transmitted over existing PC-based local area networks (LANs). Remote access technology now makes this task far more economical.

Court events, conferences, and meetings using videoconferencing are similar to those conducted when all parties are in the same room. As fewer defendants are transported to the courthouse, security is enhanced for the law enforcement agency, and risk is reduced for the court. Holding areas, reserved for prisoners, can be used for other purposes. Of course, the fees attorneys charge for travel and waiting time are reduced because attorneys can attend from their offices. The business case is pretty strong. (See Fig. 2.2.)

2.2.1.3 FILE TRANSFER APPLICATION. This is a generic application that enables mobile computers to transfer a variety of files. These files may contain word processing documents attached to an e-mail message, batches of sales orders for the day from salespersons, local database

Figure 2.2
Videoconferencing
across the nation.

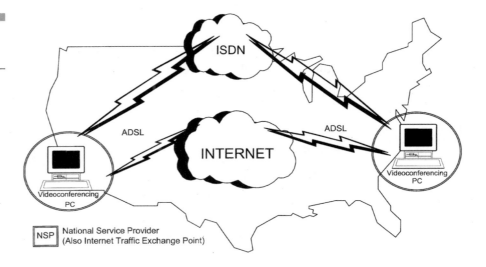

updates, changes to company policies and procedures documents sent to remote offices, sales presentations, and performance reports from the field. One common requirement that sales professionals have is to be able to update price files on a regular basis.

We should remember that in spite of improvements in compression technology, it takes a long time to transmit long multimedia files over remote access networks using 28,800 bps dial-up links. ISDN, ADSL, and cable modems are much faster and more economical. In the case of wireless networks, where you have to pay on the basis of connection time or amount of data transferred, it may cost a lot more with respect to time and money. As an example, it may cost $6.00 to $8.00 to transfer a 1-megabyte Powerpoint presentation file on a cellular network. (Note that effective throughput on cellular networks is generally 600 characters per second after you allow for call setup, protocol, and turnaround delays.) This option may still be more economical when you consider that you have to pay over $10.00 to UPS or FedEx for transferring important documents and diskettes. There are three other advantages with electronic file transfer:

1. Information is received by the ultimate user almost instantly as compared to overnight transfer involving many intermediaries inside and outside the organization.

2. The information is in digital form and can be incorporated into other documents without any transformation.

3. There are no paper- or media-handling delays.

On the other hand, you cannot indiscriminately transfer files over remote access networks because you may tie up ports at the remote access server. However, there are many situations where you can economically justify the file transfer of large files. The following examples can be easily justified even on wireless networks:

■ The transmission of news reports/photographs in time to meet printing deadlines (see Fig. 2.3)

■ The transmission of emergency diagnostic information to medical specialists (e.g., electrocardiogram information transmitted from an ambulance en route to a hospital)

■ The transmission of identification material to/from police on patrol

In other, noncritical situations, remote users should await the opportunity to access a wired switched PSTN connection before making major transfers.

Figure 2.3
File transfer remote
news coverage.
(Source: Ericsson
Mobile Data News.)

There are numerous communications software packages that provide file transfer capabilities. Most e-mail packages have built-in file transfer features. Important considerations in selecting file transfer software are compression capabilities, checkpoint restart capabilities, and support for PSTN and wireless networks of choice. If you plan to use a specific network, such as CDPD for a file transfer application, an appropriate software driver for the network would also be required.

2.2.1.4 SOFTWARE DISTRIBUTION.

The need to update application software in mobile computers is an important requirement in distributed and decentralized organizations. Therefore, electronic software distribution is quickly becoming a large business. It is different from file transfer application in several ways:

1. The software distribution process must be automatic because updating software in a large network of thousands of computers cannot be handled manually.
2. You must be able to schedule this distribution in a systematic fashion with positive confirmation of a successful transmission logged into a database.

One possible strategy is to send large software files on diskettes or CD-ROMs, and send smaller and more frequent updates by way of remote access networks. Apart from specialized software distribution packages, such as NETVIEW Distribution Manager, Xcellenet's RemoteWare can also be used for this purpose for smaller networks.

2.2.1.5 FIELD AND SALES FORCE AUTOMATION. Besides e-mail, file transfer, and database access applications, field and sales force automation (FSFA) is perhaps the next most important mobile application that is fueling the installation of remote access technology. Unlike the support and administrative of applications discussed so far, FSFA is an operational application that continues to draw the increasing attention of senior executives. In a competitive world, anything that supports sales activity gets the attention and the funding from executive management. We support this emphasis by senior management, since FSFA impacts an organization's bottom line. Indeed, FSFA is a mission-critical application. IDC, as quoted in a special mobile computing issue of *Communications Week* in 1995, expects FSFA to grow by 30 percent, from $180 million in 1994 to $435 million in 1999 (see Fig. 2.4).

Many of the FSFA applications started out as client-tracking applications. Gradually, these applications have been enhanced to allow customer service, sales analysis, sales order entry, and communications. Today, many of them contain price lists. Many of these applications are being integrated with back-end operational applications residing on LAN servers, minicomputers, and mainframe legacy systems.

Since many of its requirements are generic across the industry, FSFA applications are generally available in shrink-wrap versions with certain customization capabilities. Important aspects of FSFA are described in Table 2.3.

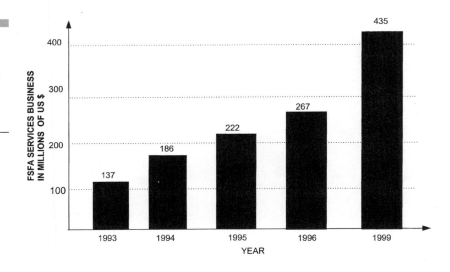

Figure 2.4
Field and sales force automation application. (Source: IDC as published in *Communications Week* [April 1995].)

TABLE 2.3

Remote Access
Applications—Field
and Sales Force
Automation

Description	■ Provides computer-based automation for sales professionals.
	■ Many of the functions of FSFA are implemented as local applications on mobile computers.
	■ Data for these applications is uploaded from central information servers.
Functions and features	■ Contact- or client-management—names, addresses, telephone #s, product portfolios, etc.
	■ E-mail and communications—enables the sending of e-mail from mobile workstations.
	■ Product catalogues and price lists.
	■ Order entry, order status inquiries, database updates.
	■ Sales reporting.
	■ Sales analysis and forecasting.
	■ Lead tracking and call tracking.
	■ Generation of custom letters to clients.
	■ Personal Information Management—expenses, commission tracking.
	■ Multimedia and sales presentations.
Vendors	■ Aurum Software's Salestrak.
	■ Brock Control's TakeControl.
	■ Modatech Maximizer.
	■ SalesKit.
	■ National Management Systems' SalesWorks.
Cost range	■ $200 to $700 per seat.
Benefits	■ Sales productivity gains.
	■ Shortened sales cycles leading to increased productivity (10—30% reported improvements).
	■ Minimization of errors as a result of order entries at source.
	■ Improvements in professional image.

TABLE 2.3 CONTINUED Remote Access Applications—Field and Sales Force Automation	Typical platforms	■ DOS, Windows 3.1, Windows 95, Windows NT, Unix, NetWare.
	Network support	■ Some vendors (e.g., Brock's TakeControl, SakesKit's SalesKit, Sales Technology's Virtual Office, Xcellenet's RemoteWare) have ARDIS, RAM, and cellular network support.
	Databases supported	■ Many single-user versions support Btrieve files or Xbase databases.
		■ Modern multiuser versions run on SQL Server, Sybase, and Oracle Informix.
	Customization	■ Most packages have capabilities to develop custom forms; some have macro capabilities.
	Compatibility with legacy systems	■ Usually capable of exchanging information with legacy systems through file transfer protocols or through database gateways or through ODBC. Internet and Java versions available with some applications.
	Special considerations	■ Client/server or workgroup versions appearing soon.
		■ FSFA applications should move as much functionality dependent on fixed data to the remote device as possible.
		■ FSFA should have interfaces into operational applications such as order entry, accounting, corporate sales analysis.
	Typical example	■ RJR Nabisco—2000 salespeople outfitted with Fujitsu Poquet PC, plus Palmtop.

SOURCE: Vendor information and author's research.

ISSUES WITH FIELD AND SALES FORCE AUTOMATION APPLICATIONS.   The following issues should be considered while selecting or developing field and sales force automation applications:

Interfaces with Back-end Operational Applications on Mainframes or Other Servers.
Note that in a typical scenario, the organization probably has previously implemented business applications that process sales data entered by inside sales staff. FSFA essentially replaces this arrangement with source order-entry by mobile salespersons. Software interfaces between the front-end FSFA applications and these backend systems should be carefully evaluated. Otherwise, software bridges with the necessary audit controls will have to be built to accept this data.

Sales-Force Automation Through Workgroup Computing Platforms Such as Lotus NOTES.
Many organizations are building FSFA applications through Lotus
NOTES. If this is a strategy adopted by your organization, you should
look into the application design issues discussed in Chap. 12.

Open Application Design.   Sales professionals are well known for their lists of
features and functions that they expect to see in future versions of business applications. More often than not, they have good reason for their
requests, based on sound business practices. Accordingly, we advise application designers to keep their functional design as open as they can, and
to choose those packages that hold greatest promise of further enrich-
ment in future versions.

AN EXAMPLE OF SALES-FORCE AUTOMATION APPLICATION.   AlliedSignal, an
industrial conglomerate located in New Jersey, has equipped its 200-plus-
strong sales force with NEC notebooks and Xcellenet's RemoteWare software. This allows for remote access of Allied's sales-assist application that
runs on Oracle database, which is linked to Sequent back-end Unix server.
The application not only assists salespeople with comparing competitive
products, but also calculates cost advantages of switching from competitive products to Allied's product line. With access to up-to-minute inventory and prices, it can respond to customer's queries, and book orders on
the spot. Sales presentations are sent to the mobile sales force on a regular
basis.

2.2.1.6 MULTIMEDIA.   Until now, sophisticated sales presentations
meant marching into boardrooms equipped with expensive projection
gadgetry. This is changing fast. Modern, high-tech sales professionals now
carry their presentations in notebooks equipped with integrated high-
performance CD-ROMs, stereo speakers, and MPEG-compatible full-
motion video. Through this new technology-in-a-briefcase, salespeople
are able to gain entrance to hard-to-get customers, such as physicians,
senior executives, and computer equipment buyers. If necessary, for bigger
audiences they can project to larger screens. With a little extra work, they
can even display up-to-the-minute information direct from the head
office, right before their customers' eyes. In our competitive world, it is the
salesperson with the sophisticated presentation that is asked to come
back—and that closes the deals. The following types of multimedia applications are being increasingly implemented on notebooks:

- Sales presentations
- Product demonstrations
- Sales catalogues

In most cases, multimedia applications are stand-alone, single-user applications that run on a new breed of mobile notebooks equipped with multimedia features. However, these sales presentations can be updated on remote access networks.

Mobile multimedia applications are becoming increasingly popular in the pharmaceutical industry. According to *PC Week*, Bristol-Myers Squibb Company of New York, has developed a multimedia application for its pharmaceutical sales staff. The application uses Oracle Corporation's Business Objects and Sybase's PowerBuilder front-end development tool. It uses a graphical user interface (GUI), featuring charts and interactive voice prompts that highlight physicians' prescription histories and provide details of customers' professional practices. Upjohn Company, in Kalamazoo, Michigan, is developing a similar application for its sales force. Shared Medical Systems Corporation of Malvern, Pennsylvania, is using a multimedia application developed by CompuDoc Inc. in Warren, New Jersey, for marketing a portable radiology machine. According to the company, as reported in *PC Week*, sales have increased 30 percent as a result.

2.2.2 Generic Horizontal Applications Requiring Extensive Customization

In this category are included those applications utilized across many different industries, but that require customization at the front end or at the back end of application systems. Database access and service representative dispatch are among the most common applications here.

2.2.2.1 DATABASE ACCESS. The ability to access information from a DBMS server on a LAN from a minicomputer or from a mainframe legacy database server is commonly requested by mobile users. Users can interact with several front-end inquiry packages (e.g., Forrest & Tree's Info-Access, Oracle's personal 2000 client software, Sybase PowerBuilder, Xcellenet's RemoteWare) to submit a query to a server database directly. If the database access protocol differs from the front-end inquiry tool supported, the query must be submitted through a gateway. The queries are custom developed for specific organizations in most cases. The back-end server software retrieves the information from the database, or acts as a gateway for the retrieval of information from a minicomputer or mainframe. (See Fig. 2.5.)

2.2.2.2 FIELD AUDIT AND INSPECTION APPLICATION.   There are many situations where head-office staff need to carry out on-site field audits. In such scenarios, field auditors can carry notebooks connected to the client-information database. The following examples illustrate this type of implementation:

- IRS and tax audits by federal or state agencies (up to 8000 agents in the United States alone are equipped with notebooks connected to IBM mainframes)
- Financial audits in the field by accounting firms such as KPMG, Deloitte & Touche, and Ernst & Young
- Site inspections by building permit inspectors
- Environmental control inspections
- Automobile insurance adjusters

At the end of day or during the day, data collected in the field is uploaded to the central computer by remote access. Figure 2.6 shows a photograph of the use of pen-based computers by an automobile insurance adjuster.

Figure 2.5
Mobile database access market growth. (Source: Yankee Group as published in *Communications Week*.)

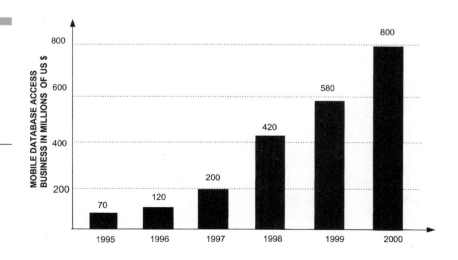

Figure 2.6
Insurance agent
application. (Source:
Telxon.)

2.2.3 Vertical Applications—Industry-Specific

In this category, we include applications that are specific to industries such as insurance, banking, government, utilities, transportation, and so forth. Usually there are business-process characteristics unique to a particular industry that make certain vertical applications inherently suitable for remote access networks. As a result, some vendors have developed turnkey solutions.

2.2.3.1 FINANCIAL INDUSTRY—INSURANCE AND FINANCIAL PLANNING. The insurance and mutual fund industries have begun equipping their sales agents with notebooks that can display and analyze current product and price information right before their customers' eyes. (See Fig. 2.7.) Sales illustrations can be created, printed, and presented to customers on the spot. With a modem connection, additional information stored in central databases can be retrieved from the client's office or home. The overall effect is a measurable increase in sales productivity.

In an extensive business-process reengineering, Merrill Lynch is spending over $500 million to equip 25,000 agents with Pentium-class multimedia notebooks connected to Unix servers running Sybase DBMS. The individual notebooks are loaded with Windows NT and a financial planning application, called *Trusted Global Advisor.* The software features new analytical and presentation tools that help customers understand different investment scenarios more easily. One of the key features of the new application is the ability to integrate customer data residing in multiple accounts into a single image and to provide reports in an ad hoc fashion as needed. The notebooks will make Merrill Lynch's financial consultants fully mobile, freeing them to visit their customers' offices or homes.

Figure 2.7
Insurance adjuster.
(Source: Telxon.)

2.2.3.2 THE RETAIL INDUSTRY. The retail industry is especially suited for remote access because of the decentralized nature of the business. Most retail stores or franchise outlets operate relatively independently without a permanent real-time connection to the central computers, except to banks for credit and debit authorization. However, they must receive pricing updates, transmit sales, and inventory data at the end of day. The following examples are illustrative of two different customers using remote access effectively:

REMOTE ACCESS FOR RETAIL STORE CONTROL.* Mervyn of California offers trend-right apparel in over 270 stores. Controlling costs is extremely important in a competitive industry like fashion apparel.

---

*Information from Xcellenet application profile.

Mervyn is using Xcellenet's RemoteWare software with satellite communication between the stores and two central servers. Each of the two servers communicates with up to 32 nodes simultaneously. Headquarters transmits 1300 to 1500 files to all stores through an average of three to five communications sessions with each store. Each session lasts 10 to 12 minutes. Significant cost savings have been achieved as a result of this implementation. RemoteWare is used for:

- Electronic mail

- Development of store displays in consultation with individual stores

- Development, distribution, and collection of survey forms

- Distribution of operating procedures and manuals to stores

REMOTE ACCESS FOR FRANCHISE OUTLET CONTROL.   Mail Boxes Etc.® franchisees receive, process, and deliver millions of packages, letters, and faxes for customers every year. Tracking all those transactions, and exchanging information between franchisees and the company was quite difficult. Now they are using remote access based on RemoteWare from Xcellenet to manage this exchange of information. (See Fig. 2.8.)

Remote access is used to transmit information for various applications, including processing packages, controlling mailbox rentals, providing centralized billing to multisite accounts, and e-mail.

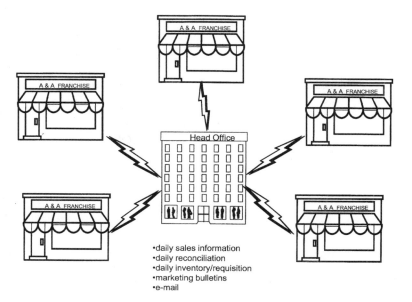

Figure 2.8
Remote access for retail franchise control.

2.2.3.3 HEALTHCARE APPLICATIONS.

Though physicians tend to have a reputation for resisting the move toward computers, they seem to be intrigued by and interested in mobile computing that involves untethered (i.e., wireless) connections and specially designed portable devices. They like the idea of calling up patient records, researching diseases, and ordering prescriptions from anywhere in the hospital or home.

Summit Health Services, in Greenburgh, Pennsylvania, has developed MobileNurse, software that enables home-care nurses to access inventory, call up patient records and Medicare information on wireless notebooks, and compare notes with other caregivers. MobileNurse uses the Mobitex Ram Mobile Data network to access a centralized database. Grace Hospital uses IBM system for a similar application.

Additionally, doctors are increasingly using handheld devices such as Newton and pen-based Palm Pads connected to hospital LANs through wireless PCMCIA cards to record patient records. As handwriting-recognition software improves, the use of this technology will grow.

Because of the high cost of physician services, business cases for the use of mobile computing and remote access are relatively easy to build. Even small increases in physician productivity or diagnostic accuracy translate into savings sufficient to justify the expense of remote access technology solutions. With the current all-pervasive squeeze on healthcare budgets everywhere, tremendous attention is being paid to the ability of technology to reduce costs.

TELEDIAGNOSTICS. In certain cases, critical patient data like EKG and MRI need to be transmitted from an ambulance to a hospital or from a small hospital to a medical specialist at a larger hospital. In one such implementation, Systems Guidance Ltd., a computer software and hardware supplier in the United Kingdom, has developed a remote patient monitoring system (RPMS). Vital information—including EKG waveform, pulse oximetry, heart rate, and blood pressure—is captured either at the scene or en route to the hospital. In a matter of seconds, it is then transmitted in real time to a hospital emergency room, over the Mobitex network. The information is monitored by the consulting doctor, who can make an immediate diagnosis and transmit advice back to the ambulance.*

EMERGENCY MEDICAL SERVICES. Quite often, computer technology needs to be able to follow a medical worker out into the emergency room or a patient's rehabilitation room after surgery. In these situations, pen-based

---

*See Ericsson's Mobile Data News (issue 5/95) for further details.

computing can be used to provide a solution which meets the stringent requirements of the medical profession. Westech Information Systems, Inc., a Vancouver, British Columbia, company has developed Emergency Medical Services (EMS), an application that covers the full gamut of activities—from the time an emergency call is dispatched through to the arrival of the patient in the emergency room. Vancouver General Hospital, British Columbia Service (BCS), and 33 other customers in North America are using Westech's application.

The Westech application has three objectives:

- To increase the accuracy and completeness of patient data
- To enable data capture at the source (that is, at the time and the location of the emergency)
- To improve the sharing of patient information among caregivers

On-the-Spot Data Capture.   The access to accurate and timely information is a critical part of delivering quality emergency care. Westech's EMS makes it possible for paramedic crews to record important trauma information at the scene, for later use by emergency room specialists. This information includes:

- Patient demographics, including name, address, birth date, and so forth
- Patient condition, including state of consciousness, central nervous system, and cardiovascular system; condition of head, neck, chest; amount of blood lost; and so forth
- Medical history in light of the current injury or illness, including past history, current medications, known allergies, and so forth
- Vital signs, including blood pressure, pulse, respiration, temperature, and so forth
- Treatment provided, including IVs, medications, oxygen administered; transportation details
- Trauma and scene details
- Information about the crew and the particular call, including information about expenses

Upon arriving at the hospital, the trauma team in the emergency room also uses a pen-based system to record vital signs and other medical information at the patient's bedside. The information from the paramedics is downloaded to an emergency room computer, which then displays both sets of data simultaneously to all members of the team via an overhead monitor.

Benefits of Westech's EMS.   It replaces several paper forms that previously made finding and entering information difficult. It takes less time to enter all the information, which is entered only once, at the source. Data can be uploaded for billing and other needs, such as the production of legal reports.

Technology Used.   The graphical user interface was developed with Pen-Right. It was originally developed on GRiD™ Convertible and PalmPad SL pen computers. Westech also supports IBM 730T and Fujitsu 325 pen computers as platforms of choice.

2.2.3.4 PUBLIC SECTOR—MISCELLANEOUS APPLICATIONS.
There are many other applications of remote access in the public sector. The following sample applications are based on pen computers: In these applications, information is uploaded to a central information server either through telephone dial-up or wireless network access.

The Department of Defense is using wireless-enhanced Apple Newton MessagePads to access patient information (medical histories and patient records), document lab test results, and schedule consultant visits.

Two New Jersey companies (CDP, Inc. and Computers at Work) have developed an application for parking control officers. The system is based on Norand PEN*KEY 6100 handheld ruggedized pen computers. Major advantages include improved timeliness and accuracy of ticket writing.

The City of Richmond Hill, in Ontario, has automated time reporting for its employees. The city's crew supervisors use Fujitsu's pen-based PoquetPads and custom forms created with the PenRight development environment.

Booz-Allen is field testing an interesting solution for U.S. Army infantry patrol soldiers. Currently, soldiers must fill out a coded situation report on paper. This has been replaced by an electronic form on Newton MessagePad 120s.

UTILITIES.   Utility companies are using remote access technology for collecting and updating inventory information about field assets such as transmission towers, transformers, and the like, which is then uploaded to head office computers. Previously, this information was brought back from field trips in paper form and only then entered into computers. On-the-spot data entry updates information in databases more quickly, reduces errors, and improves productivity.

As well, field technicians and project staff are now able to stay in touch with head offices through e-mail and messaging applications, while CAD applications are used to dispatch service representatives to repair sites.

A good example of a utility company exploiting remote access technology is the New Jersey utility, PSE&G. They expect to save millions of dollars by equipping 750 field-service workers with pen-based computers that send and receive information via a wireless network. The new network allows field workers to receive real-time information on customers' orders, repair requests, and existing maintenance contracts. Once a job is completed, details are entered into a pen-based computer, and fed in real time to the main computer. The dispatch system is based on a client/server paradigm.

Boston Edition's sales force has implemented e-mail that allows use from anywhere in the metropolitan area, thereby improving the efficiency and timeliness of communication.

TRANSPORTATION.   The transportation industry was an early adopter of remote access technology. Due to the high capital and operating costs of long-haul delivery and pickup trucks, the industry was easily able to justify the costs of new technology and wireless networks, as long as they were able to show that their implementation would result in even a small increase in productivity. Planners needed only to point out that CAD information about emergency drop shipments or en route pickups could achieve significant savings to warrant the implementation of remote access technology solutions. Additionally, the ability to indicate exact delivery times made for excellent customer service that enabled appreciative customers to better plan their day.

Special ruggedized terminals have been developed by companies such as QUALCOMM/CANCOM group (operating in the United States and Canada). These terminals, which utilize satellite-based communication, can send and receive messages to and from long-haul trucks on the highway. According to the company, dozens of freight carriers have installed the equipment. For a fee, CANCOM provides a bundled one-stop service for a fixed number of messages per month. (See Fig. 2.9.)

J.B. Hunt Transport Inc., in the United States, has already installed a customized tablet computer, built by IBM with a touch-screen interface and satellite communication. The mobile tablets are connected to 13 Intel-based servers at regional sites. These, in turn, are connected to an IBM 9000 mainframe. QUALCOMM Inc. and Rockwell Corp. were also involved in this project. Each computer contains a number of databases that include locations of fuel stops, safety information, and other travel-related information. Engine information regarding idle time, overspeed, overwind, and other features that require routine adjustments are also recorded. The project is already showing major returns. The company has

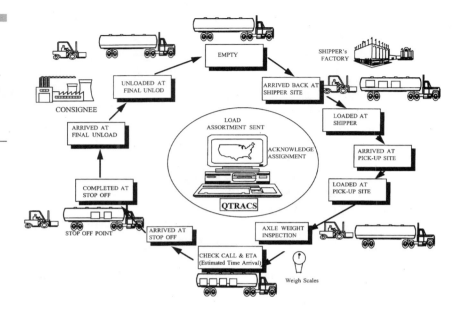

Figure 2.9
QUALCOMM's QTRACS application for fleet communication. (Source: QUALCOMM software description brochure.)

achieved improved customer service, more miles per day per truck, more satisfied drivers, and, most important, more efficient fleet managers and more efficiently managed fleets. In the past, fleet managers used to spend most of their time in just phoning truck drivers.

2.2.3.5 THE HOSPITALITY INDUSTRY. All large hotel chains have teams of salespersons on the road selling meeting and convention space. A few technology-innovative chains have adopted remote access technology solutions to help smaller sales forces sell more space. IT Sheraton is one such company. About 100 sales staff received 75-MHz 486-based AT&T Globalyst notebooks, with 340 MB or 540 MB hard drives loaded with a Windows 3.1 application suite that includes WordPerfect 6.1, Lotus 1-2-3, and FreeLance Graphics. Communication software is based on pcAnywhere from Symantec Corporation. Also included is Delphi hotel reservation software. Each salesperson's home office also was supplied with an HP OfficeJet—a multifunction device with the capabilities of a printer, a fax, and a copier combined in one unit. Thus equipped, the sales force can now dial into the Delphi system at the main office and make reservations directly from the field. This arrangement incorporates elements of telecommuting, working from remote offices, working from home offices, and remote access technology. Among the benefits, ITT reports costs savings, productivity gains, and better customer service.

Summary

In this chapter, we reviewed some of the major applications of remote access technology. Both horizontal and vertical industry applications were discussed. E-mail continues to be the most prevalent horizontal remote access application because it is an essential requirement of a mobile worker's life. File transfer is a common requirement for remote access, whether it is for software updates or for transferring sales information at the end of the day. Videoconferencing is emerging as another application that will become affordable and viable with new ADSL and cable modems. Sales force automation and database inquiry are growing as remote access applications. On the other hand, vertical applications such as those built by QualComm for the transportation industry and Telxon in the manufacturing arena have been extremely successful because vendors have provided complete solutions to users.

The variety of applications described in this chapter demonstrates the reach of remote access into almost every industry. Business depends on timely information reaching the field personnel. That creates a natural demand for remote access. With tomorrow's high-bandwidth networks, Internet, and groupware applications, remoteness will not be based on the physical distance but on the inverse ratio of the speed and capacity of remote access links. Then all desktop applications will become remote access applications.

CHAPTER 3

The Business Case for Remote Access

An emerging technology may have a lot of value for the scientist who invents it, but it has very little value for the business person, unless its economic benefits are more than the costs associated with its implementation.

—*Chander Dhawan*

About This Chapter

Our discussion in Chap. 2 gave us an idea of the type of business applications that early implementors of remote access have implemented. In our discussion of these applications, we emphasized that there was a strong demand from the user community for remote access of desktop applications. Some organizations have gone ahead with remote access purely on the basis of user demand; others have done some analysis of benefits and costs. This chapter is about the formal methodology of developing a business case. We believe the analysis of a project is a fundamental part of justifying the project and proceeding with it, on the assumption that senior management supports it.

Information in this chapter will help us in building a preliminary business case for an application of our own. First, however, we need to ascertain the preliminary costs and benefits of the application; since the estimates are preliminary in nature, they will have to be revised as more accurate data becomes available.

3.1 Methodology for Developing a Business Case

Figure 3.1 shows a flowchart of the various steps involved in developing a business case. This flowchart is based on the premise that the development of a business case is an iterative process, with each subsequent iteration producing a more refined scenario. The preliminary case can be based on a rough technical design, and rule-of-thumb costs and benefits, obtained from user executives and business process analysts. A pilot project should be created to simulate the new mobile environment, so that measurements can be made for improved productivity. The data should then be modified as more precise cost information is made available by vendors and the application's environment is implemented. At each stage, a go/no-go decision can be made that results either in revision or proceeding on to the next step.

3.1.1 Quantifying Unquantifiable Benefits

Traditional business school return on investment (ROI) analysis can deal with many of the tangible costs and benefits (e.g., productivity improvements). ROI analysis does not, however, deal effectively with intangibles, such as risk assessment, and benefits, such as improved business processes,

Figure 3.1
Methodology for
developing business
case for remote
access.

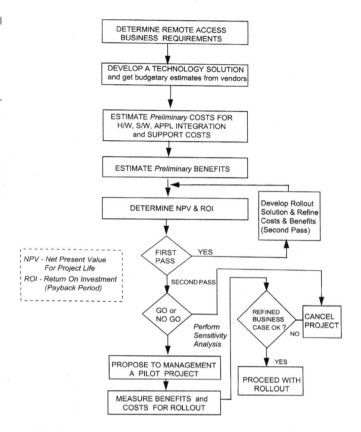

superior customer service, or increased competitiveness. Increases in market share as a result of such benefits are, therefore, quite difficult to predict.

We intend to propose a methodology—based on the emerging field of information economics—which attempts to quantify, by a group consensus process and the assignment of probabilities, many of the more intangible costs and benefits. *We strongly believe that quantification estimates by experienced business executives lead to decisions that are inherently better than those based purely on the intuition of a decision maker.*

The discipline of information economics should be used to identify intangible costs (or risks) and benefits. This is done by assigning weights and then adjusting total cost according to these weights.

3.1.2 Enhanced Weight Assignment Scheme

The analysis can be further enhanced by a weight multiplier which takes into account the degree of support of a specific factor being analyzed.

Several management consulting companies have devised different methodologies for this type of analysis. They use the same principles, but have implemented them differently. The author has used a scheme based on the conversion of weighted scores into dollar values. This is done by determining the relative weights of tangible and intangible benefits, and then, once the dollar values of the tangible benefits are known, using the relative weighting to determine the dollar values of the intangible benefits. Oracle Corporation has developed a similar scheme in its CB-90 financial evaluation model. The reader may obtain the CB-90 model from Oracle Corporation for this enhanced analysis.

3.1.3 Overall Process for Evaluating Tangible and Nontangible Costs and Benefits

1. Divide the costs and benefits into the following categories:
   - Tangible costs
   - Intangible costs/risks
   - Tangible benefits
   - Intangible benefits

2. Determine both one-time (i.e., capital) costs, as well as ongoing (i.e., operational) costs for the entire life cycle of the project. Calculate NPV (net present value) for the entire project.

3. Define the groups of major stakeholders in the project (e.g., business-user managers, IT managers, accounting staff, corporate financial managers, etc.), who would be affected by an investment in remote access technology.

4. Have the groups assign weights to each of the categories, with a total for all the categories not exceeding 100. These weights should be discussed, debated, and agreed upon by the groups. Once agreement has been reached, it will reflect the combined wisdom and judgment of the group.

5. Come to a consensus on intangible benefits and risks. Ask the groups to assign scores between 0 and 10 which indicate their estimation of the degree to which these risks and benefits will be realized.

6. Multiply each score by the previously assigned weight, sum the resulting numbers, and divide the total by 10. This will give a relative cost or benefit number on a scale of 100.

7. Determine the return on investment (ROI) by using conventional means acceptable to the organization, but only after adjusting the tangible costs and benefits by the weights assigned by the group to the intangible factors.

Figure 3.2 shows a sample spreadsheet based on the preceding process.

3.2 Costs

We must consider tangible costs, as well as intangible costs. Tangible costs can be identified, quantified, and estimated more easily. The intangible costs are more difficult to quantify in economic terms. As a result, project sponsors tend not to pay adequate attention to them. Nevertheless, we strongly recommend that intangible costs be evaluated carefully. It is precisely because they require so much more effort to analyze that they, in particular, tend to concern senior management.

Figure 3.2

Remote access business case.

	WEIGHT	SCORE (Consensus)	Weighted Score	Year 1	Year 2	Year 3	Year 4	Year 5
1. COSTS								
1.1 Tangible Costs	60							
1.1.1 One-time Capital								
- Hardware,Software				$500,000	$250,000			
- Development				$350,000	$100,000			
- Training				$50,000	$100,000			
- Miscellaneous				$250,000	$100,000			
1.2 On-going								
- Network				$25,000	$100,000			
- Maintenance				$25,000	$50,000	$100,000	$120,000	$120,000
- Miscellaneous				$25,000	$60,000	$60,000	$60,000	$60,000
Total Tangible Costs				$1,225,000	$760,000	$160,000	$180,000	$180,000
1.3 Intangible Costs/Risks								
Resistence to Change by Users	10	6	6					
Bleeding Edge Technology	20	5	10					
Poor response Time On Road	10	4	4					
Total Assigned Score	40		20					
Costs Assigned to Intangible Factors				$408,333	$253,333	$53,333	$60,000	$60,000
Total adjusted Costs				$1,633,333	$1,013,333	$213,333	$240,000	$240,000
2.0 BENEFITS								
2.1 Tangible Benefits	35							
- Headcount savings				$0	$200,000	$400,000	$600,000	$600,000
- Other Tangible savings				$0	$100,000	$300,000	$300,000	$300,000
- Total Tangible Benefis				$0	$300,000	$700,000	$900,000	$900,000
2.2 Intangible Benefits								
- Increase in Marketshare	20	8	16					
- Improved Customer Service	15	9	13.5					
- Future Competiveness	20	6	12					
- Misc Benefits	10	8	8					
Weighted Value of Intangible Benefits	65		49.5					
Total Adjusted Benefits				$0	$424,286	$990,000	$1,272,857	$1,272,857
Net Benefits over Costs				$0	$724,286	$1,690,000	$2,172,857	$2,172,857
				($1,633,333)	($289,048)	$1,476,667	$1,932,857	$1,932,857
NPV of Costs over 5 years	$2,795,539							
NPV of Benefits over 5 years	$4,701,569							
ROI								
EXPLANATION								
1. Weighted Score=(Weight*Score/10)								
2. In this example,intangible benefits (wt=65) are more important than tangible (wt=35) benefits.								

3.2.1 Tangible Costs

Although these costs are relatively easy to quantify, several design options should be considered to find an optimal solution. The spreadsheet in Fig. 3.3 shows component costs and benefits in a typical business case of an information technology project. This can be customized for a remote access project. We would like to recommend that the following pointers be considered while undertaking a high-level analysis of this kind.

1. In many remote access projects, network costs are among the most important cost items. Appropriate time and attention should be paid to the type of network selected. Also, reliable estimates should be made of the usage patterns and the number of data packets. Remember: if the application is properly designed, network traffic might ultimately be higher than estimated in the initial phases. Second, the user transaction packets, as well as the protocol packets and the system overhead must be counted. The network provider can assist you in this estimation process.

2. Depending on the area of coverage required, you should consider both public shared and private network options, especially if you are in a relatively small geographic area. If your coverage area is small, your data traffic is on the high side, and you already have a spectrum license, a private network may be cheaper if capitalized over a longer life cycle.

3. Do not ignore the cost of making changes to applications to make them mobile-aware.

4. Do not underestimate user-training requirements, especially if the users are not familiar with the application and the equipment.

5. Allow sufficient resources for comprehensive technical support. Remember that mobile users tend to always be in a hurry, and therefore, are not willing to tolerate long delays or to hold patiently for the *next available representative*.

6. Provide adequate estimates for systems-integration costs, both internal and external. The business case is normally built before signing contracts with a vendor. Get budgetary estimates and do not be surprised when you get higher cost estimates at the time of actual proposal or response to an RFP. Vendors are known for giving lower preliminary estimates.

7. Logistics costs for the permanent installation of ruggedized notebook computers in vehicles are quite significant. They can be as high as $500 to $1000 per vehicle, depending on the type of mount, any special cabinetry required to accommodate dual-air-bag constraints, and other health and safety considerations, especially in police vehicles. Include these costs after discussion with the appropriate installation contractors.

REMOTE ACCESS PROJECT - Sales Force Automation	ASSUMPTIONS FOR COSTING	1997/98 PLAN $	1998/99 PLAN $	1999/2000 PLAN $	2000/2001 PLAN $	TOTAL
Staff Expenses - Salaries and Expenses	Internal Human Resource Costs (Provide details in a separate sheet - categories numbers and $$)	240,000	540,000	430,000	350,000	1,560,000
Staff Benefits	Assume 25 % of the above	60,000	135,000	107,500	87,500	390,000
Total Additional Payroll		300,000	675,000	537,500	437,500	1,950,000
	FOR PHARMACEUTICAL DISTRIBUTION COMPANY (1000 Mobile Sales Professionals) *(Gradual phased implementation with a 3 year roll-out)*					
Capital Expenditure						
A.1 Hardware						
- Mobile Workstation Hardware Acquisitions	# of Notebook Computers=200, 400 and 400 in 3 years; 200 in pilot	100,000	800,000	1,400,000	1,200,000	3,500,000
- Radio Modems	Decreasing Costs For Hardware assumed over 3 years		100,000	160,000	140,000	400,000
- Remote Access Server hardware	High-end Multimedia Computers - 4000, 3500, 3000, PSTN/Wireless Modems @ 500, 400,350		650,000	125,000	125,000	900,000
A.2 Software						
- Client workstation software	Workstation software like Windows, TCP/IP, ARDIS drivers, etc.@ 300 per workstation		60,000	120,000	120,000	300,000
- Software for RAS, network management, etc.						
A.3 Software Development						
- Application Development	Developing Sales Force Automation Custom Interface	300,000	400,000	100,000	100,000	900,000
- Communications Software Changes	Changes to Communications interface for Wireless Network - TCP/IP Interface					
- Application Software Enhancements	Backend System Changes plus making Applications "Mobile-aware"					
A.4 External Consulting	Specialized technical expertise, Remote and Mobile Computing architecture Development	200,000	200,000	200,000	0	600,000
A.5 Training	$ 400 per professional; plus initial training development		400,000	0	100,000	500,000
A.6 Installation						
- Installation and user setup	$ 300 per sales professional decreasing to 200 ; includes customization of individual workstation		500,000	225,000	225,000	950,000
A.7 Miscellaneous - Describe	For errors and ommissions & misc. items e.g. Server HW/SW upgrade		60,000	120,000	100,000	280,000
A.8 Contingency	$ 150,000 Specialized Services + 100000 For HW+100000 For installation		500,000	250,000	250,000	1,000,000
A.9 Initial R&D - Pilot Installation -20 Workstations	No allowance for Enhancing Servers in Back and Mainframes & Minis etc.		350,000	0	0	350,000
Sub-total one time		600,000	4,020,000	2,700,000	2,360,000	9,680,000
B. Ongoing Operations						
- Telecommunications Network Services	200/user/month going down to 150 in 2nd & 100 in 3rd year		480,000	1,080,000	1,200,000	2,760,000
- Computer processing	Net additional costs only included;		200,000	500,000	700,000	1,400,000
- Maintenance (Hardware/software/equipment)	$300/in Car WS/yr on-site, 100K/yr for MCSS - 12% of HW (assume 50 % only for current year because of gradual installation)		210,000	430,000	550,000	1,190,000
- Technical Support (outsourced from vendor)	Outsourced from vendor		120,000	360,000	600,000	1,080,000
- Ommissions, Errors, Contingency	Assumed at 15%		178,235	418,235	538,235	1,134,706
Sub-total on-going		0	1,188,235	2,788,235	3,588,235	7,564,706
Total Capital & Ongoing (A+B)		600,000	5,208,235	5,488,235	5,948,235	17,244,706
TOTAL PROJECT		900,000	5,883,235	6,025,735	6,385,735	19,194,706

Figure 3.3

Sample cost spreadsheet for a remote access project (sales force automation application).

8. Include an appropriate amount for contingency costs. A figure in the range of 20 to 25 percent of the initial budget is not uncommon to take unsophisticated cost estimates into account, especially in the first phase of the business case.

3.2.2 Intangible Costs

Because remote access is an emerging technology, it is subject to the risks inherent in all new technologies. The business case should include the following intangible costs, which are either difficult to quantify or are non-quantifiable:

The multiple network hops and slower supported speeds, associated with switched wired networks and wireless networks, result in a generally poorer performance than that obtained from private leased-line WANs or LANs. This can give rise to a degree of dissatisfaction on the part of users who expect the same level of performance outside the office as they get inside the office.

Mobile workers interface with many different application-servers on a multiplicity of platforms. The overall cost of interfacing in such environments is difficult to assess, especially in the early phases of a business case development. We suggest that the business case should be increasingly refined as more accurate information becomes available.

The cost of converting from existing business processes to reengineered ones is also difficult to estimate. The method suggested in Sec. 3.1 should be used to estimate these costs.

3.3 Benefits

The benefits can be classified into two categories—tangible benefits and intangible benefits. Tangible benefits can be quantified more easily than the intangible variety. We discuss both types briefly.

3.3.1 Tangible Benefits

The saving that results from staff reductions is probably the most obvious economic benefit associated with a remote access solution. Remote access can lead to increased individual productivity, increased sales per salesperson, more service calls per repair person, less time spent by professionals on administrative work, and decreased travel, all of which can ultimately translate into a reduction in total staff required.

However, there are several other tangible benefits associated with remote access solutions. A higher order-fill ratio as a result of accessing real-time inventory information at the time an order is submitted can translate into reduced inventory costs. On-the-spot invoice production in service vehicles can lead to shorter payment cycles and better cash flow. The electronic citation/ticketing applications with credit card payment of traffic violations, that public safety agencies are experimenting with, can lead to a higher ratio of paid fines.

Table 3.1 shows a range of potential benefits that can all be used to justify remote access applications.

TABLE 3.1

Nature of Benefits from Remote Access

Remote Access Application	Type of Benefits (Tangible and Intangible)
Formula for quantification	$ Savings = Number of users * $ bundled wages * % productivity improvement + $ increase in sales revenue * profit margin/unit of sales revenue (as a result of superior customer service) + decrease in inventory costs * interest rate
Sales automation	▪ Shorter sales cycle; increased sales per sales person—reduced head count ▪ Elimination of order-entry staff; more accurate data in database ▪ Better customer service—reduced merchandise return ▪ Lower inventory costs ▪ Increased market share
Electronic mail	▪ Less time spent calling the office for mail messages
Computer aided dispatch for service representatives	▪ Increased number of service calls per day ▪ On-the-spot invoicing—faster payment cycle ▪ Electronic dispatch of parts—reduced administrative and inventory costs ▪ Improved customer service
Health care industry	▪ Less administration for doctors ▪ Better patient care ▪ Telediagnostics—faster diagnosis and reduced physician costs
Accident/collision data systems	▪ Real-time entry of accident data avoids subsequent duplication of effort ▪ Information available to police and insurance simultaneously

Remote Access Application	Type of Benefits (Tangible and Intangible)
Taxi dispatch	■ Less cruise time and more trips ■ Credit card authorization leads to improved service and more customers ■ Reduced dispatch personnel costs
Financial industry (insurance)	■ Reduced selling cycle ■ Superior customer follow-up and customer retention ■ Higher dollar sales per sales presentation
Financial industry—(stock trading)	■ Faster trades—more trades per hour ■ More accurate trading
Retail industry	■ Faster service during seasonal sales ■ Reduced electrical and wiring costs ■ Improved customer service ■ Automated vending machines with credit authorization lead to higher sales
Airline industry	■ Better customer service leads to higher sales ■ Fewer missing baggage complaints—reduced tracing costs ■ More accurate maintenance data leads to safer plane flights ■ Electronic dispatch of parts results in less time in grounding of planes under service
Manufacturing	■ Lower wiring costs as a result of wireless LANs ■ Accurate inventory control ■ More accurate production tracking
General office applications (public sector)	■ Better document control ■ Reduced administrative staff requirements ■ Timely access to bills under debate in the legislature
News communications (sports, conferences)	■ Less travel and higher productivity ■ Real-time scores fed to media—score competitive advantage

3.3.2 Intangible Benefits

Many of the applications of remote access involve automating sales, improving customer service, or gaining a competitive advantage—all benefits that tend to be difficult to quantify. The project team should outline these benefits with as much detail and as specifically as possible. The group should then translate these benefits into percent increases in sales, market shares, and productivity improvements. If the benefit cannot be quantified with a high degree of reliability (we suggest above 70 percent), it should be handled with weights and scores as described in Sec. 3.1.

3.4 Return on Investments

Once costs and benefits have been identified, return on investment should be calculated using any method that the organization condones or uses. Popular spreadsheet packages like Lotus 1-2-3 and Microsoft Excel have formulae for this purpose.

3.5 Industry Experience of Return on Investment

Management consulting companies, such as Gartner, have analyzed business cases for remote access for several different applications. Their research studies indicate that the payback period with remote access applications can be as low as 18 months. Customers, such as Liebert and Bank of America, have indicated that they have been able to achieve up to 20 percent productivity improvement with computer-aided dispatch of service representatives.

Summary

In this chapter, we studied a methodology for developing a business case for a remote access project. We discussed the many nonquantifiable benefits of remote access that are either difficult or outright impossible to recognize by conventional cost/benefit analysis, and proposed a methodology to deal with them based on the assignment of weights for these factors. We looked at cost components and used a spreadsheet example for illustrating these components. Finally, some of the common tangible benefits that may be derived from remote access applications were enumerated in a table. Information such as this should give you ideas for listing the costs and identifying the benefits of a specific application, and can lead to the building of a business case for a project.

PART 2

Remote Network Access Technology Solutions and Architecture

4

How Does Remote Access Work?—Different Approaches

The current incarnation of remote access technology simply extends a LAN to remote devices through PSTN, ADSL, cable, or the Internet. It achieves this through software tricks and firmware wizardry in a remote access server. Their objective is to provide transparency to LAN applications from remote devices. They succeed to a reasonable extent.

—Chander Dhawan

About This Part and Chapter

In Part 1 of this book we reviewed at a high level the remote access scene, the overwhelming demand for remote access solutions, business applications, and a methodology for developing a business case. So far, we have not explained how this technology works. After all, why should anyone worry about a piece of technology that does not help us in our jobs and has no economic justification in our business? We shall leave that worry to the research and development community. As technology practitioners, we are interested in utilizing this technology to improve our competitiveness.

Assuming that there is a business case, we should try to understand how this technology works, and what kind of architectural framework your solution should have—this is what we intend to do in Part 2 of the book. First, we need to understand remote access requirements. Then we review four basic approaches to remote access solutions. We explain the remote bulletin board system (BBS) or terminal emulation. After understanding BBS, we next describe remote control, remote node, and mail gateways. We also compare these approaches. Finally, we describe some product examples from major vendors in this industry.

4.1 Basic Technology Behind Remote Access

In this book, we have used the term *remote access* to define a physical and logical connectivity session between an end-user device and an information server, so that the remote user can either retrieve information from the information server or send information to it. While the end-user device usually initiates the connection, it is possible for the central information server to initiate the call, as well. Of course, the remote device must be either powered on or have a remote power-on capability (like IBM's APTIVA line of PCs) to complete the connection. While a majority of remote access connections are between a stand-alone PC and a LAN-based information server or a mail server, the following scenarios constitute other common variations of remote access:

■ An end user wants to communicate with an application or DBMS server on a minicomputer that is either resident on the same LAN as the communications server or another LAN, the two LANs being interconnected through a wide area network.

- An end user wants to connect to an application or DBMS server on a mainframe that is remote from it. Since this mainframe may not be on the LAN, and have a different communications and/or application presentation protocol, we would need a gateway.
- An end user on a LAN wants to dial out to an external network service (e.g., Internet)
- An end user utilizes his/her connection to Internet to access an information server that resides on the Internet or is a part of an Intranet configuration belonging to the organization.

In all of these scenarios, except the last one, we need to connect a remote computer to a LAN. These scenarios are shown in Fig. 4.1.

4.1.1 Understanding the Basic Remote Access Problem and Solution

Connecting a remote terminal or computer to another computer was a common requirement for networking in a Unix setting. Multiple terminals were connected to a terminal server, which in turn was connected to the Unix computer, either directly or through an Ethernet connection. When PCs were used as terminals, they had to emulate dumb terminal devices. (What an old-style, undemocratic thinking—the master does not

Figure 4.1
(a) Basic remote access scenario—remote client to LAN information server; (b) Remote access scenario for mini- or mainframe access.

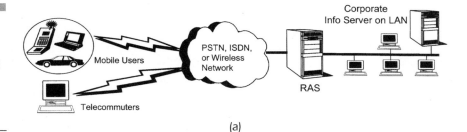

(a)

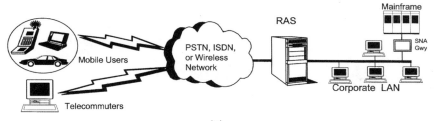

(b)

Figure 4.1
Continued
(c) Remote user
accessing Internet
from corporate LAN;
(d) Remote users
using Internet to
access Intranet server
on LAN.

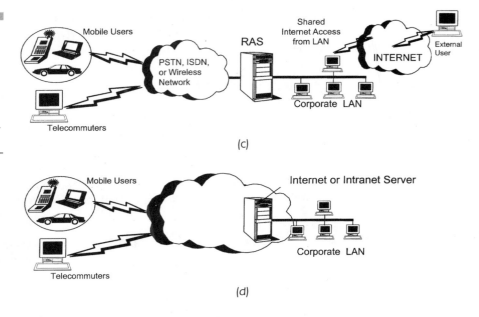

know how to deal with intelligent subordinates and starts treating them as dumb?) However, for logical connectivity purposes, the Unix world invented point-to-point protocol (PPP) and serial line Internet protocol (SLIP). Most Unix workstations and intelligent terminals used this method for connecting to a multiuser server in a star arrangement where terminal servers were at the outer end of a star, and the Unix multiuser computer was at the center. A LAN with intelligent PCs as clients and server, where files, applications, and data are stored, works somewhat differently. It uses the same physical medium for transport as all other client workstations. What the remote access industry has done is to retrofit a Unix terminal connectivity protocol to remote access between remote PCs and LAN-based servers. So what began in the mid-1980s as a way for dumb terminals to share a single Ethernet terminal LAN interface has now evolved into a feature-rich set of products that connects a remote user to an organization's LAN backbone. What had been a fixed function terminal server has become a more functional communications server of the remote access world. That is a simplistic view of this RNA world.

4.1.1.1 CLARIFYING THE TERMINOLOGY—RACS, RAS, OR MCSS. In this book, we call this server by various acronyms—RACS for remote access communications server (our preferred terminology) or RAS for remote access server (industry or trade publications' common terminology). We shall use RACS and RAS interchangeably. In the wireless world, the server may do switching, as well. Therefore we call it MCSS (mobile communications server switch).

The most common requirement for remote access is that remote users have to dial in and connect to various information servers on a LAN, such as Novell NetWare or NT server. Ideally, they would like to use the same graphical user interface as if they were attached to the LAN. Internet access for e-mail and Web browsing are other common requirements.

In order to achieve this remote connection, the end user needs some sort of client software that uses PPP or SLIP as the core communications protocol. At the remote access communications server (RACS) end (i.e., on the LAN), you need the following additional functionality:

- Provide corresponding PPP or SLIP support on the LAN
- Manage a modem pool that supports multiple serial ports
- Allocate and de-allocate ports to different remote users
- Security
- Support for multiple network links, such as ISDN, ADSL, or cable modem
- Buffering between slow-speed serial link (typically at 28,800 bps) and faster-speed LAN (10/100 Mbps for Ethernet or 16 Mbps Token Ring)
- Establish communication with application and database servers on behalf of the remote user
- Emulate a remote workstation device as a locally attached LAN workstation device for interactive applications that assume a LAN-attached device
- Operate under the control of the LAN network operating system (NOS)

This is precisely what RACS does, as a minimum. We see in the next section how various components of hardware and software provide the required functionality.

4.2 Remote Access Solution Components and Their Individual Roles

A remote access solution consists of the following components (see Fig. 4.2):

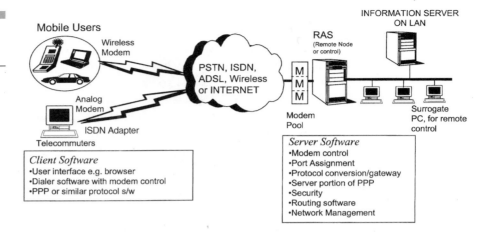

Figure 4.2
Remote access
components.

4.2.1 Remote Components

- A remote client PC—typically a desktop PC, a PC notebook, diskless network computer, PDA, a two-way pager, a personal communicator à la Nokia's PCS 1900, or in the Apple world, a MAC Powerbook
- A PSTN modem, ISDN adapter, a similar device for ADSL connection or a cable modem
- Communications software residing in the remote client device (provided by the remote access vendor)
- Modern operating systems, such as Windows 95, Windows NT, or OS/2 have a built in remote access functionality which many RACS/RAS vendors support

4.2.2 Network Connection

- A dial-up public switched telephone analog, an ISDN/ADSL digital network connection, or a cable modem

4.2.3 At Communications Server End

- A modem pool or equivalent digital ISDN hardware at the communications server (RACS/RAS) end

■ Host PC or communications-server hardware at the location where information resides (software supplied by the remote access vendor)

■ Communications-server software (typically provided by the remote access vendor) that handles the server portion of PPP or similar protocol

■ Routing software (supplied by the remote access vendor) to direct information to the correct application server or information server

■ Security software (supplied by the remote access vendor or by a specialized security software vendor)

■ Protocol-conversion or mainframe-gateway software (typically supplied by a third party), for example, SNA server for IBM mainframe connection

■ Network management software

■ Backbone WAN connecting communications server to mainframe or other application servers in the network

4.3 Remote Access Approaches

The following four generic approaches are used to provide network access to remote users, mobile workers or telecommuters:

■ Bulletin board system (BBS)—style terminal emulation

■ Remote control

■ Remote node

■ Mail gateways

We describe each of these approaches briefly, in order to clarify the differences.

4.3.1 Bulletin Board System—BBS (Also Called Remote Terminal Emulation Mode)

In this implementation, a terminal emulation software runs in a remote computer that is connected through a public switched network to a server on a LAN, as if the remote device were locally attached. There are a number of terminal emulation software choices available, depending on the communications protocol of the host information server. These range from specialized terminals, such as TN3270 for IBM mainframes and DEC VT220 for DEC-based servers, to the ASCII terminals implemented on CompuServe and LAN-based information services. On the host end, there is a terminal emulation server software or a network bulletin board software, such as Wildkat BBS from Mustang software. With this approach, a LAN manager can provide access to LAN files, network e-mail, and some LAN applications. One of the most common uses of BBS is to provide a file repository service from which location files can be uploaded or downloaded by remote users. Access to e-mail is provided through an e-mail gateway. (See Figs. 4.3 and 4.4.)

4.3.1.1 HOW DO YOU START UP A BBS SESSION?   Setting up a BBS session is very easy. The remote user starts up a terminal emulation program in his/her PC or MAC workstation. Most of these programs support an address book that stores several connection settings such as telephone number, number of data bits, number of stop bits, any parity requirements, and so forth. The terminal software calls the BBS. The BBS modem answers, and BBS software sends a terminal prompt that identifies the BBS. The BBS may ask for the user's name and any other information for keeping track of calls. It may ask for a password. After validating user identification and password, if required, BBS may send a menu selection indicating the services that the user may access.

Figure 4.3
BBS for remote access (also called remote terminal emulation). (Source: *Communications Week*, September 1995.)

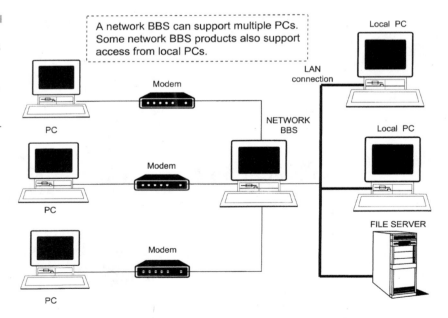

Figure 4.4
Mail gateway for remote access. (Source: Adopted from *Communications Week*, September 1995.)

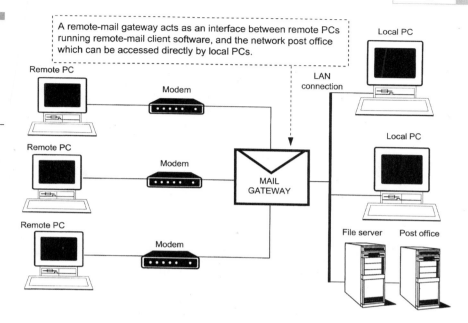

A remote-mail gateway acts as an interface between remote PCs running remote-mail client software, and the network post office which can be accessed directly by local PCs.

4.3.1.2 BBS FROM AN INTERNAL ARCHITECTURE PERSPECTIVE.

BBS is a specialized remote access software that typically includes both communications software functionality, as well as application modules that interface with the file system, management subsystem, menu subsystem, e-mail subsystem, e-mail gateway, and user interface. Early versions of BBS products based on DOS were implemented as multitasking applications with system and communications functionality provided by BBS software. Newer versions based on OS/2, Windows NT, and Unix utilize the system facilities provided by the operating system. Figure 4.5

Figure 4.5
BBS architectural components,

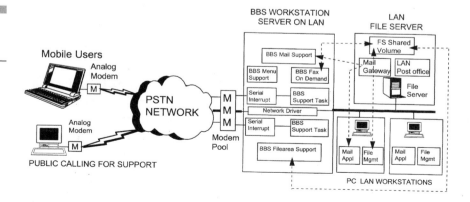

shows a schematic of the architecture of BBS from a functional software perspective.

4.3.1.3 BBS'S USER INTERFACE.   Most bulletin board or terminal servers operate in archaic character mode, though the interface can be upgraded to graphical user interface with the use of terminal emulation software that supports remote imaging protocol (RIP). Mustang Software's *Qmodem* software is one such example. RIP allows lines, bitmaps, and color. Menu selection can be made using the mouse or keyboard. In such an implementation, terminal software must support mouse-based input.

4.3.1.4 BBS SUPPORT OF LAN APPLICATIONS.   Several BBS products support direct LAN connection, that is, it can access LAN files or LAN applications. A remote BBS user can download a file (receive from the BBS) or upload a file (send it to BBS). Many common file transfer protocols, such as XMODEM and YMODEM, are supported. Some BBS products support e-mail from remote users. Through an e-mail gateway, e-mail can be exchanged with LAN users, as well.

4.3.1.5 BBS USES BEFORE AND AFTER THE INTERNET WAVE. Before the recent evolution (or, perhaps, the revolution) of the Internet, during the mid-1990s, BBS was the most popular method of posting notices, updates, problem descriptions, software fixes, frequently asked questions with answers, and upgrades to products after they had been shipped to customers—a general form of technical and field support. The BBS software was quite inexpensive, and existing hardware, such as PCs, MAC, or Unix workstations could be used for hosting technical support information. Business justification of BBS was very easy because you could reduce the number of technical support personnel. As long as the BBS telephone number was well publicized through manuals or other means, it made the task of sending fixes and answering most frequently asked questions (FAQs) much easier. BBS also provided an inexpensive electronic mail system for a limited number of internal and external users, before the current suite of LAN mail systems such as cc:Mail or MS Mail came into vogue. However, the recent Internet wave and standardization of internal LAN-based messaging have almost completely stopped the growth of BBS as a method of remote access. Nonetheless, BBS continues to be a legacy application in certain organizations, and BBS vendors are adopting their software to allow remote access through the Internet.

4.3.1.6 BBS INSTALLATION AND MANAGEMENT.   From a network manager's perspective, management of a BBS is similar to that of a

LAN. Initial installation can be quite simple and may take from half a day (for a simple BBS with a couple of lines) to several days, depending on the amount of customization and configuration setup requirements. Of course, any new computer installation calls for extensive testing—do not underestimate the requirements for a complete installation and testing. It is a common occurrence that what starts as a small experiment with a few users grows mushroomlike, such that management requirements increase significantly. Quite often, requirements increase gradually.

Apart from hardware and software installation, it is the design and customization of the menu that may take up the maximum amount of the BBS administrator's time. You also need to set up the e-mail configuration, file area definitions, user area configuration, and any public forum configurations that you may define. From a user registration perspective, you can set up BBS for automatic user registration for technical support, file transfer of software fixes, and marketing information. In other cases, you may opt for manual registration of remote users who may be allowed to access various LAN applications. If the number of remote users is large, you may prefer to automate the registration process, just as most Internet providers do.

You may want to set up audit trails on the use of BBS by different users. Detailed audit trails can be set up if required—as a minimum, initial login, use, and logoff times. You may track use of various applications by remote users.

4.3.1.7 PROS AND CONS OF BBS. The major advantage of this approach is that it puts minimal demand on remote client nodes. No specialized software is required. Therefore, BBSs are more geared to accepting new users without any approval or configuration matching between the remote end and BBS. All that you need are a modem and terminal emulation software. Most common terminal settings are 8, 1, N for 8 data bits, one stop bit, and no parity. Windows 3.1, Windows 95, and OS/2 all provide terminal emulation software as an integral component of their communications driver suites. BBS accepts new users without any problem. Therefore, BBS is the ideal mechanism for technical support. It is also among the cheapest implementations for the limited functionality that it provides to external users. Accordingly, most platforms such as PC, MAC, and Unix are supported.

The disadvantages are the limitations inherent in an old terminal interface, and participation on the LAN as something decidedly less than a peer—a fact that will doubtless not be appreciated by your users. Not all applications on LAN can be accessed by remote users. More important, the Internet phenomenon has overtaken BBS technology completely.

Thus, this approach does not have much of a future. It is a short-term arrangement.

4.3.2 Remote Control Mode[*]

This scheme involves two PCs—the mobile user's remote PC notebook and the host (surrogate) PC at the LAN. The remote user dials into the LAN and takes control of the host PC. With both PCs running matching software, every piece of information accessed by the surrogate PC on the LAN appears on the client notebook PC. Any keyboard or mouse action on the remote PC is sent down to the host PC which simulates the remote actions, as if those were done on its own keyboard or mouse. The host PC runs the actual application which the remote PC sees and interacts with. You may note that the host PC may or may not have a keyboard or for that matter a monitor. In this manner, the remote user can run LAN applications as if he/she were on the office LAN. More sophisticated remote control programs allow the remote and host PCs to share each other's hard disk. They also provide file-copying facilities to transfer files between the two PCs. Once disconnected, after a logoff command, the remote user can do whatever local stand-alone processing that it may want.

You should be aware that with this approach, the actual application processing takes place at the host PC on the LAN, not on the remote PC, which sends only keyboard commands or mouse clicks. An interesting scenario for remote control may be one where a LAN administrator can perform tape backup at night, after all the users have logged off. As long as the tape has been inserted into the tape drive, he/she can start up the backup operation from home and monitor it from there. The administrator may even disconnect while the backup is going on and do some other work. He/she can log in again, later, to ensure that the operation did take place successfully.

The remote screens may be updated synchronously or asynchronously. In the first mode, changes to the screen are sent as they occur. Therefore, you would see the remote screen updated in the same manner as the host PC's screen, albeit after a slight time delay (that is, the time taken to transfer this screen data on the dial-up link). In the asynchronous mode, changes may be batched, and the latest screen image is sent across.

---

[*]"Using Remote Node with Remote Control"—a white paper by Shiva Corporation, an RNA vendor.

Products such as CarbonCopy, pcAnywhere, and ReachOut are among the most well known examples of remote control implementation. Remote control software, in fact, uses the client/server design paradigm, insofar as the host/surrogate PC runs the application and the client notebook displays the information. However, this implementation should not be confused with true client/server business applications that partition overall processing of a unit of work across a client workstation (handling presentation services and processing client-unique data locally) and a server that stores shared multiuser databases or mail post offices on the LAN.

This approach was originally devised as a solution, so that software technicians could provide remote support working from home or diagnose remote system problems from their offices. Since they were solving LAN-based problems, they just wanted to see what a local user would see. It was important that the problem was not made more complicated by processing in the remote computer. It did not take long for other mobile users to discover the benefits of this approach, however, and as it was very easy to implement, it grew rapidly in popularity. Figure 4.6 shows how remote control works.

4.3.2.1 HOW DO YOU START A REMOTE CONTROL SESSION?

Remote control programs provide an address book functionality of some kind for remote hosts, similar to BBS systems. This function is used to store the telephone number of the dial-up line that is connected to the modem of the host PC. Both the remote and host PCs must be running the remote control program, such as pcAnywhere or CarbonCopy, before a physical connection can be invoked. The host PC may load the remote control software at the initial power-on and boot-up sequence, or it may be done by the LAN administrator as part of the start-up operation. The actual connection takes place when the remote PC starts up the dial-in screen and makes the call. Thereafter, you can start a session on the host PC from the remote site. Figure 4.7 shows a remote user accessing a contact database application.

After the physical connection is made, the two PCs synchronize their screens. Once this happens, the remote user can start typing in any command to start an application in the same way as if he/she were attached to the LAN. One difference that a remote user will notice is the longer response time than that observed on the LAN. The actual response time difference will depend on the link speed and the amount of data displayed on the screen. The difference in response time is more pronounced with graphical screens, where a large amount of data is sent across the link to the remote PC.

Figure 4.6
(a) Remote control concept in remote access architecture; (b) Remote control concept for remote access. (Source: pcANYWHERE brochure.)

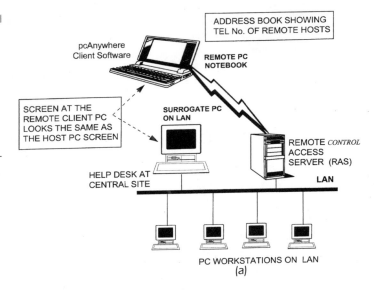

(a)

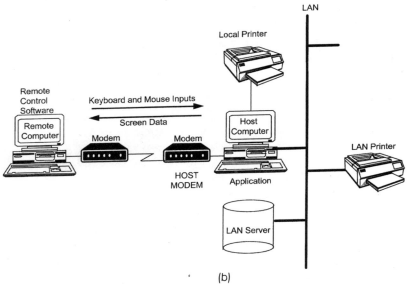

(b)

The remote user may use the remote control session in two modes—the interactive session mode or the specialized remote function mode. In the first mode, the user interacts with an application running on the host PC by using the remote PC's keyboard, mouse, and monitor. Besides this interactive mode, the remote user may invoke special functions such as file transfer or chat mode. In the chat mode, the remote and host PCs can exchange messages. (See Figs. 4.8 and 4.9.)

Figure 4.7
pcANYWHERE icon
for host PC.

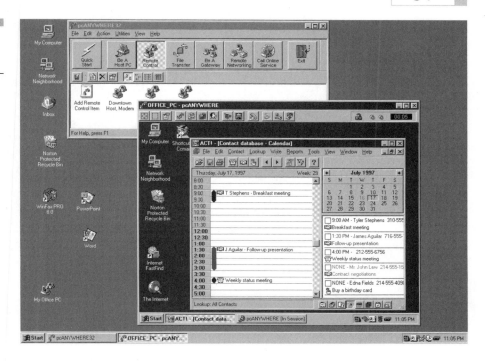

Figure 4.8
pcANYWHERE file
transfer mode.

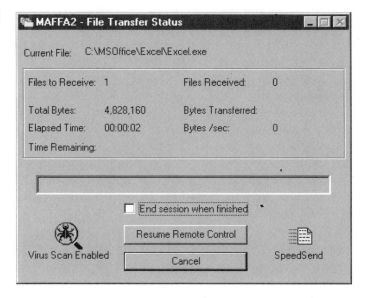

Figure 4.9
pcANYWHERE chat
mode.

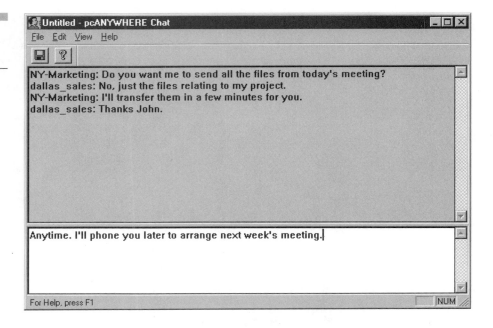

4.3.2.2 SECURITY LEAK WITH REMOTE CONTROL.   We men-
tioned that the host PC can carry on the task, even after the remote user
got disconnected because of network errors. Meanwhile, another user
could dial in, and get connected, assuming the previous user's identification and password. In fact, the new user could change the password of
the original user, with the result that the first user could never log in
again, without an intervention of the systems administrator. This is a
serious security leak that you should be aware of; you should check with
the vendor regarding any available solutions for this problem. If you use
LAN network operating system (NOS) security, you may be exposed.
However, many remote control products provide security features to prevent this type of problem. The remote control host server may implement advanced security based on password authentication, dial call-back,
or token exchange. (See Fig. 4.10.)

4.3.2.3 SINGLE HOST PC FOR MULTIPLE REMOTE CLIENTS.
One of the complaints against the remote control approach has been that
you need as many dedicated host PCs as the number of active remote
clients. This is no longer a restriction with implementations, such as
Symantec's pcAnywhere remote control server that runs on a single-host
PC, and supports up to eight independent remote control sessions. This

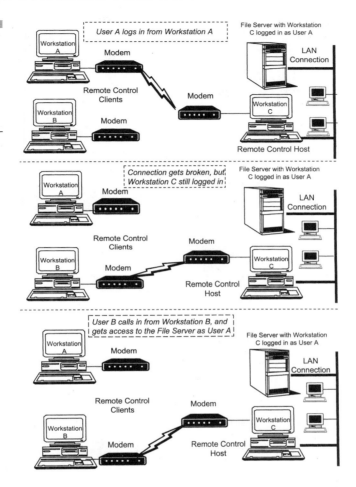

Figure 4.10
Security leak as a result of link failure. (Source: Adopted from *Remote LAN Connections* by Wong.)

capability is achieved through multitasking operating systems, such as OS/2 or Windows NT. Other vendors have taken other approaches to solve this problem. Several vendors have started designing servers with multiple PCs on a card, in the same box. You do not really need peripherals, such as the keyboard, monitor, or mouse at the host PC. Therefore, you can reduce the cost further. (See Fig. 4.11.)

4.3.2.4 MUST WE MATCH REMOTE CONTROL CLIENT PLATFORM WITH THE HOST PLATFORM?

We have talked about Intel PC as a generic remote and host device. While this is the most common and widely used platform, in reality, you may have to support multiple remote devices, including Mac and Unix workstations. Newer implemen-

Figure 4.11
(a) Remote control
gateway; (b) Remote
control server—
multiple concurrent
sessions. (Source:
Symantec.)

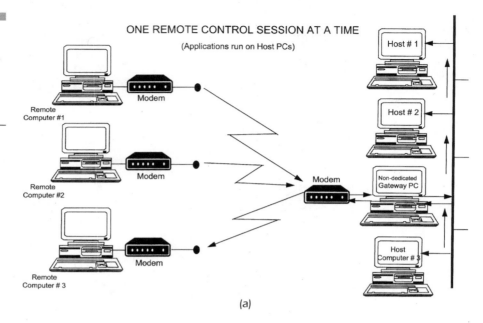

ONE REMOTE CONTROL SESSION AT A TIME

(Applications run on Host PCs)

(a)

Remote Control Server

(Applications run on the Host Remote Control server)

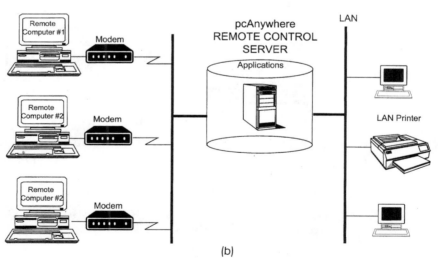

(b)

tations of remote access support multiple remote clients with the same host server platform; for example, Windows NT server as host may support PC, Macintosh, and Unix workstations as remote control clients.

4.3.2.5 APPLICATIONS SUITABLE FOR REMOTE CONTROL APPROACH. The following types of applications benefit from remote control:

1. Flat file database applications, based on FoxPro, dBase, or Paradox, that require the transmission of a large amount of data from the file server to the workstation (the surrogate or host PC) before a user can access needed information. With such applications, it is necessary for the remote control to achieve certain minimal performance levels across the phone connection.

2. Older e-mail applications.

3. Any application whose license prohibits its transfer to a remote PC. In this scenario, remote users are saved the expense of purchasing additional licenses.

4.3.2.6 PROS AND CONS OF REMOTE CONTROL. An advantage of the remote control mode is its ability to access programs from the hard disk, and records from the databases at LAN speeds because of the fact that the host or surrogate PC actually does reside on the LAN, even though it is controlled remotely. Only displayable screen data is sent across the modem dial-up link. Therefore, it is quite practical to use a word processing package on a remote control solution because the actual program code is not sent across the communications link. On the down side, as the remote node camp has been accustomed to point out, more often than not, as many surrogate PCs have been needed as the number of remote concurrent users. To counter this criticism, remote control vendors are now offering solutions, such as the Symantec's (pcAnywhere) remote control server that runs on a single-host PC, and supports up to eight independent remote control sessions. We have already discussed security leak possibilities with remote control. We summarize the pros and cons of remote control in our comparison table in a later section of this chapter.

4.3.3 Remote Node Mode

Remote node (also called remote workstation, in some books) refers to the ability of a remote PC user to dial into a LAN-attached communications server and function as a full node (or a peer) on the network—thus permitting remote users the same privileges and level of access to LAN-based resources as their office-based colleagues. In this mode, packets are transmitted to and from the remote PC over the telephone lines (POTS or ISDN), directly or through Internet. Therefore, one major difference between a remote device and a locally attached LAN device is that of performance. Although the locally attached devices operate at LAN speed (10 or 100 Mbps for Ethernet and 4 or 16 Mbps for Token Ring network), remote node devices operate at network speed (28,800 to 56,000 bps for PSTN and 128 Kbps for ISDN). This may have an important bearing on the initial reaction to remote access from mobile users. An application that works very fast on a LAN may appear very slow on a remote device. Such is the nature of remote telephone links. ISDN-based connection may be much faster, but ADSL connection will be much closer to the performance of a workstation directly attached to the LAN.

4.3.3.1 HOW DOES REMOTE NODE WORK?   Let us understand the difference between a device that is locally attached and a device that is remotely connected to a LAN. Instead of a network interface card (NIC), you need a modem and a client communications software that manages the remote node. This is depicted in Fig. 4.12. On the LAN, you need a remote access communications server (RACS). This server is essentially a special implementation of a traditional router in the LAN-WAN internetworking context; very appropriately, it could be called a remote access

Figure 4.12
Remote node concept—replace NIC with modem.

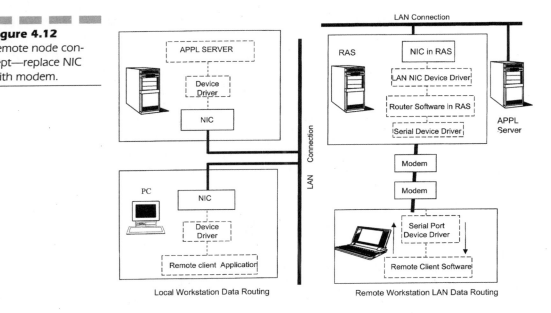

Local Workstation Data Routing

Remote Workstation LAN Data Routing

router. RACS or the remote access router receives packets from the remote access link and routes them to an appropriate device on the LAN.

RACS may be implemented as a hardware/software combination or a software-only implementation on a network operating system (NOS), such as Novell NetWare or Windows NT. In Novell terminology, it may be NetWare load module (NLM). In the first category, it may appear as a standard PC workstation with a keyboard and a monitor, or it may be housed in a box, without a keyboard or monitor. In this form, it is best located in a computer closet or a rack. Many different implementations are available, as you can see in App. A.

The remote node essentially uses the same network software as that of a locally attached PC; however, there is one main difference: it has a remote modem device driver, whereas a LAN-attached node has an NIC (network interface card) device driver. Data from an application on a LAN-attached workstation goes through a software device driver and NIC itself. The data from a remote node is sent through a modem device driver, remote modem, local modem, and conventional NIC device driver. This is shown in Fig. 4.13.

4.3.3.2 HOW DOES A REMOTE USER SET UP AND TERMINATE THE CONNECTION?
Typically a remote user starts up a remote access session. The remote node client software presents a connection screen that asks for the name of the RACS server and telephone number to call. Figure 4.14 shows a typical remote connection setup screen. Secu-

Figure 4.13
Remote node mode.

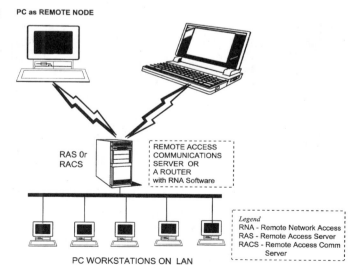

PC as REMOTE NODE

RAS 0r
RACS

REMOTE ACCESS
COMMUNICATIONS
SERVER OR
A ROUTER
with RNA Software

PC WORKSTATIONS ON LAN

Legend
RNA - Remote Network Access
RAS - Remote Access Server
RACS - Remote Access Comm
Server

Figure 4.14
Windows screen shots
of Shiva NetRover
using remote node.

(a)

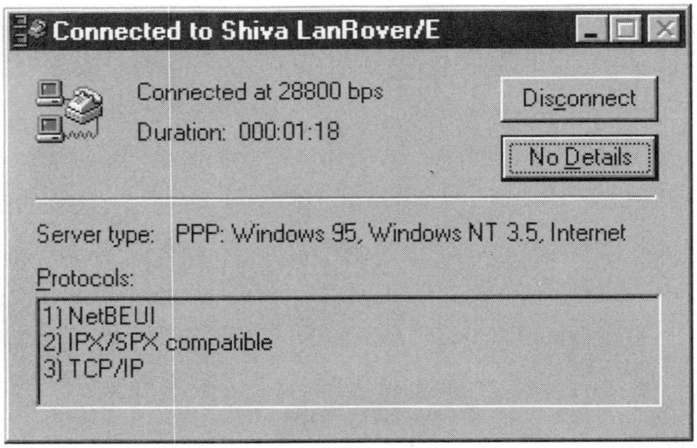

(b)

rity software may require a userid and password. Once you are connected, the user may log in and log out from the remote session. Thereafter, you can use all the applications that may be available on the LAN. While the most common applications used to be file and print services in the past, more common present-day applications are e-mail, sales force automation, and many other business applications. These applications may reside on other application or database servers to which RACS communicates over the LAN in the same manner as it does for locally attached devices.

4.3.3.3 DIFFERENCES IN LOCALLY ATTACHED AND REMOTELY ATTACHED DEVICES—USER'S PERSPECTIVE.

While both groups of users have access to the same set of applications, there are a number of differences that result from the slower speed of the remote link, when compared to the LAN speed. Everything happens in slow motion—file manager commands, file transfer, and database commands. Better implementations of file transfer and file copy commands in the Windows environment show a visual indicator of the progress of the operation as a visual feedback. This may satisfy the user to some extent, but the problem may cause frustration when the remote user is accessing a database application designed for LAN. As an example, take the case of the Lotus organizer application that has a large contact database. When you start the application from a remote site, the application loads the entire database before it can display the information. Similarly all the Windows DLLs (dynamic link libraries) are transferred over the remote link, causing slow response time to the extent that the remote user may either get used to the price to pay for mobility or get frustrated.

Slow response time may also lead to an incorrect perception in the mind of the remote user that nothing is happening, even though considerable data transfer of application DLLs and databases is progressing in the background. Unfortunately, it is difficult to provide visual feedback of this background activity to the user. This is so because of two reasons. First, it means that you need hooks into the LAN application, implying some modification to the application. If the application had been designed with the remote user in mind—a feature of mobile-aware design—then this visual feedback should have been built in to start with. Such is not the case with most LAN applications. Second, as a result of these modifications, access to the LAN applications is no longer transparent.

When there is no visible response on the remote user's screen, the user may resort to reloading the program, rebooting, or attempting a reconnection. These acts may lead to more serious problems, such as corrupted files or databases. This highlights the need for the careful evaluation of applications that should be offered for remote access. Users should be given adequate training and technical support, besides setting their expectations right from remote node on slow PSTN links.

These are the problems inherent in those applications that are run on the network infrastructure for which they were not initially designed. As compared to remote access solutions, you may look at Internet applications, such as Netscape, that do provide continuous feedback of data transfer. This causes relatively less frustration, even though it may take a long time to complete the operation.

4.3.3.4 NETWORK MANAGER'S VIEWPOINT.

Network managers like the remote node option because it is easy to manage and extend. You can increase the number of ports by adding modems and multiport serial boards. Of course, you may have to upgrade to a higher capacity RACS to handle more users.

Remote node has the great advantage over remote control of full security. As an example, several advanced products integrate with Novell's NetWare bindery software. Some products also integrate third-party hardware security products that employ handheld devices or *tokens* with remote node—based communications servers. (Refer to Chap. 13 for an analysis of security.)

While remote node is the method most preferred by network managers, its speed is a major drawback. Typical dial-up speeds of 28.8 Kbps are slow in comparison to the 10 Mbps of a LAN. Even with compression factors included, modem connections are several times slower than LANs. This results in noticeable delays for remote users while accessing LAN file servers. Even with an ISDN connection at 64 Kbps or 128 Kbps, it is unrealistic, we believe, for remote mobile users to expect the same performance levels from remote connections as they expect from locally attached workstations on the LAN.

Accordingly, remote node mobile applications should be designed differently from local LAN applications, with as many functions as possible being performed on the remote notebooks. Host information servers should be used to access database information, and only the results of the queries should be transferred across dial-up links. It is important to appreciate that there are many concepts in older mainframe-based systems that are relevant in modern remote access technology.

Newer client/server applications that involve file transfers or access to SQL databases are more suitable for remote node operations. Many applications are now using the client/server design. These include Lotus NOTES, cc:Mail Mobile, Microsoft Mail Remote, and the like, as well as popular relational databases, such as Informix, Oracle, Ingress, and Sybase. These applications operate well over remote node, and it is expected that more such applications will be developed, while keeping the remote user in mind.

What we have stated in this section suggests that network managers should first understand the nature and extent of slow response time. Then they can try to solve it by any of the following methods:

- Evaluating business applications' suitability for remote access
- Upgrading to faster links, such as ISDN
- Loading the applications on the remote PCs

- Using remote access solutions that utilize caching software, like Airsoft's Powerburst
- Making the applications mobile-aware

Even though applications may be transparent, remote users present slightly different support requirements than locally attached LAN users. We discuss these requirements in greater detail in Chaps. 15 and 17. Suffice it to say here that remote users have a greater number of problems that are typically more difficult to solve because it is not easy to send a support technician to the site in a hurry.

4.3.4 E-Mail Gateways

This is a special application-specific remote access communications server that was invented when e-mail became quite popular. E-mail, like real mail, is sorted at and stored in post offices that may reside on a LAN. An e-mail post office may be considered as a specialized application server that manages e-mail users and the actual electronic mail. It may be a network load module on a Novell NetWare file server, Microsoft NT Server, or OS/2 LAN Server. A separate client software program handles the user interface in terms of composing, sending, retrieving, and filing e-mail messages on the remote workstation—whether it is a PC, Powerbook, PDA, or a host of new PCS telephone devices. Most of the popular operating environments, such as Windows (3.11, 95, or NT), OS/2, MacOs, Netscape, and Pegasus (new OS for PDAs from Microsoft) have incorporated mail interface of one kind or another.

An e-mail gateway is the interface between a remote workstation and an e-mail post office server configuration within the organization or the Internet public e-mail service. The gateway provides the following functions:

- Communications interface, that is, RACS functionality, similar to remote node and remote control solutions offer. This handles the modem interface directly from the remote workstation or through Internet.

- Simultaneous handling of multiple local and remote users.

- Post office functionality, described previously, may be integrated within the gateway, as well.

- Translation of different e-mail formats, for example, between cc:Mail and MS Exchange.

- As an option, interface to other mail applications.

This is depicted in Figs. 4.15 and 4.16.

Figure 4.15
E-mail systems for
remote access.

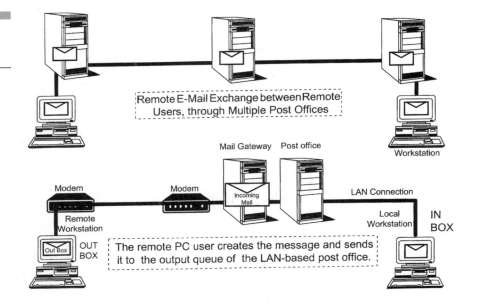

Figure 4.15
E-mail systems for
remote access.

Mail gateways from various vendors have different user interfaces and support different document format standards. Most modern gateways support attached documents following one or more of the three document standards—Mail Interchange Multimedia Extensions (MIME), Common Object Model/Object linking and Embedding (COM/OLE), and OpenDoc.

Figure 4.16
Mobile users sending
e-mail to internal and
external users
through mail
gateway.

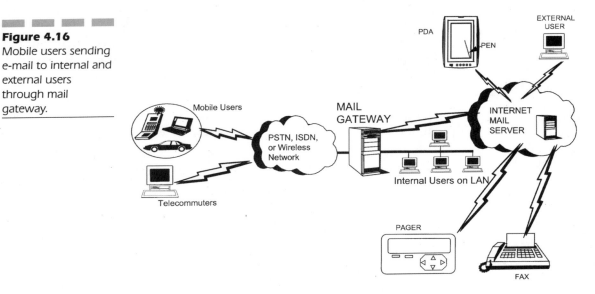

Those who have used Eudora mail package on Internet will find the MIME standard quite familiar. The documents may contain multiple types of data, including graphics, sound, video, besides text formatting information, such as fonts, columns, and the like. While these document standards are general data exchange standards, they do meet the needs of mail exchange across disparate platforms. Document viewers provided with different products allow these documents to be viewed on-line or off-line.

4.3.4.1 MAIL GATEWAY AND POST OFFICE. These are two distinct functions. Mail gateway is a software/hardware configuration that interfaces with remote users from a communications perspective, providing transport services, as well as dealing with external mail systems that do not follow the same messaging protocol as the sender of a mail message. Mail gateway may also manage a modem pool for multiple-port configurations. On the other hand, the post office handles non-transport-related functions, in terms of storage, address resolution, sorting, and batching of messages.

A mail gateway sends an initializing string to the modem or modems and enables them to receive incoming calls from remote users in auto-answer mode. It may perform initial handshaking with the remote user's modem to set up the logical connection with the e-mail software on the remote client workstation. After validating the user by authenticating his/her userid and password, it assumes the role of a transport engine between the remote user and the post office.

4.3.4.2 USER INTERFACE ON MAIL GATEWAYS. Installation of the user interface software is a relatively straightforward process. Since e-mail is becoming a generic requirement, several vendors have started embedding a basic e-mail interface into the operating system package. Such is the case with the Microsoft exchange interface with Windows 95. However, specialized e-mail packages provide additional functionality, including file attachment, file transfer, and multiple document protocols. Also, typical mail gateways support a diversity of OS environments. In fact, this is an important consideration because you will find several flavors of workstations among the user population. (See the example in Fig. 4.17.)

Typically, the user invokes e-mail directly from Windows Program Manager, or start-up menu, or from a word processor such as WORD 6.0, Corel WordPerfect, or Lotus AmiPro in the case of integrated desktop suites. First you create the message, attach additional documents, if necessary, and then use the transport function of e-mail software to ship the message to the gateway on the LAN or the mail server of an Internet Ser-

Figure 4.17
Screen shot of e-mail interface (Eudora Professional).

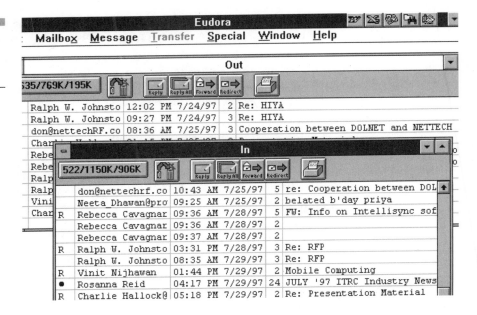

vice Provider (ISP). At this stage, the message may still sit in the out mailbox of the remote workstation. Actual transmission takes place when the *Send* function is invoked by the user. Depending on the configuration option selected, the physical connection between the modems may have been completed before this process or may happen now.

At the e-mail server, the messages will be sorted according to the destination address, based on the name and address list that maps logical addresses to physical addresses on the same LAN, another LAN connected through a router-based WAN, or an external user through the Internet connection from the LAN. The messages are stored on the post office until a finite time interval (typically a selectable configuration parameter) elapses. This allows multiple messages from different users to be sent to the recipient just like a real postal system. Also, you may configure your server to take advantage of cheaper long-distance charges by delaying the transmission to nonprime time.

4.3.5 Hybrid Implementations—Remote Node and Remote Control

Figure 4.18 shows a hybrid implementation of both remote control and remote node in the same RACS.

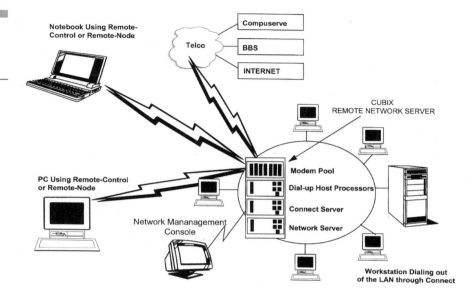

Figure 4.18
Hybrid solution—
remote control or
remote node.

4.3.6 Comparison of Remote Access Options

Even though we have discussed four approaches in this chapter, essentially
there are two technical approaches in remote access technology—remote
control and remote node. BBS and mail gateways are simply application-
specific implementations that got extended to provide access to LAN
applications. Nonetheless, in Table 4.1 we summarize some of the advantages and disadvantages of the four approaches to remote access.

A remote node connection can take a number of different forms,
depending on whether it is incorporated into a file server or it is connected to a router or a stand-alone PC.

4.4 Remote Access Communications Servers (RACS and RAS)

There are a number of remote access communications servers available in
the market, from a variety of vendors ranging from internetworking vendors (such as Cisco, 3COM, Ascend, IBM Cabletron, Bay Networks) to specialized remote access server vendors (such as Shiva). In this section, we
review the type of servers, their functionality, important features, and eval-

TABLE 4.1
Comparison of
Four Remote
Access Approaches

Comparison Criteria	BBS	Remote Control	Remote Node	Mail Gateways
Description	Bulletin Board System for technical support information	Remote access system where a surrogate PC LAN is controlled by key strokes and mouse clicks from a remote workstation	Remote workstation acts as a peer on a LAN by switched connection through a communications server.	A specialized mail server with both remote communications and mail post office functions built into one logical configuration
Application support	As a bulletin board for technical support and software updates	Almost all LAN applications, but application compatibility must be tested	All LAN applications	Essentially e-mail; some LAN applications
User interface	Generally character-based	Slightly different interface from LAN	Same interface as on LAN	
Network performance	Acceptable since BBS uses older but more efficient terminal interface	Only key strokes transferred on the network	Lot of network traffic Suitable for client server design	
Scalability	Supports relatively fewer users	Not as scalable because surrogate PC requirements	Most scalable (Few servers support >1000 users)	Same level as Remote Node (Not for >>500—1000 users)

Security	Weak; open to public users / Central security not always enforced	Weaker than remote node; easier to break-in / Easier to change applicationsoftware on the LAN	Better than remote control; should implement third party s/w / More difficult to manage remote application software	Better than remote control; may have own authentication / Easier to manage
Network management cost	Relatively inexpensive way of providing support information	RACS hardware more expensive / Software license costs lower	RACS hardware is less expensive / Software license costs are higher	Cheaper for mail-only application
Pros	Minimal requirements for remote client software / Simple conventional terminal emulation interface / Generally no user registration / Some BBSs allow access to other applications	Network traffic is less / Ability to continue with application if connection broken / Ability to run application if remote user disconnected	Transparency to LAN applications / RAS hardware cheaper / More scalable	Simple user interface / Support popular e-mail systems / Connection time is minimized / Optimized for e-mail application
Cons	No access to LAN applications without violating security	User interface not always same / More expensive hardware / Security is weak / Manual connect and disconnect	Too much network traffic / Software license costs higher	Limited or no access to LAN applications

SOURCE: Author.

uation criteria. The reader is reminded that in many trade publications, RAS is more commonly used as an acronym for remote access servers. We shall use RACS and RAS interchangeably, even though the former acronym is more descriptive and covers more implementations, including wireless-network-based mobile communications server switch (MCSS), defined in the author's book, *Mobile Computing—A Systems Integrator's Handbook,* cited previously.

4.4.1 Types of Remote Access Servers

Remote networking comprises a number of technologies, including LAN-to-LAN switched connections, where small LANs at the remote branch office can dial into corporate LANs. However, in this book we concentrate our discussion to remote PCs, Macintosh computers, PDAs, and similar devices, which can use one of the many approaches described in this chapter to connect to LAN-based information servers through remote access servers. The emphasis here is on single remote users à la mobile workers. RACS/RAS may be categorized into any of the following types of remote access servers:

4.4.1.1 FIXED PORT REMOTE ACCESS COMMUNICATIONS SERVERS. This category is the most common implementation, and consists of RACS technology platforms that have a fixed number of ports—4, 8, 16, 32, 64, and more. Typically, the number of ports is fixed, even though they are housed in a variable capacity chassis for upgrade purposes. What is important in this designation is that there be one-to-one correspondence between the ports and communications lines coming in from the users. Many of the older-style terminal servers that provide remote access are being replaced by general purpose RACS.

4.4.1.2 REMOTE ACCESS SERVER (RAS) COMMUNICATIONS CONCENTRATORS.˙ In this category, we include internetworking products that are primarily used for leased-line connections at T1 and ISDN/BRI speed, but have been extended to provide remote dial-in capabilities. A RAS concentrator takes lower-speed incoming analog, or ISDN-based signals (thin pipe) from off-LAN telecommuters and remote users, and concentrates them into higher-speed T1 or ISDN-based primary rate

---

˙Information based on IDC report on RAS Servers (June 1996).

(thick pipe) interface. This concentration is achieved through multiplexing techniques, as explained here:

Analog dial-up calls come into a concentrator box, located at customer premises, or at a Telco site, where they are processed through a T1 card and its CSU/DSU. ISDN calls are transported over a digital network to Telco's central office switch, where a corporation or an ISP terminates a T1 or ISDN PRI connection. A T1 card in the RAS concentrator performs any protocol processing of packets, such as PPP. The packets are then sent through the remote access communications server's dial-up router component. Then the packets merge into an Ethernet, Token Ring, or frame-relay network. One T1 pipe can handle 24 channels, each one representing a dial-up port. This concentration is important because it is more efficient to have one T1 termination rather than 24 lower-speed terminations at the customer's premises. In the case of ISDN, one PRI can handle 23 BRI sessions, each of them equivalent to one dial-up port. The following remote access communications servers are examples of the concentration technique:

- 3COM's AccessBuilder 5000
- Ascend's Max, Pipeline 400T, and Pipeline 400B
- Bay Networks (Xylogics') Remote Annex 6100, 6300
- Cisco's AS5200
- Shiva's LanRover Access Switch
- Telebit's Mica Blazer
- US Robotics' Total Control Enterprise Network Hub

4.4.1.3 DYNAMIC SWITCHING. Many of these products perform dynamic switching, where all calls are dynamically switched for either analog, frame-relay, Switched 56, T1, or E1 type of access. The concept of switching in remote access will become more prevalent when switching (of connections and therefore of the traffic) at the LAN level becomes a more common internetworking solution. We refer to this topic in Sec. 4.5 of this chapter, but discuss it in detail in Chap. 11.

4.4.1.4 SMALL OFFICE, HOME OFFICE (SOHO) TYPE REMOTE ACCESS SERVERS. As a result of the popularity of telecommuting and home office phenomenon, remote access vendors have started bringing into the market low-end RAS products that are getting cheaper in cost and easier to install. The target market is the small office and the home office. Many of these products support both PSTN and ISDN connections.

4.4.2 Remote Access Servers—Some Vendor Product Examples

We now describe several product implementations of remote access server technology, in order to give you a flavor of what is out there in the marketplace. No doubt, some of these products described here will get replaced by newer models as the book goes to press or as time passes by. You are well advised to surf the World Wide Web on the Internet, obtain preliminary information on current products, and then ask the vendors for a more detailed explanation of product features. Some products are hardware-independent but software-specific; others use proprietary hardware and software. We describe the following products:

- Operating software—based—Microsoft's remote access server (RAS)
- Network software—based—Novell's NetWare Connect 2
- RAS hardware—based servers—Shiva's LanRover/NetModem
- Internetworking hardware—based servers—CISCO's 5200 Universal Server
- Internetworking hardware—based servers—3COM's AccessBuilder family
- Application software—based servers—Xcellenet's RemoteWare

4.4.2.1 OPERATING SOFTWARE—BASED—MICROSOFT'S RE-MOTE ACCESS SERVER (RAS).   Microsoft provides the remote access server as a software implementation under Windows 95, Windows for Workgroups (WFW), and a bundled operating software service under Windows NT server.

RAS (à la Microsoft implementation) is a communication software service that runs on Intel or any other supported hardware under Windows NT operating system. It can utilize file server facilities or application services on the LAN under TCP/IP or Netbios. TCP/IP is preferred if the remote workstation runs SNMP communications software. In terms of network connections, Microsoft's RAS supports dial-up, X.25 packet switched links, and ISDN. The initial implementation supported 64 workstation connections per server. Compression employed is generally based on the modem hardware. Clients supported include DOS, Windows 3.1, Windows 95, and Windows NT workstations. (See Fig. 4.19.)

RAS supports NT's security system—that means authentication using the normal name and password combination. The initial connection

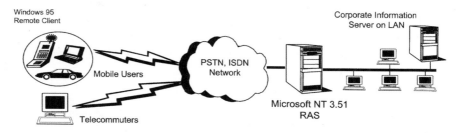

Figure 4.19
Microsoft NT 3.51
remote access server
architecture.

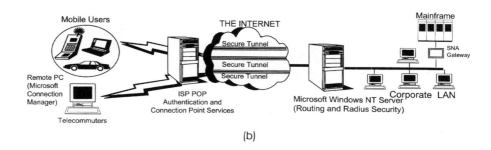

handshake procedure uses encryption techniques. You can also implement callback from the server. Under this implementation, the client workstation authenticates the user; but before any application activity can begin, the server hangs up the connection. Then the server calls the remote client at the telephone number that has been stored for that user, and sets up the real session.

4.4.2.2 NETWORK SOFTWARE—BASED—NOVELL'S NETWARE CONNECT 2.
NetWare® Connect™ 2 provides a comprehensive suite of remote access solutions on a single integrated platform built on NetWare. Combining management, security, and NetWare Directory Services™ (NDS™) support, NetWare Connect 2 provides a NOS-based platform for remote access to LAN resources. In NetWare terms, it is implemented as an NLM (NetWare load module).

NetWare Connect 2 enables communications equipment to be shared by a suite of remote access modules in a dynamically allocated, bidirectional manner. Up to 128 remote PC and Apple Macintosh users can dial in and access network resources, including files, databases, applications, printers, electronic mail, and host computer services. Local PC and Macintosh workstations can dial out through NetWare Connect to access on-line ser-

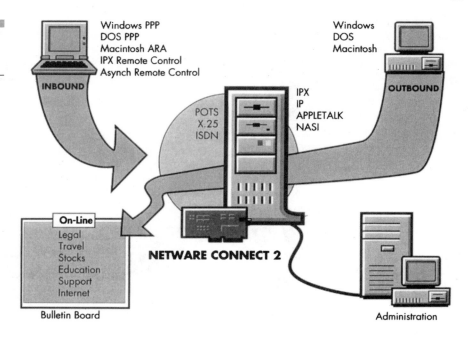

Figure 4.20
NetWare Connect 2.
(Source: Novell.)

vices, such as bulletin boards and the Internet. (See Fig. 4.20) Novell's Net-Ware Connect provides the following features and capabilities:

COMMUNICATIONS RESOURCE SHARING.   Cross-platform sharing of Novell and third-party communications services means higher resource utilization, lower communications costs, and centralized management. This resource sharing eliminates the need to dedicate equipment to individual users, thereby reducing costs, improving security, and simplifying management.

NetWare Connect intelligently separates its hardware interfaces from its suite of communications services. This separation allows any service access to an available communications port on an as-needed basis. A universal pool of modems can be dynamically used by Novell or third-party applications for IP, IPX™, AppleTalk, fax, paging, remote control, dial-out, and host access.

SUPPORT FOR BOTH REMOTE NODE AND REMOTE CONTROL CONNECTIONS.   Net-Ware Connect 2 supports both connection methods. Remote node connections provide remote users with a familiar work environment. NetWare Connect 2 provides DOS and Windows PPP dialers for IP or IPX connections. These dialers can be distributed freely to remote users and provide time-saving features, such as a phone book, compression support, and call scripting. Macintosh AppleTalk users can dial into a network via NetWare Connect using Apple's ARA 2.0. Once connected, users simply click on Apple's Chooser to access network servers from their remote desktops.

Third-party remote control software users can dial in and take over a workstation on the LAN via NetWare Connect's pool of modems. Unlike solutions that provide remote control access only over IP or IPX connections, NetWare Connect can also host a remote control connection using third-party packages' highly streamlined asynchronous mode for the best possible performance.

NetWare Connect provides a security firewall, and acts as a communications front-end to remote control users. It presents a dynamic list of available remote control host workstations to inbound users, and directs them to the appropriate station.

Third-party remote control applications, such as Symantec's pcAnywhere, Triton's Co/Session, Microcom's CarbonCopy, Stac's ReachOut, and Citrix Systems' WinView, can all use NetWare Connect ports.

SUPPORTS OUTBOUND COMMUNICATIONS ACCESS.   Adding a modem pool to a network can be the single biggest network cost savings. NetWare Connect extends the benefits of dial-out modem pooling to PC or Macintosh applications. NetWare Connect provides modem independence, so a pool of dissimilar modems can be easily accessed by dial-out applications without reconfiguration by users.

Dial-out access can be secured by requiring each user or group of users to have rights on the NetWare Connect server. A comprehensive audit log is maintained, so dial-out usage can be tracked. Additionally, administrators can set up dial-out number filters to control costs or restrict access to particular dial-up services.

NetWare Connect allows terminal emulation packages to access minicomputers, including models from DEC, Prime, Data General, and Hewlett-Packard, through its pool of modems.

FLEXIBILITY TO SUPPORT MIXED NETWORK MEDIA.   NetWare Connect manages access to switched telephone lines, X.25, ISDN, and other services from a central point. It creates an environment where modems, fax units, PAD devices, or ISDN terminal adapters can be shared among multiple dial-up users, thereby lowering the overall equipment and monthly service costs.

ENHANCED SECURITY BUILT ON NETWARE.   NetWare Connect offers global and service-specific security options along with dial-back and dial-out call restrictions. Users are secured individually or as members of a group.

They can be passed directly to NetWare or authenticated first by Net-Ware Connect. PAP authentication can be used to give users different login IDs on the NetWare Connect server than on the network, and they can use third-party challenge-and-response security, such as Enigma Logic's SafeWord.

NetWare Connect leverages NetWare's security features, including NDS, login, password, and audit capabilities. Centralizing communications activities into a server gives administrators a single point for controlling an organization's remote access.

NETWARE DIRECTORY SERVICES INTEGRATION.   Users have remote connectivity attributes defined for them in the schema of the NetWare directory. As users are added to or removed from the directory, those actions are reflected in their remote access rights, too.

NETWORK MANAGEMENT.   NetWare Connect provides a comprehensive set of management tools for your enterprise-wide remote access needs. Net-Ware Connect includes a redesigned setup utility, with autodetection of ports and modems, which simplifies installation. Configuration summary information can be printed for reference. An enhanced audit trail displays port and user activity, line conditions, connections durations, as well as system alerts. NetWare Connect can generate Simple Network Management Protocol (SNMP) and ManageWise alerts, and send them to NetView and other management consoles.

With ConnectView, network administrators can get a global view of communications resources and activities from a Windows workstation or in conjunction with ManageWise™.

SCALABLE CONFIGURATION.   Each NetWare Connect server can be scaled to support up to 128 ports. The product is available in packages supporting 2-, 8-, and 32-port licensing. The number of users supported can be many times this port count—typically it can be 10 to 12 times as a rule of thumb for an average remote access user profile.

4.4.2.3 RAS   HARDWARE—BASED   SERVERS—SHIVA'S   LAN-ROVER/NETMODEM.   Shiva is one of the leaders in the remote access market, with a very focused product strategy exclusively devoted to this segment. They are able to achieve superior end-to-end optimization as compared to those vendors who have enhanced their networking products, such as routers, to support remote access. This is achieved by incorporating analog V.34 modem cards, ISDN Basic Rate (BRI) cards, and asynchronous serial cards with multiprotocol signaling software directly

Figure 4.21
Shiva's NetModem.
(Source: Shiva.)

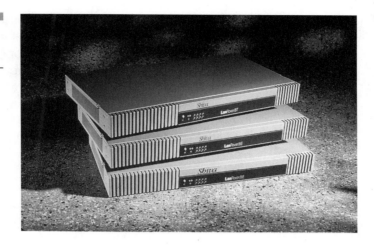

into Shiva's LanRover/NetModem remote access servers, and ShivaRemote client software. As a result, according to the vendor, worldwide there were more than two million users of Shiva remote access servers. (See Fig. 4.21.)

Shiva remote access servers offer Microsoft Windows 95 dial-in support, along with dial-in support for DOS, Windows 3.X, Windows NT, MAC OS, and OS/2. Shiva's LanRover/NetModem utilizes STAC data compression technology that delivers significant performance improvements during dial-in and LAN-to-LAN operation. Innovative dial-in and authentication APIs for Windows enhance your programming capabilities to integrate remote access with client/server applications. We consider this an important feature in order to develop mobile-aware applications or modify existing client-server applications to make them mobile-aware.

WEB-ENABLED REMOTE ACCESS. Like many other solutions, Shiva servers can provide easy Internet connections from either the office or on the road.

MANAGEMENT AND SECURITY. Shiva servers provide a high level of security and network management. Security features include integration with security devices, such as Radius, TACACS+, centralized user lists, password aging, Security Dynamics, Digital Pathways, and other options. To make network managers' jobs easier, Shiva solutions feature DHCP support for centralized TCP/IP addressing, terminal server support, and SNMP-based management.

FAX. Shiva servers allow you to add outbound fax capability, and save users time and effort.

END-TO-END OPTIMIZATION.   Performance for remote access applications depends on the number of factors and components involved. Shiva has analyzed the contribution of several different components, besides server capacity and link speed, toward end-to-end throughput. As a result, they have optimized some of these components. A performance-optimized 32-bit VxD implementation of ShivaRemote for Windows offers support for Multilink PPP MLP, virtual connections, client event logging, power switching, and STAC data compression all contribute to this overall improvement. Integrated client and server support for MLP and client PC ISDN cards enable remote dial-in users to send data over both 64 Kbps ISDN BRI B-channels for aggregated throughput four times faster than a 28.8 Kbps connection. Performance in dial-in, dial-out, and LAN-to-LAN modes is further enhanced by innovative remote-adapted routing, data and header compression, broadcast filtering, packet fragmentation, and delta technologies.

TRANSPARENT CONNECTIONS.   Users do not want to change the way they work when they move locations. They would like to perform the same actions regardless of whether the user is at the office, working from home, or in a hotel room. Virtual Connections provide the ability to automatically suspend and resume a physical analog or digital connection while spoofing network protocols, routing, and applications. In an ISDN environment, Virtual Connections are somewhat transparent to the user due to the relatively faster digital call setup times.

EASE OF INSTALLATION.   Shiva is providing an intelligent wizard to automate and simplify the installation process for the most common configurations. Autodetection of components, on-line help, and client installation scripting assist the novice network administrator in these tasks.

TARIFF MANAGEMENT.   Tariff Management offers a unique set of technologies designed to allow companies to control telecommunications costs. Shiva Tariff Management technologies fall into three areas—bandwidth, connection, and data control. The combination of these technologies gives network managers a robust tool set for managing variable network expenses, which can account for as much as 80 percent of the overall cost of operating a network.

4.4.2.4 INTERNETWORKING HARDWARE—BASED SERVERS— CISCO'S 5200 UNIVERSAL SERVER.   Cisco is a leader in the LAN-WAN internetworking market, especially with its router product line. In the past, Cisco provided limited remote access capabilities in its routers.

However, during 1996—1997 the company has significantly enhanced its capabilities for the remote access market. This seems to fit well into many large organizations' networking infrastructure plans because they already have Cisco products. Cisco's AS5200 is its first entry into a new product line of universal integrated access servers. AS5200 provides dial access to enterprise and service provider networks from single users or from remote networks by using either asynchronous modems or digital Integrated Services Digital Network (ISDN) technology.

The AS5200 has become quite popular because it integrates the functions of stand-alone channel service units (CSUs), channel banks, modems, communication servers, switches, and routers into a single, stackable chassis. The AS5200 contains digital ISDN and asynchronous modem access server functionality, ideal for the mixed-media dial environments that are becoming more prevalent every day.

By terminating both analog and digital calls on the same chassis, from the same trunk line, the AS5200 gives customers an easy migration path from today's predominantly analog dial services to tomorrow's principally digital ISDN dial services. (See Fig. 4.22.)

Table 4.2 presents the components, features, and benefits of the AS5200. We elaborate further on the following key features of AS5200:

UNIVERSAL ACCESS THROUGH INTEGRATION WITH INTERNETWORKING SOLUTIONS. The AS5200 provides an integrated solution for routing and internetworking. It allows ISDN and analog modem callers to dial into the same chassis, using the same telephone number. This feature enables customers to save money by using one trunk line instead of two for all calls, thus reducing the number of system components, which reduces operational costs. The AS5200 also supports multiple networking and routing proto-

Figure 4.22
Cisco's AS52000.
(Source: Cisco.)

TABLE 4.2

Description of
Cisco AS5200

Feature/Component	Description of Feature	Benefit
Packaging	Two PRI/T1 lines with integrated CSUs	Up to 48 simultaneous callers terminated in one server
	Up to 60 integrated fully managed modems	Full network management of server modems
	Ethernet Port	Supports local LAN attachment
	Two synchronous serial ports for back-haul to the corporate hosts	Distributed network design
Software support	Cisco IOS	Reduced training requirements and common management in predominantly Cisco infrastructure
Remote access support	Both remote node, remote control and terminal services (Telnet, Rlogin, TN3270)	
Multiple terminal support	Telnet, Rlogin, TN3270, XRemote, X.28, X.25 PAD, protocol translation, LAT, Reverse Telnet	Provides support for legacy applications and allows these applications to migrate to LAN remote node applications
Router support	Extensive routing support RIP, RIP2, SAP, IGRP, EIGRP, NHRP, PIM, OSPF, IS-IS, BGP4, EGP, IDRP	Multiprotocol routing support eliminates need for external router
WAN optimization	Modem compression Link sharing for asynchronous and synchronous connections Dial-on-demand routing Bandwidth-on-demand routing	Reduced operating costs
Network management	SNMP, MIB II, enterprise MIB, SNMP2, RMON, ISDN MIB, Modem MIB, CiscoWorks, and CiscoView GUI applet	Full-function network management for monitoring and problem isolation
Security	TACACS+, RADIUS, PAP< CHAP, token cards, filtering, Kerbros V, PPP, Callback, per user accounting, access list violation	Can be used as a firewall

SOURCE: Cisco product catalogue on Cisco's Web site www.cisco.com.

cols available in the industry. Deploying the AS5200 provides central site administrators the ability to connect users with a single, easy-to-use platform. The AS5200 is designed to be flexible enough to fit in many different network designs.

COMMON SOFTWARE FOR INTERNETWORKING AND REMOTE ACCESS. Since the Cisco IOS is the software running on many different internetworks (both on Cisco routers and other platforms), AS5200 offers compatibility to those networks. AS5200 supports both remote-node and remote-LAN dial protocols, besides the full suite of routing protocols. Cisco customers find this environment familiar since it reduces their training requirements for installing, configuring, and managing their networks.

INTEGRATED MODEM TECHNOLOGY. As many network administrators have learned over the years, all modems are not created equal. Connection rates, throughput, reliability, and manageability are all important factors for modems deployed in a central site. The AS5200 integrates modem technology from Microcom, a leader in the central site modem environment, with the CSU, modems, router, and access server components integrated into one chassis. AS5200 avoids incompatibility concerns among multivendor installations.

SCALABILITY. The AS5200 supports ISDN B-channel aggregation over multiple chassis. Using Multichassis Multilink Point-to-Point Protocol (MMP) across several AS5200 chassis. AS5200 allows the aggregation of multiple calls terminated on multiple servers. Network managers have the ability to stack multiple AS5200s for high-density access server systems. The modular design allows implementation of future technology.

SUITABILITY FOR DIFFERENT APPLICATIONS. The AS5200 can be used for a variety of different applications as follows:

- Service provider POP
- Centralized corporate dial access site
- Mixed-media enterprise dial

The AS5200 can be used as a distributed analog/digital dial platform in networks with multiple, geographically dispersed points of presence (POPs). The AS5200 accepts calls from service users over a variety of protocols, and allows ISDN users to dial into the same platform as analog modem users, using a single ISDN PRI trunk. The AS5200 is fully manageable, using both a Command Line Interface and CiscoWorks™. The AS5200 has many remote diagnostic features, specifically for the demands of a service provider network.

The AS5200 is designed with high-density, centralized environments in mind. The AS5200 uses only two rack units of real estate, allowing for high concentrations of modem and ISDN dial users to be serviced in less physical space than multibox, multivendor solutions. The AS5200 can aggregate calls using MMP across multiple chassis. This feature allows the AS5200 to scale to many more PRI lines than a single chassis. This capability also enables a resilient network implementation where a CPU or power supply failure impacts only a small part of the access solution.

The AS5200 also operates very effectively in the mixed-media enterprise environment. The AS5200 supports the needs of mobile users, who typically use modems to dial in from disparate locations, and telecommuters, who are increasingly using high-speed ISDN Basic Rate Interface (BRI) technology to access the enterprise network. Both types of users have specific needs when dialing in, and the AS5200 services them both in a single, easy-to-manage platform. By deploying the AS5200, network managers can leverage training that their people have on other Cisco equipment.

4.4.2.5 INTERNETWORKING HARDWARE—BASED SERVERS—3COM'S ACCESSBUILDER FAMILY.

3COM is another major internetworking vendor that offers a family of products for remote access servers from the low end to the high end. Remote users dial into the AccessBuilder over analog or digital (ISDN) telephone lines, achieving direct, transparent links to Ethernet and Token Ring LANs—just as if these users were locally attached. Some of the key features and benefits of the AccessBuilder family are:

- Dial-in/dial-out connections for geographically dispersed employees, business travelers, telecommuters, and remote office workers.
- Multiprotocol bridging and routing for client-to-LAN connections.
- Support for most popular desktop operating systems, such as Windows 3.1, Windows 95, Windows NT, MacOS, Unix workstations, and so forth.
- 3COM's Transcend™ graphical SNMP management software for IP and IPX environments. This can be integrated with HP's OpenView®, Sun Microsystem's SunNet™, and IBM's NetView®/6000.
- Advanced suite of built-in security features coupled with comprehensive NOS and third-party security products. The AccessBuilder built-in

security prevents unauthorized network access through password protection, standard authentication schemes, such as PAP, and CHAP, automatic call-back. CallerID validation denies access to unauthorized ISDN calls even before the call is answered. In AppleTalk environments, AppleTalk zone filtering hides sensitive AppleTalk zone resources from remote users. AccessBuilder servers automatically log and report all unauthorized access attempts to the management station.

- AccessBuilder servers also work with hardware security systems from third-party vendors, including Digital pathways, LeeMah, and Security Dynamics.

- Simultaneous analog (modem) and digital (ISDN) connectivity on a single-server platform.

- Flexible, dynamic bandwidth management to reduce dial-up line charges. Dial-on-demand keeps line connections open only when remote access connectivity is required. To save line charges, connections are torn down when network requests have been satisfied. With 3COM Impact ISDN adapter, IP and IPX spoofing rebuilds the connection for end-user data only. Batched data transfer can be scheduled after hours during less-expensive rate periods.

There are a number of products in the AccessBuilder family. The SuperStack II AB2000 serves the needs of remote office and workgroup environments, with fixed WAN ports for Ethernet configurations. The multiprotocol AB4000 serves the needs of the enterprise marketplace, where integrating disparate network environments is a key consideration. AB4000 supports both analog and ISDN links.

4.4.2.6 APPLICATION SOFTWARE—BASED SERVERS—XCELLE-NET'S REMOTEWARE. Our objective in this book is to provide information on end-to-end logical connectivity. Remote access servers described so far provide physical network communications between remote users and a LAN. A major objective here is to make the remote network connection transparent to the user—inside and outside the office. In reality, the remote users want many value-added services that are unique to the needs of the mobile user. To provide this logical connectivity—between the client applications on the end-user devices and back-end application or database servers—another remote access application server, such as Xcellenet's RemoteWare is required. Xcellenet's Remote-Ware is based on a client-agent-server architecture. It uses a sophisticated agent technology to provide mobile-user-specific application functions, such as the following:

- Provides field professionals with current information from central headquarters, such as the latest product, service, and pricing changes
- Delivers software applications and updates to the field
- Assembles and republishes information collected from on-line services
- Delivers mainframe reports and multimedia presentations electronically

In such an implementation, remote access servers, such as Shiva's Lan-Rover or Microsoft NT Remote Access Server may sit in front of the Xcellenet RemoteWare application server, as shown in Fig. 4.23.

This combination through two separate logical processors provides a powerful turnkey application-oriented solution for remote access needs.

4.5 Remote Access and LAN Bridges, Routers, and Switches

As you may have observed from the material covered in this chapter so far, remote access functionality can be provided by software or hardware products that had their heritage rooted in bridges, routers, LAN switches, Internet Web servers, or specialized remote access servers. One of the questions network designers must answer is: *which is a better approach?* There is

Figure 4.23
Xcellenet's Remote-Ware remote access application server.

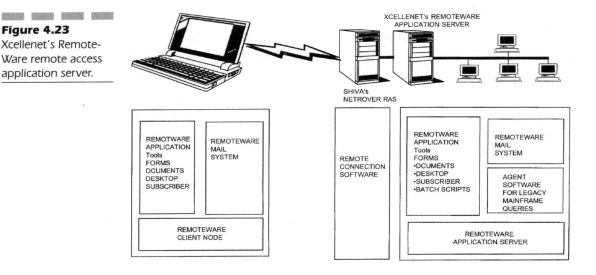

TABLE 4.3

Comparison of Remote Access Server Approaches

	OS-based RAS (MS Windows NT RAS)	NOS-based Server (Novell Connect)	Specialized RAS (Shiva LanRover)	Remote Control (pcAnywhere)	Bridge/Router/ Switch (Cisco, 3COM)	Application-Oriented (Xcellenet RemoteWare)
Remote access functionality	Medium	High level	Richest functionality from RNA perspective	High level of transparency	High functionality	High application level capability
Support for various network links	Limited	High—most links supported (analog, digital now)	Highest support, including support for emerging links (e.g., ADSL)	Limited to medium	High	Depends on which network server is selected—Look for one supported by the vendor
Security and network management	Medium—Third parties gradually building interfaces Familiar management, limited network trouble shooting	High (Novell bindery services)	High (Integration with third-party security products)	Low	Medium	Basic—medium (Security not inherent capability of RemoteWare)
Internet support	Not inherent; available through other Web servers from MS	Yes, through Web services	Yes	Not at the time of publication	Yes	Yes, through ISCOCOR messaging gateway
Scalability	Medium	Medium	Low to medium high	Low	Highest	Medium
Cost	Low	Medium	Generally high	Relatively high	Low	High for server, though low when applications are included
Suitability	Low cost, adequate function for certain situations	Best fit for NetWare-centric configuration	Good overall solution for general-purpose remote access	Easy-to-implement solution	High network optimization; superior integration with network infrastructure	Superior business solution, preferred by users

113

no simple answer to this question. It should be analyzed from architectural, functional, cost, and management perspectives. Although we discuss design-related issues and considerations in Chap. 13, the comparison presented in Table 4.3 summarizes some of the key points. We would like to warn the reader that the remarks are very general. The reader should do a more detailed analysis after the research of current products than what this table provides.

Summary

We discussed how remote access works in this chapter. After describing the components of a remote access solution, we investigated four basic approaches: terminal emulation, remote control, remote node, and mail gateways. The first and the last mode of remote access are application-dependent implementations with limited future enhancements. Therefore, remote control and remote node are the two essential modes of implementing remote access. Remote control implementations, such as pcAnywhere and Reachout are popular among smaller organizations. Remote node is preferred by medium and larger organizations because of cost and security. In fact, there are several hybrid implementations allowing both modes. To further understand remote access, we surveyed server product implementations by companies, such as Microsoft, Novell, Shiva, Cisco, and Xcellenet. The RemoteWare product from Xcellenet is a unique application-specific remote access product that enhances the physical communications capability of remote access servers.

5

Remote Access Technology Architecture

Remote access products were created for helping technical support personnel but found a fertile home for a bigger need in the market. It is not surprising, therefore, that remote access does not have a unified architecture. It just extends the LAN architecture through a remote access server. Now that it has acquired a critical mass, it is time for us to sit back and see how it fits into the rest of its surroundings—general-purpose mobile computing arena.

—Chander Dhawan

About This Chapter

Remote access solutions take on many different forms. We discussed detailed implementation of these methods in Chap. 4. Here, we want to raise certain architectural issues. We discuss these issues in the context of a generic mobile computing architecture that can extend current implementation of remote access solutions to address needs of wireless networks as well. This discussion comprises the following sections:

- Impact of business users' information access needs on technology architecture
- Hardware architecture
- Network architecture
- Software architecture components
- Logical (application level) architecture
- Interoperability considerations from an architectural perspective
- Architectural issues
- Technology principles for an integrated architectural framework
- Methodology for developing a custom architecture

Simple to More Complex Configuration with Enterprise Deployment.   While many early implementations of remote access projects were often relatively simple and employed only one specific application and one network and few servers on a LAN, we illustrate a more complex configuration that extends the solution to legacy servers as well. Our discussion of the configuration's architecture highlights the technical design issues that arise as such solutions are rolled out across organizations and are integrated with other business applications.

5.1 Impact of Business Users' Information Access Needs on Technology Architecture

We reviewed in Chap. 1 how the increasingly competitive nature of our business has led to extensive mobility and scattering of the workforce. Traditional information technology solutions have catered to the needs of fixed-site workers inside the organization during the last few decades. However, the new business paradigm of the 1990s requires that we provide connectivity to a new breed of workforce consisting of mobile workers, telecommuters, and workers in remote regional offices, who might be in fixed locations but are not connected permanently. Controlled access must also be given to business partners, suppliers, and customers. These users are extremely critical to the success of the business because they deal with your revenue-generating customers.

The term *mobile user* now includes professional executives making business decisions away from home offices, salespeople closing deals, service representatives on emergency repair calls, suppliers' shippers advising shop floor supervisors of exact times of arrival of just-in-time inventory items, and business partners accessing databases of technical support material or sales presentations.

Figure 5.1 represents some of the more important factors determining mobile computing technology architecture (types of remote users, types of business applications, amount of data transferred, currency of data, and geographical territory that mobile users travel).

A typical remote access solution being implemented today and its technology architecture are based on the following assumptions:

1. A large number of mobile users (several hundred to several thousand) need to access information from or send information to one or more application or database servers. Many of these users have access to public switched telephone networks (PSTN) or the Internet, which provide inexpensive and reliable communications. Those users who

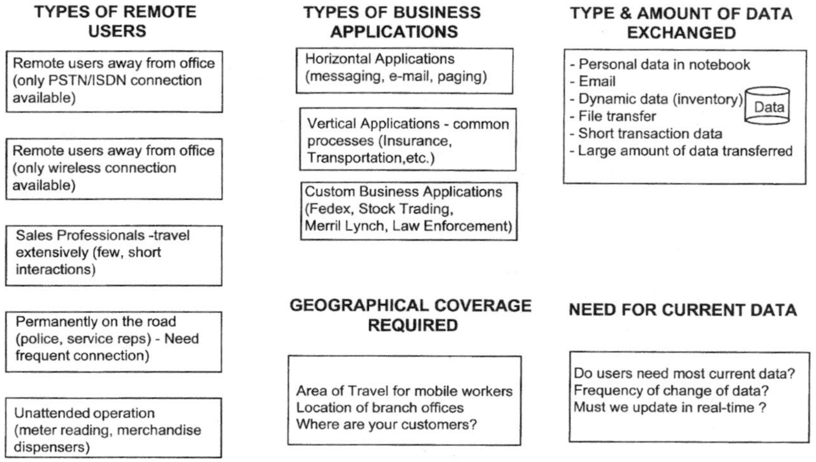

Figure 5.1

Business factors influencing technology architecture.

TYPES OF REMOTE USERS

Remote users away from office (only PSTN/ISDN connection available)

Remote users away from office (only wireless connection available)

Sales Professionals -travel extensively (few, short interactions)

Permanently on the road (police, service reps) - Need frequent connection)

Unattended operation (meter reading, merchandise dispensers)

TYPES OF BUSINESS APPLICATIONS

Horizontal Applications (messaging, e-mail, paging)

Vertical Applications - common processes (Insurance, Transportation,etc.)

Custom Business Applications (Fedex, Stock Trading, Merril Lynch, Law Enforcement)

GEOGRAPHICAL COVERAGE REQUIRED

Area of Travel for mobile workers
Location of branch offices
Where are your customers?

TYPE & AMOUNT OF DATA EXCHANGED

- Personal data in notebook
- Email
- Dynamic data (inventory)
- File transfer
- Short transaction data
- Large amount of data transferred

Data

NEED FOR CURRENT DATA

Do users need most current data?
Frequency of change of data?
Must we update in real-time ?

spend a lot of time on the road (road warriors) and cannot find a telephone jack nearby require wireless connection.

2. The types of business applications may include:

   ■ Store-and-forward messaging applications such as electronic mail and paging (one-way or two-way)

   ■ Updates of local operational data files (e.g., price list, service schedules for the day) or downloads of data (e.g., sales summaries, presentation materials)

   ■ Retrieval of operational data such as orders, audits, insurance appraisals, traffic citations, and the like, is handled several times during the day or at the end of day

   ■ Internet access for external e-mail and the new breed of electronic commerce applications

3. The data transferred by remote users may consist of:

   ■ File transfers of data to synchronize local files on notebooks

   ■ Records transfer of transaction data (typically message-based) stored in databases on regional or central servers; elements of data required to complete a unit of work may reside on disparate databases and platforms

4. Frequency of connections by remote users and the amount of information transferred between remote users and homes/offices during a working day varies with the type of application used. This may range from occasional use of networks (i.e., once or twice a day) to almost continuous connections for several hours (e.g., as in the case of telecommuters).

5. The number of physical network connections (implying the number of physical ports required at communications servers) is typically a fraction of actual numbers of remote users. This is analogous to a telephone network where the number of trunk lines is always less than the number of telephone sets in the organization.

5.2 Hardware Technology Architecture

Hardware involved in a remote access configuration consists of one or more of the following components:

■ A mobile computer. Although PC notebook continues to be the end-user device of choice today, it may not retain the supreme position in the future. PC might be supplemented by several of its incarnations or variations, such as a pen-based portable computer, application-specific handheld computer, new Java-capable network computer for Internet browsing, a PDA (personal digital assistant—e.g., Newton or Simon), a palm pad (e.g., HP200LX or Zenith Cruise Pad), or Macintosh-based devices. New two-way pagers introduced by Research in Motion in 1996 or Motorola PageWriter being introduced in 1997 may also provide limited messaging and Internet browsing capabilities in the future. Typically, a variety of hardware platforms are involved, ranging from the most common Intel to Power PC and RISC microprocessors.

■ A suitably configured modem or an appropriate digital network interface device. This might be a switched network modem, a digital adapter for ISDN, a dial-tone generation adapter for cellular connections, or a specialized radio modem for wireless networks such as ARDIS, RAM, or CDPD (e.g., ERICSSON Mobidem 901 for RAM, IBM's CDPD modem). A single universal modem that supports all connection scenarios does not currently exist, though PC Card—based Ethernet/fax/cellular modems are now available in the market. Where wireless networks are concerned, modems can work only with mobile computers that have radio transmitter equipment installed. In cellular connections, cellular telephones have this functionality built in.

■ A communications server and/or a wireless network switch/gateway (e.g., Shiva's NetRover for asynchronous remote network access; IBM's ARTour for wireless networks). This type of communications server/switch provides the following functionality:

—Asynchronous wireline session connection and disconnection services

—Session router management across large numbers of remote mobiles

—Mobile identification

—Network login (i.e., security management)

—Asynchronous wireless network connection and disconnection services

—Protocol conversion (i.e., a gateway function) in some cases

■ An application and/or database server on a LAN, a minicomputer, or a mainframe where information resides. Normally there is a LAN (Ethernet or Token Ring) connection between the communications server/switch and the hardware platform on which information servers reside. This connection can also be accomplished either through high-speed buses, channel connections, or wide area network private lines.

■ An electronic mail server on an appropriate platform, typically a LAN.

A basic hardware configuration for remote access computing is reflected in the schematic in Fig. 5.2.

As remote access requirements grow in terms of numbers of users, types of networks supported, and application processing platforms accessed, hardware configuration becomes more complex, as depicted in Fig. 5.3.

5.2.1 RAS or MCSS

The author has proposed the concept of a general-purpose mobile communications server switch called MCSS in his previous book, *Mobile Computing—A Systems Integrator's Handbook,* cited earlier. We believe that RACS/RAS is a special form of MCSS designed specifically for remote access, primarily with port management and LAN device emulation. From an architectural perspective, it serves the same functions.

Figure 5.2
Hardware architecture for a basic remote access configuration.

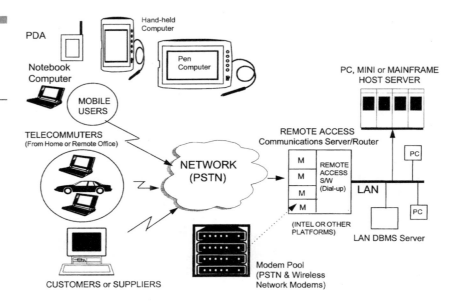

Figure 5.3
More complex remote access/mobile computing configuration.

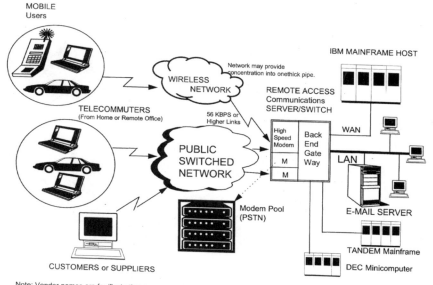

Note: Vendor names are for illustration purposes only. It does not imply an exclusive role of named vendors in architecture.

5.3 Network Architecture for Remote Access

In the following chapters, we review the types of network connections involved in a remote access solution. Figure 5.4 reflects the variety of possible choices available to a network designer.

The following points are worth noting in terms of network architecture of a remote access solution:

1. The needs of users are sufficiently different to warrant support of multiple network links. There is no single network that can satisfy the requirements of all the applications at an affordable price.

2. While ISDN coverage is well entrenched in Europe and is getting better in most states in the United States and Canada, ADSL coverage will continue to be spotty in different parts of the country for several years.

3. Wireless networks are generally slower and more expensive than their wired counterparts. Moreover, most wireless networks do not use LAN extension as a physical connectivity mechanism.

4. The gap between remote access links and LAN speeds will continue to exist for the foreseeable future.

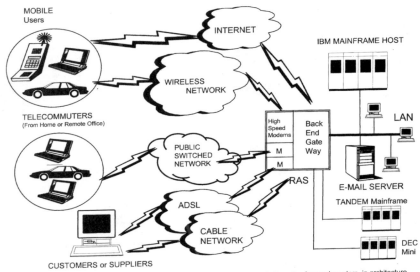

Note: Vendor names are for illustration purposes only.   It does not imply an exclusive role of named vendors in architecture.

5. Information resides not only on LAN-based servers but on mini-
    computers and mainframes. There may be another backbone net-
    work to deal with to reach the ultimate information server.

5.4 Software Technology Architecture

Having discussed the hardware and network architecture, we now discuss
software architecture of remote access solutions. We start with software
components that comprise this architecture and then see how these components are glued together to construct an end-to-end solution.

5.4.1 Software Components

Here we concern ourselves with mobile client software and the systems
software in the communications server. We briefly mention the application and DBMS servers residing on various platforms. Then we discuss the
role of an agent software that will enhance the client-server model for
mobile computing.

5.4.1.1 REMOTE CLIENT SOFTWARE. Figure 5.5 shows software components from a mobile client's perspective.

The client software suite in mobile client nodes provides the following functionality:

- Remote mobile workstation operating software (DOS/Windows 3.1, Windows 95, OS/2, Mac OS, Solaris Unix) along with network software drivers controls hardware devices and peripherals (display, pen, voice input, PCMCIA cards, etc.).

- A user interface or shell. This might be included in the operating system itself, as in the case of Windows 3.1 and Windows 95, or it may be distinct from the operating system, as is the case of Unix with MOTIF.

- The client portion of the communication server software (e.g., Windows 95's RAS client, Netware Connect, Novell's asynchronous remote access software—which provides connectivity through PSTN as well as cellular circuit switched network). Almost all remote access software vendors have a proprietary client component that runs on a laptop.

- Communications transport layer software like TCP/IP. TCP/IP has become the most common transport protocol for LAN/WAN internetworking.

- Optionally, middleware software that sits between business applications and transport layers like TCP/IP.

- Software drivers for specific networks (e.g., PPP or SLIP for PSTN, RAM Mobitex, ARDIS, CDPD, or the emerging PCS network).

Figure 5.5
Software architecture of the remote workstation.

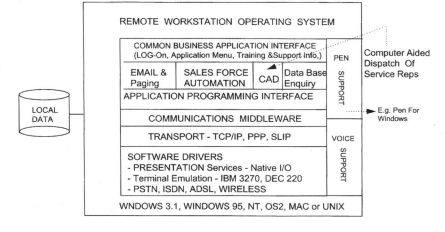

■ Pen and voice recognition interfaces are becoming fashionable in new mobile applications. Pen not only facilitates the inputting of data, but the interface can also be used as a mouse replacement.

5.4.1.2 REMOTE ACCESS COMMUNICATIONS SERVER (RACS/RAS) SOFTWARE.   Communications software that is the brain of RACS/RAS residing on office LAN where information resides, provides the following functionality:

■ Network connection assignments to remote users

■ Communication port or modem pool management

■ Security verification

■ Multithreading of physical connections from different users on multiple ports on a single logical communications channel like COM1, COM2, and so forth

■ Communications protocol handling

■ Logical connections to back-end application processors or DBMS systems

5.4.1.3 APPLICATION AND DBMS SERVERS.   Application and database server software suites provide business application functionality. This software may reside on one or more of the following three platform categories:

■ LAN-based database servers such as Sybase, Oracle, SQL server implemented under OS/2, Windows NT, or Unix operating systems on Intel, IMPS or similar hardware platforms

■ Minicomputer-based servers such as DEC VAX or AS/400

■ Mainframe supersavers, including IBM, VMS, UNISYS, and Tandem Cyclone Guardian

Communications support for applications on these platforms may dictate a need for protocol conversions or gateways, or an equivalent function, unless middleware software is implemented with embedded protocol conversion.

5.4.1.4 HOW DOES IT FIT TOGETHER?   A schematic of major software components for client and server implementations is shown in Fig. 5.6.

Figure 5.6
End-to-end software architecture for a remote access solution.

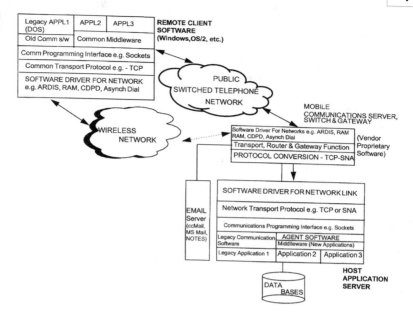

5.4.1.5 TWO MODERN PARADIGMS FROM AN APPLICATION DEVELOPMENT PERSPECTIVE

THE CLIENT/SERVER MODEL.   The client/server model provides the flexibility required in mobile and remote access applications. This model has in fact become the de facto standard in the application development paradigm. In mobile computing, it is very important to give the maximum amount of business functionality and a replicated copy of relevant data to remote users without compromising good information resource management (IRM) principles and security. This does mean, however, that a common secure data repository should be on the LAN server behind the firewall.

THE AGENT-BASED CLIENT/SERVER MODEL.   This is a paradigm in which the basic client/server model is enhanced with the "intelligent agent" software concept. The agent portion of the software works on behalf of remote clients while they tend to other matters, such as dealing with customers. The agent organizes incoming information in the background (inquiry responses, e-mail messages, etc.) and forwards it to the remote user whenever a connection is made.

The agent-based client/server model has the following advantages when it comes to mobile computing:

■ It makes possible single automated communications sessions that cater to all information needs at any given time.

- It reduces costs and the need for resources, since sessions do not have to be maintained while remote users are inactive.
- It boosts user productivity. With the agent working in the background, users are free to pursue other matters.
- It takes advantage of both client/server and store-and-forward technologies.
- Because the server portion of the agent can be centralized under the management of the IS organizations, it shares control responsibility between IS and the user organization.

5.4.2 Agent-Based Client/Server Software Example

Since this concept is important for making existing LAN remote access applications partially mobile-aware without a lot of application changes, we describe in detail an agent-based client/server implementation, using as an example a product called RemoteWare, developed and marketed by XcelleNet, an Atlanta-based company.

RemoteWare is essentially an application engine for mobile communications with a few interesting enhancements over a straightforward application gateway. RemoteWare addresses the client application portion of the client/server paradigm, interface between a remote access communications server (RACS) and an application processor, and e-mail server functionality. It uses an existing RACS such as Shiva's NetRover.

Simply put, RemoteWare is a 32-bit OS/2 platform (NT version became available in 1996) that lets mobile users send messages, transfer files, and query remote databases. The software comprises e-mail and forms applications, OS/2-based server management, and client software that runs on Windows, OS/2, Macintosh, and other environments.

5.4.3 Agent-Based Software Technology

Agent-based design is a combination of real-time interactions with batched store-and-forward transmissions. The store-and-forward implementation is appropriate because mobile workers are more often than not interrupted in their work sessions by other activities and can seldom sit down for long interactive sessions. We look at several fundamental concepts in an agent-managed remote work session as follows:

■ Communication session

■ Staging of session work

■ Session scheduling

■ Agent processes

5.4.3.1 COMMUNICATION SESSION. The communication session, in this context, may consist of the following:

■ Security validation at the start of each session

■ Communication tasks, such as file transfers initiated by the client application, file synchronization, operating system commands, or processes invoked as a result of central server commands

■ Application tasks, such as servicing of specific remote client application-directed activities (e.g., posting of a sales order into a sales order database)

■ Electronic software distribution under the direction of central server

■ Resource monitoring and diagnostic tasks (disk space and memory consumption)

With script capability available in application engines such as Remote-Ware, these tasks can be completed automatically using integral run-time data compression and checkpoint restart techniques where appropriate—as if there was an agent supervising the completion of these tasks as one set.

5.4.3.2 STAGING OF SESSION WORK. *Staging* is a process by which activities to be accomplished (session work) during subsequent communications sessions are queued up at the central server as well as on remote notebooks. In this context, staging can be considered as batching of work to be handled at a later point. For example, when a remote salesperson submits an inquiry, the central server needs to access information from several distributed databases. In agent-based implementation, the salesperson can book an order off-line; whenever the next communications session takes place, the agent will both update the data and provide answers to any previous inquiries. (See Fig. 5.7.)

5.4.3.3 SESSION SCHEDULING. *Session scheduling* is the process by which session work is queued and initiated at a central RemoteWare server. In this context, the RemoteWare server can intelligently stage session work for thousands of remote notebooks. For example, if information is to be retrieved from 100 notebooks and there are only 10 modems in the

Figure 5.7
Client-agent-server
model for remote
access technology
architecture. (a) Ses-
sion work staged at
SERVER or CLIENT;
(b) Agent Process on
Server and Client.
(Source: Xcellenet
RemoteWare Technol-
ogy Overview White
Paper.)

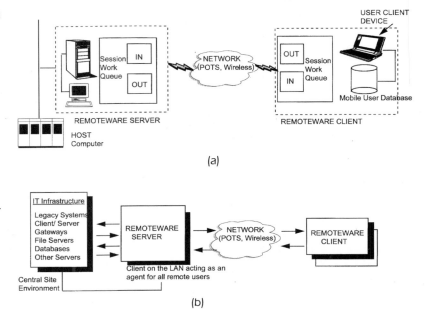

modem pool, RemoteWare schedules 100 sessions—launching the first 10 and holding the remaining in a queue. The scheduling of sessions and their execution can be controlled by time of day, type of connection (e.g., use only PSTN and not wireless network for file transfers), or the availability of resources (disk space in the remote notebook).

5.4.3.4 AGENT PROCESSES.

Agent process is simply a process that does work on behalf of a user. Whereas users interact with an application program with keystrokes, mouse clicks, and pen- or voice-activated commands, agent processes do not require user interactions. With RemoteWare, agent processes themselves interact with other program processes on behalf of the remote user in order to accomplish specific session activities. For example, an agent process may post an order into a central DB2 database on behalf of the remote user, if certain conditions (e.g., credit verification) are met.

Agent processes can be designed to handle both communications and application tasks. RemoteWare applications have agent processes automatically built into the system (e.g., querying from an ODBC-compliant database). Agent processes may reside at the central server, at the remote client workstation, or at both places.

The preceding concepts of sessions, staging, session work, session scheduling, and agent processes provide a context for implementing extensions to the traditional client/server model. The RemoteWare server acts as a central agent capable of performing complex functions on behalf of the large number of remote clients. With continuous connection to a central IT infrastructure, the RemoteWare server buffers remote clients from the complexities of connection to a number of distributed computers. It improves the overall system integrity and performance by acting as a central information wall à la a firewall in the Internet context. (See Fig. 5.8a and b.)

The agent/server model does not require continuous connection, which fits in nicely with an important characteristic of the mobile workers' lifestyle. All work takes place off-line and is staged for the next communications session. When a connection takes place with a specific RemoteWare client, the RemoteWare server takes over control and performs all the session work that needs to be done. It automatically synchronizes all session work and drops the connection after completing the tasks. This allows for all staged tasks for each user to occur in a single, automated communications session.

5.4.4 Intelligent Agent Functionality

Intelligent agent in the server handles the following functions:

5.4.4.1 SESSION WORK QUEUING AND SCHEDULING. In order to use the wireless link more efficiently, message traffic should be inserted into multiple queues with different priorities (e.g., one for immediate transmission and the second for a later transmission when the user has access to a cheaper cost link).

5.4.4.2 MULTIPLE MESSAGES PER SESSION OR TRANSMISSION. To minimize the protocol overhead, multiple messages could be batched during a single transmission.

5.4.4.3 RETRIEVAL OF MAIL MESSAGES BASED ON PREDEFINED CRITERIA. For example, retrieval includes receiving headers of messages from the boss and the last three customers (through an intelligent mail search engine).

5.4.4.4 INTELLIGENT MESSAGE CODIFICATION. This codification provides the ability to define a limited number of codes with predefined meanings.

5.4.4.5 DEFAULT MESSAGE PROCESSING BASED ON LINK TYPE. Send only urgent messages to wireless connected device.

Figure 5.8
(a) Agent-based client
server example—
general purpose;
(b) Agent-based client
server architecture for
wireless networks
(messaging example).
(Source: Xcellenet
RemoteWare Technol-
ogy Overview White
Paper.)

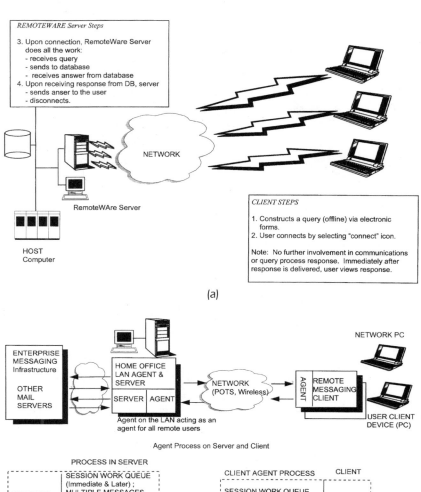

REMOTEWARE Server Steps

3. Upon connection, RemoteWare Server
   does all the work:
   - receives query
   - sends to database
   - receives answer from database
4. Upon receiving response from DB, server
   - sends anser to the user
   - disconnects.

NETWORK

RemoteWAre Server

HOST
Computer

CLIENT STEPS

1. Constructs a query (offline) via electronic
   forms.
2. User connects by selecting "connect" icon.

Note: No further involvement in communications
or query process response. Immediately after
response is delivered, user views response.

(a)

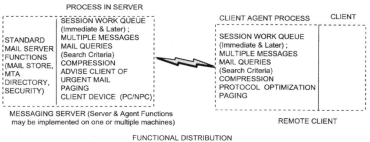

NETWORK PC

ENTERPRISE
MESSAGING
Infrastructure

OTHER
MAIL
SERVERS

HOME OFFICE
LAN AGENT &
SERVER

SERVER | AGENT

NETWORK
(POTS, Wireless)

AGENT

REMOTE
MESSAGING
CLIENT

USER CLIENT
DEVICE (PC)

Agent on the LAN acting as an
agent for all remote users

Agent Process on Server and Client

PROCESS IN SERVER

CLIENT AGENT PROCESS   CLIENT

SESSION WORK QUEUE
(Immediate & Later) ;
MULTIPLE MESSAGES
MAIL QUERIES
(Search Criteria)
COMPRESSION
ADVISE CLIENT OF
URGENT MAIL
PAGING
CLIENT DEVICE (PC/NPC)

STANDARD
MAIL SERVER
FUNCTIONS
(MAIL STORE,
MTA
DIRECTORY,
SECURITY)

SESSION WORK QUEUE
(Immediate & Later) ;
MULTIPLE MESSAGES
MAIL QUERIES
(Search Criteria)
COMPRESSION
PROTOCOL OPTIMIZATION
PAGING

MESSAGING SERVER (Server & Agent Functions
may be implemented on one or multiple machines)

REMOTE CLIENT

FUNCTIONAL DISTRIBUTION

(b)

5.4.4.6 PROVIDE PAGING FUNCTION.   The paging function is provided if requested by sender and acceptable to the recipient.

5.4.4.7 HANDLE DEVICE DIFFERENCES.   Full-function PC and
network PC workstations may have different functional capabilities.

5.4.5 Are There Any Implementations of Client-Agent-Server Architecture?

As far as we know, there is no true implementation of client-agent-server architecture of remote access solutions. However, some vendors such as Xcellenet of Atlanta have a general-purpose mobile application and development environment based on this design. While it fits well in a wireline remote access environment, it has limited support for wireless networks, except for circuit-switched cellular and CDPD. Oracle's Mobile Agents and IBM's ARTour product provide mobile-aware application development design capabilities.

5.5 Interoperability Considerations from an Architectural Perspective

One of the greatest challenges in networking across heterogeneous platforms is to find out what works with what in different components. Figure 5.9 shows a variety of client operating systems, end-user interfaces, physical network links, LAN operating software, transport layers, and information server platforms. You must select a remote access solution that supports your preferred choice in each of these categories. Only then, an end-to-end connectivity can be assured.

In spite of vendor claims, system designers must watch for all the disclaimers, exceptions, and other constraints which may be imposed by nonstandard combinations of these layers. Finally, they should pilot and test the precise hardware, network, and software combinations they intend to implement.

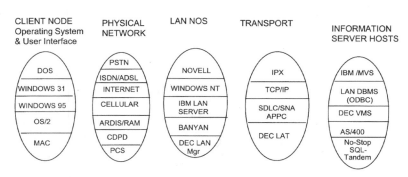

Figure 5.9
Interoperability considerations for remote access solutions.

CLIENT NODE Operating System & User Interface	PHYSICAL NETWORK	LAN NOS	TRANSPORT	INFORMATION SERVER HOSTS
DOS	PSTN	NOVELL	IPX	IBM /MVS
WINDOWS 31	ISDN/ADSL	WINDOWS NT	TCP/IP	LAN DBMS (ODBC)
WINDOWS 95	INTERNET	IBM LAN SERVER	SDLC/SNA APPC	DEC VMS
OS/2	CELLULAR	BANYAN		AS/400
MAC	ARDIS/RAM	DEC LAN Mgr	DEC LAT	No-Stop SQL-Tandem
	CDPD			
	PCS			

5.6 Architectural Issues in Remote Access

Based on the discussion so far, we can list a number of architectural issues that arise. These issues should be analyzed in detail before selecting a comprehensive remote access strategy for the organization.

5.6.1 Application Suite Beyond LAN Applications

While the initial suite of remote access applications may be LAN-based, you should consider operational applications that your users will demand in order to make them truly mobile. E-mail and personal productivity applications just quench the initial thirst of a free-spirited sales force. After getting comfortable with this initial suite, they will ask for more and more of the applications that they can access inside the office. Many of these applications may be legacy applications on mini- and mainframe computers. This, in turn, will lead to protocol conversion and database middleware issues such as TCP/IP to SNA conversion. While SNA Server from IBM or Microsoft could satisfy the first requirement, EDA/SQL from Information Builders may be used for multivendor database integration.

5.6.2 Making LAN Applications Mobile-Aware Eventually

We have repeatedly stated the differences between LAN and remote access WAN speeds. While some of the LAN applications could be used on remote access solutions without change as they are, sooner or later, the application designers would have to face up to the desirability of redesigning the architecture of LAN applications to make them mobile-aware. Client-agent-server technology could be employed to make existing applications mobile-aware.

5.6.3 Selecting a Remote Operating System and Client Software

Although Microsoft Windows (3.1 and 95) has held sway as the leading client operating software so far, it is not certain that it will continue to be the dominant environment in the future. The Internet revolution and emerging JAVA-capable browsers are certainly creating an environment where remote access server solutions would be expected to support a hybrid environment consisting of both user interfaces. The best bet for the future is to cover most of your bases and be able to support multiple environments, even if you select one platform as your standard environment.

Besides the operating system (Windows 95 and NT do provide basic remote access client and server support), you may have to select a remote access client software from vendors such as Symantec, Traveling Software, Artisoft/Tritone, or Shiva. A specialized remote client software of this type provides functions over and above baseline functionality provided by Windows 95. You should consider the ease of use, functionality provided, and level of integration with the end-user operating environment while selecting the client software.

5.6.4 Single Communications Server for Wireless and Switched Wireline Networks

While the same user would like to use wireline networks such as PSTN, ISDN, or Internet when a telephone jack (analog or digital) is available, he/she may want to access a wireless network at other times whenever such facilities are not available. It would be nice if the same solution could address both these needs from the same client software. In order to meet this requirement, you will need support for wireless networks in your traditional remote access server.

5.6.5 Single versus Multiple Communications Servers

Whether you should employ a single large remote access server at a central site or several smaller distributed servers at multiple sites is a question that you should analyze on the basis of costs, availability of local technical support, centralized network management, security, and redundancy capabilities. Vendors such as Cisco and 3COM are enhancing their routers to provide remote access capability. If your organization already has this type of router configuration, you should consider upgraded router configuration as one of the solution scenarios. During the last year, the vendor community has introduced higher-capacity models that can meet the needs of thousands of remote users from a single server.

5.6.6 Hybrid Communications Servers with Remote Node versus Remote Control

We reviewed remote node and remote control implementation in remote access servers. There are pros and cons of these approaches—each approach has distinct features that are suitable for specific types of applications. Since it is difficult to modify existing applications, we should select communications servers with both capabilities. It should be easy to switch from one mode to another.

5.6.7 Scalability and Upgradability of Solution

Many remote access solutions have been implemented by users in a small group here and another group there. Now that users like what they get, they are asking network managers to roll out these solutions across the organization. It is important to select technology that is scalable and upgradable, even if initial pilots are small. Too many pilots become default solutions for the enterprise because technical people get accustomed to it and do not want to start the learning cycle all over again.

5.6.8 Internet as VPN for Remote Access

Internet is fast becoming an inexpensive mechanism for remote access of information by external users—customers, suppliers, and business partners. Now network planners are finding that they can create virtual private networks for internal users by using Internet as the front-end for all electronic communication by internal users as well. You need to explore this option in a serious fashion.

5.6.9 Integration and Compatibility with Internetworking Products

During the 1980s, LANs created islands of information for different departments. During the early 1990s, these LANS were connected by bridges and routers. During the last couple of years, local-premises wiring and WAN-based internetworking products such as hubs, switches, and ATM have introduced another level of complexity. Many of these products have limited remote access functionality. Network designers must create a network blueprint for the organization and continue to update it. Remote access solutions can be introduced into this grand scheme either as stand-alone solutions without introducing complexity into already complex network configuration, or we can upgrade these internetworking configurations with their own remote access capabilities.

5.6.10 Lack of Standards

Many of the remote access solutions are based on proprietary implementations by different vendors. You need not wait for the standards to be established and products based on these standards to emerge. In this type of environment, you can utilize proprietary hardware platforms from vendors that have long staying power in the market, so long as you select de facto standards in software, such as TCP/IP for transmission layer. You should still ask yourself the question: How would I migrate from the chosen solution if a standards-based environment came along during the next few years?

5.7 Generalized Mobile Computing Architecture in Remote Access Context

Mobile computing technology is replete with unique and proprietary products. The author, in his previous book, *Mobile Computing—A Systems Integrator's Handbook,* has proposed a general-purpose technology architecture for mobile computing. This architecture is based on a mobile communications server switch from communications perspective and utilizes a client-agent-server paradigm from an application partitioning perspective.
   Essential attributes of this architecture are briefly listed here:

■ Client-agent-server model from application partitioning point of view

■ Multiplatform and open design based on standards (de facto or de jure, where they exist)

■ Support for multiple applications, including messaging

■ Applications written to standard APIs that recognize mobile-aware design concepts described in Chap. 13

- Support of multiple wireline and wireless networks by these APIs

- Mobile-aware middleware such as Oracle's Mobile Agents, RACOTEK's KeyWare, or IBM's Webexpress on ARTour

- Multiple wireless network support to allow migration to higher performance, cheaper, and more reliable wireless networks of the future

- Mobile TCP/IP as a transport

5.8 Technology Principles for an Integrated Architectural Framework

Because the remote access industry has grown in an unplanned manner without the presence of any standards, we cannot define a universal technology architecture for remote access. On the other hand, the absence of standards has given vendors an opportunity to respond quickly to an emerging need by providing proprietary solutions. No single de facto standard has emerged, although vendors such as Cisco, Shiva, and 3COM have been successful in creating strong acceptance of their design approaches and forging some level of third-party vendor following—necessary ingredients for acceptance as de facto standards. Here, we can enunciate certain basic technology principles that you should consider in developing an architecture for remote access. These principles, combined with technology components described in this chapter, can assist users in developing architectures for specific organizations.

1. Technology architecture for remote access must be driven by business requirements. It should be compatible with application and data architecture of those business processes which are expected to be available for remote access during the planning period, typically five to seven years.

2. Remote workstations should match the business requirements of their users. Multiple devices should be used only if they meet specific technical or business needs of the user community. Cost savings achieved by using cheaper and less functional workstations should be weighed against the costs of multiple application-development efforts, higher technical support costs, more complex network management, and additional training requirements.

3. Remote workstations should retain their functionality in offices, homes, hotels, and vehicles. They should support LAN connections, wireline (switched) connections, or wireless connections. Client software should recognize modes of connection and load appropriate drivers accordingly.

4. Response times observed by remote clients should be analyzed with an eye to optimizing network performance from an end-to-end perspective. Wireless network usage should be optimized both for communication costs and for user productivity.

5. As much application logic and data should be moved to mobile devices as is consistent with security guidelines and data architecture principles. This will give users maximum mobility and flexibility.

6. Client and remote access servers should support multiple networks available today. Support for future networks such as ADSL should be investigated as well.

7. Remote access security considerations must meet corporate standards. There should be a firewall.

8. Wherever feasible, applications should be made mobile-aware.

9. Technical support and network management should primarily be centralized, but allowance should also be made for the distribution of those management functions that may benefit from regional support (e.g., call a service person from a local area for mobile hardware repair).

10. The architecture should be open and based on industry-supported standards (either de facto or de jure). Proprietary vendor-specific implementations may be acceptable, provided there is a strong user community with a strong voice. In a similar vein, de jure standards should not be followed if the customer community does not support these standards. You may recall OSI experience—a perfectly well-intentioned de jure standards effort that was rejected by the user community for a variety of reasons.

5.9 Methodology for Developing a Technology Architecture

The flowchart in Fig. 5.10 shows a simple methodology for developing a technology architecture for remote access—related networking.

Figure 5.10
Methodology for developing technology architecture.

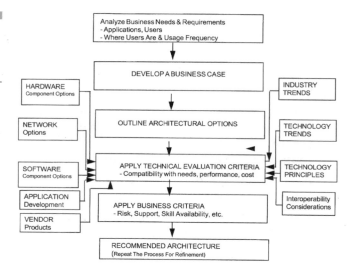

You should use this methodology as a guideline and adopt it to suit your organizational needs. If you are using a formal application development methodology, you should incorporate this flowchart as one of the submethodologies into the formal methodology.

Summary

In this chapter, we have covered an overview of remote access technology architecture. This overview included an analysis of users' business requirements and their effects on an organization's remote access solution. We discussed each of the architectural components—hardware, network, and software. Application software and systems control software architectures were illustrated graphically. Finally, important technology architectural principles were outlined briefly. The information contained in this chapter should lay the groundwork for users to develop custom technology architectures for their organizations. The methodology illustrated in the last section will help in this exercise.

PART 3

Remote Access Network Options

6

Remote Access Network Options—PSTN

Bandwidth of switched networks continues to be a major impediment in the widespread adoption of remote access applications designed for high-speed LANs and emerging multimedia applications.

—Chander Dhawan

About This Part and Chapter

We learned in Part 2 of this book how remote access works. We also developed an understanding of a general-purpose remote access architectural framework. In this part, we look at various network options available to us for satisfying our users' needs. PSTN, ISDN, ADSL, the Internet, and wireless networks are among the most common options available today as well as in the near horizon. Without any doubt, dial-in is the most common method of accessing corporate information and is the substance of the discussion in this chapter.

6.1 Public Switched Telephone Network—Mainstay of Remote Access

A telecommunications network is an important component of a remote access solution. In fact, it may represent the second most significant operating cost item, after technical support cost. Even though the cost of network usage is lower than that of technical support (in the context of total cost of ownership), network analysis attracts more attention from IT professionals. Moreover, the performance of a remote access solution depends a great deal on the type of network selected. Public switched telephone network (PSTN) that denotes traditional dial-in service using voice network is the most common and popular method of connecting remote users to information servers. It is also the least-cost entry point and the easiest method of connection.

PSTN provides a temporary method of providing a physical connection between two computing devices, called DTE (an acronym for data terminal equipment). It is like a voice telephone connection that gets established when you dial a number. At that moment, the telephone network consisting of telephone switches and central office equipment configures a specific path between the two parties. This is what the term *switched* connotes, because instead of a dedicated path, it provides a temporary path between various telephone switches. (See Fig. 6.1*a* and *b*.)

Since telephone networks have been traditionally designed for voice communication, using analog technology, they require a device between

Figure 6.1
(a) PSTN call using telecommunications switched network; (b) PSTN call using telecommunications digital network.

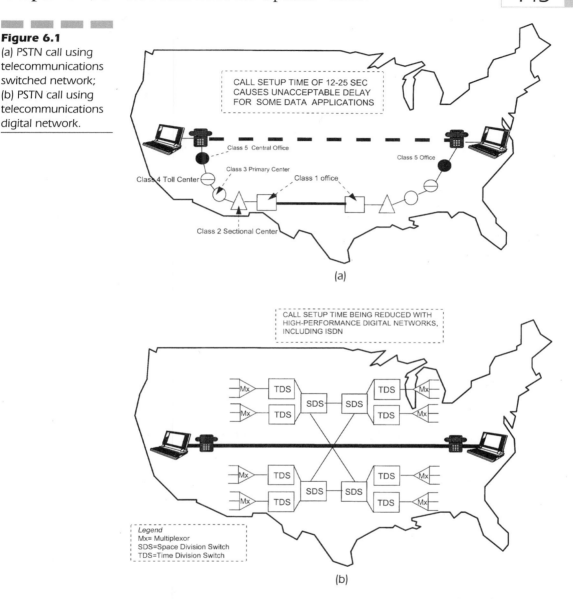

CALL SETUP TIME OF 12-25 SEC CAUSES UNACCEPTABLE DELAY FOR SOME DATA APPLICATIONS

Class 5 Central Office

Class 5 Office

Class 4 Toll Center

Class 3 Primary Center

Class 1 office

Class 2 Sectional Center

(a)

CALL SETUP TIME BEING REDUCED WITH HIGH-PERFORMANCE DIGITAL NETWORKS, INCLUDING ISDN

Mx   TDS   SDS   SDS   TDS   Mx
Mx   TDS   TDS   Mx

Mx   TDS   SDS   SDS   TDS   Mx
Mx   TDS   TDS   Mx

Legend
Mx= Multiplexor
SDS=Space Division Switch
TDS=Time Division Switch

(b)

a digital DTE and analog network that handles voice communication. This device is called DCE (data circuit-terminating equipment) and converts the DTE output to a signal suitable for transmission over standard voice circuits over a wide area. DCE ranges from line drivers to complex modulation/demodulation devices.

6.2 PSTN Modems

The word *modem* is derived from two words, *mo*dulator + *dem*odulator, that represent techniques to convey digital information on analog signals by modifying this signal in a variety of ways. These techniques are described in many telecommunications textbooks; you should refer to these books for explanation. Suffice it to say here that digital information is superimposed on a constant carrier that has a certain frequency. It is this intelligent signal that is transmitted across a network. (See Fig. 6.2.)

There are a number of techniques by which the original constant carrier signal is modified to carry information. The key differences in these techniques are in how the signal is modified by shifting amplitude, frequency, or phase of a sinusoidal wave carrier signal. Some of the popular techniques are:

- Amplitude shift keying (ASK)

- Frequency shift keying (FSK)

- Phase shift keying (PSK)

- A combination of these techniques

The network manager and professional need not analyze the underlying modulation techniques employed by vendors as long as a modem can perform to a higher level of performance characteristics than it must meet. We describe these characteristics in this section.

Figure 6.2
Modulation
techniques.

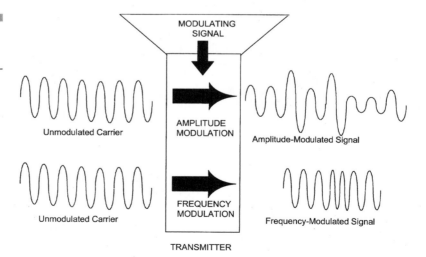

6.2.1 PSTN Modem Standards

There are a number of standards to which modems adhere. We briefly mention the older standards but describe the current and emerging standards for modems presented in Table 6.1.

TABLE 6.1

Modems Standards

Standard Designation	Description
Bell 103	Asynchronous rates up to 300 baud using FSK, allows full-duplex operation on a two-wire switched voice circuit.
V.21 (ITU-T, -ex CCITT)	Same as Bell 103. For a carrier frequency of 1080, space and mark transmitted by +100 Hz and −100 Hz, respectively.
Bell 202	Asynchronous rates of 1200 baud; employs FSK. Also species slow reverse channel for control, etc.
Bell 201	Synchronous data up to 2400 bps; full-duplex is available, though it is half-duplex typically. Modulation technique is differential phase shift. Actual modulation is 1200 bauds, with each baud capable of carrying two data bits.
V.26 (ITU-T, -ex CCITT)	Similar to Bell 201 except phase shift values are different slightly. Two bits (dibit) are carried by single phase change, e.g., $00 = 0°$ $10 = 90°$, $11 = 180°$, $01 = 270°$.
Bell 212	Supports either synchronous or asynchronous data rates up to 1200 bps. Modulation method is DPSK.
V.22 (also US Fed STD 1008)	Synchronous and asynchronous operation at 1200 bps.
V.22bis	Full-duplex 2400 synchronous or asynchronous. Modulation technique is quadrature amplitude modulation (QAM). Modulation at 600, but each baud carries four data bits.
V.27	Up to 4800 bps, full-duplex synchronous data or half-duplex on two-wire, switched, voice circuits. Modulation rate is 1600 bauds, each baud representing three data bits.
V.29	Up to 9600 bps full-duplex synchronous data. Modulation rate is 2400, with each baud representing four data bits. Incorporates fallback to 4800 bps.

	Standard Designation	Description
TABLE 6.1 CONTINUED Modems Standards	V.32	Supports up to 9600 bps asynchronous or synchronous data. Modulation is quadrature amplitude modulation (QAM), or QAM with Trellis coding—a forward error correction (FEC) scheme. Echo cancellation scheme allows same carrier (1800 Hz) to be used at each end.
	V.32 bis	Same as above but up to 14,400 bps.
	V.FC	Nonstandard V.34 (28.8 Kbps) modulation scheme, developed by Rockwell.
	V.34	Supports up to 28,800 bps. Describes a line probing process to set up the modem optimally for any type of line connection.
	New V.34 (under review by ITU-T)	Increases modem speed to 33,600 bps. Primarily pushed by US Robotics.
	Nonstandard x2 Modems by USR	This is a brand new 56 Kbps modem being marketed by US Robotics. (You may read about it further in this chapter.)
	V.42 and MNP2-4	Error correction protocol.
	V.42bis and MNP hardware-based compression	Data compression standard.

6.2.2 Shannon's Law and Maximum Data Rate Possible on PSTN

Since the modem industry is constantly upgrading the modem throughput, in terms of the bits per second (bps) rate, it is important to understand the theoretical limits to this technology. Shannon's law defines the limit as the maximum data rate possible on a voice grade circuit, in terms of its bandwidth and signal-to-noise ratio. It states that the maximum possible number of bits per second is given by the following formula:

Maximum theoretical capacity of a circuit = $W \log_2 (1 + S/N)$,

where W is the bandwidth and S/N is the signal-to-noise ratio. Assuming a typical S/N ratio of 30 dB (i.e., a signal 1000 times stronger than noise), and a usable bandwidth of 2800 bps, the theoretical maximum capacity can be calculated around 27,900 bps. An S/N ratio of 38 dB can raise this to 33.6 Kbps.

In spite of this theoretical limit, modem manufacturers keep on extending modem capacity to the limit in ingenious ways. US Robotics' announcements of 33.6 Kbps modems and x2 modem in late 1996 (available in 1997) are indicators of such efforts. These modems do not defy Shannon's law but do provide solutions for improved effective throughput by indirect means. (Read the x2 description in the following subsection.)

6.2.3 US Robotics' x2 Modem Technology (56 Kbps) for Faster Remote Access[*]

US Robotics x2 technology allows modems to receive data at up to 56 Kbps over the standard, public switched telephone network. x2 takes advantage of the fact that most remote access applications are asymmetric in nature, that is, remote users receive more data than what they transmit. x2 also exploits the fact that equipment at the central site server end may have digital connection instead of analog. This is true with many Internet service providers, on-line service providers, and Intranet host sites.

Typically, the only analog portion of the telephone network is the phone line between your home to the telephone company's central office (CO). Over the past two decades, the telephone companies have been replacing portions of their analog networks with digital circuits. However, the slowest portion of the network is the portion between your house and CO—likely to continue as analog for a few more years.

x2 may not require any changes to the wiring and equipment at the remote user end, provided you have an x2 modem or a compatible V.34 modem (from US Robotics or Courier) upgraded with an x2 firmware upgrade. In other cases, you may need to replace your older modems with x2 modems. However, there is a change required at the central site. Depending on the equipment at the central site, a software or hardware upgrade may be required at the service provider end.

The reader should note that the public switched telephone network was designed for voice communications. By artificially limiting the sound spectrum to just those frequencies relevant to human speech, network engineers found that they could reduce the bandwidth needed per call and therefore increase the number of simultaneous calls in a given

---

[*]*Source:* US Robotics' technical brief document on the Web.

spectrum. While this works well for voice, it imposes limits on data communications.

6.2.3.1 NOISE INTRODUCED BY QUANTIZATION OF ANALOG SIGNALS.

Analog information must be transformed to binary digits to be sent over the PSTN. The incoming analog waveform is sampled 8000 times per second, and each time its amplitude is recorded as a PCM code. The sampling system uses 256 discrete 8-bit PCM (pulse code modulation) codes.

Because analog waveforms are continuous and binary numbers are discrete, the digits that are sent across the PSTN and reconstructed at the other end approximate the original analog waveform. The difference between the original waveform and the reconstructed quantized waveform is called *quantization noise,* which limits modem speed.

Quantization noise limits the communications channel to about 35 Kbps. But quantization noise affects only analog-to-digital conversion—not digital-to-analog. This is the key to x2: if there are no analog-to-digital conversions between the x2 server modem and the PSTN, and if this digitally connected transmitter uses only the 255 discrete signal levels available on the digital portion of the phone network, then this exact digital information reaches the client modem's receiver, and no information is lost in the conversion processes.

6.2.3.2 SIGNAL-TO-NOISE RATIO (SNR).

The *signal-to-noise ratio* is a measure of link performance; it is arrived at by dividing signal power by noise power. The higher the ratio, the clearer the connection, and the more data can be passed across it. Even under the best conditions, when a signal undergoes analog-to-digital conversion, there's a 38 to 39 dB signal-to-noise ratio that limits the practical V.34 speeds to 33.6 Kbps.

Let's see how this happens:

1. The server connects, in effect digitally, to the telephone company trunk.

2. The server signaling is such that the encoding process uses only the 256 PCM codes, used in the digital portion of the phone network. In other words, there is no quantization noise associated with converting analog-type signals to discrete valued PCM codes.

3. These PCM codes are converted to corresponding discrete analog voltages and sent to the client modem via an analog loop circuit. There is no information loss.

4. The client receiver reconstructs the discrete network PCM codes from the analog signals it received, decoding what the transmitter sent.

6.2.3.3 UPSTREAM AND DOWNSTREAM CHANNELS: ASYMMETRIC OPERATION.

x2 connections employ one bidirectional channel, upstream and downstream. The x2 client modem's downstream (receive) channel is capable of higher speeds because no information is lost in the digital-to-analog conversion. The x2 client modem's upstream (send) channel goes through an analog-to-digital conversion, which limits it to V.34 speeds.

As discussed previously, data is sent from the x2 server modem over the PSTN as binary numbers. However, to meet the conditions of point 2 preceding, the x2 server modem transmits data (8 bits at a time) to the client end's DAC (digital-to-analog converter) at the same rate as the telephone network (8000 Hz). This means that the modem's sampling rate must equal the phone network's sample rate.

During the training sequence, x2 modems probe the line to determine whether any downstream analog-to-digital conversions have taken place. If the x2 modems detect any analog-to-digital conversions, they will simply connect as V.34. The x2 client modem also attempts a V.34 connection if the remote modem does not support x2. The x2 client modem's task is to discriminate among the 256 potential voltages to recover 8000 PCM codes per second. If it could do this, then the download speed would be nearly 64 Kbps (8000×8 bits per code). But, it turns out, several problems slow things down slightly.

First, even though the network quantization noise floor problem is removed, a second, much lower noise floor is imposed by the network DAC equipment and the local loop service to the client's premises. This noise arises from various nonlinear distortions and circuit cross talk. Second, network DACs are not linear converters but follow a conversion rule (mu-law in North America and A-law in many other places). As a result, not all 256 discrete codes can be used, because DAC output voltage levels near zero are too closely spaced to accurately represent data on a noisy loop. For example, the most robust 128 levels are used for 56 Kbps, 92 levels to send 52 Kbps, and so forth. Using fewer levels provides more robust operation but at a lower data rate.

Exploitation of x2 technology requires the following:

1. *Digital at one end.* One end of an x2 connection must terminate at a digital circuit, meaning trunk-side channelized T1, ISDN PRI (over a

T1 or E1 line), or ISDN BRI. Line-side T1 will not work because additional analog-to-digital and digital-to-analog conversions are added. In a trunk-side configuration, once the user's analog call is converted to digital and sent through the carrier network, the call stays digital until it reaches a US Robotics server modem through a T1, PRI, or BRI circuit.

2. *x2 support at both ends.*   x2 must be supported on both ends of the connection by the client modem, as well as by the remote access server or modem pool at the host end.

3. *One analog-to-digital conversion.*   There can be only one analog-to-digital conversion in the phone network along the path of the call between the x2 server modem and the client modem. If the line is a channelized T1, it must be trunk-side and not line-side. With line-side service from the phone company, there typically is an additional analog-to-digital conversion.

6.2.4 PSTN Modem Features

The following features are important in comparing modem capabilities:

- *Modulation speed (baud) and data rate (bps).*   It is important to distinguish between the baud rate and bps (bits per second). The *baud rate* is the number of times a modem's signal changes. Due to the limitations in standard phone lines, it is difficult to get baud rates higher than 2400. The *bps,* on the other hand, is the measure of how many bits per second the modem transmits. As an example, a 28,800 bps modem transmits 12 bits per baud.

   It is equally important to understand the difference between the raw data rate of a modem and its effective throughput capacity. Sometimes the vendors give optimal throughput capacity of a modem, after assuming the highest level of compression, as the data rate of a modem. This can be misleading. The raw data rate and compression efficiency should be evaluated separately, because not all data can be compressed, and the efficiency of compression depends on both the compression algorithms implemented and the compressibility of data. We revisit end-to-end optimization for remote access later in this chapter.

- *Compliance to Bell and ITU-T standards.*   Ensures interoperability across products from different vendors.

- *Compatibility with class 1 and 2 fax machines and voice mailboxes.*   Ensures that the modem supports fax and voice mail applications.

■ *Error correction protocols.* Line noise can cause significant errors in modem connections. The high speeds of modems exacerbate this problem. Error-correction protocols combat the line noise problem by packetizing the data, and providing checksums to help determine if the data has been corrupted during transmission. Several error-correction protocols are available; the two most popular ones are MNP-4 and ITU-T V.42 LAP-M.

■ *Data compression capabilities.* In their quest to improve the effective throughput on the voice-grade lines, modem vendors have been developing data compression protocols for serial transmission. These protocols remove blanks and use other techniques to compress the data. Two popular methods of data compression are MNP-5, with a maximum compression ratio of 2-to-1, and ITU-T V.42 bis, with a maximum compression rate of 4-to-1.

■ *Hardware flow control.* This is important for high-speed modems. The RTS (Request To Send) and CTS (Clear To Send) pins are used to make sure that the receiving end of the connection has enough resources to process the data that is being sent to it.

■ *Support for Hayes AT command set.* This ensures that many of the communications software applications utilizing the AT command set will work satisfactorily.

6.3 Technical Issues in Using PSTN for Remote Access

There are a number of technical and business issues that you must consider in using PSTN for remote access. We describe these considerations here:

6.3.1 Suitability of PSTN Network to Remote Access Applications

PSTN is suitable for only those remote access applications where the slow connect time of switched networks is acceptable from performance and cost points of view. It takes a finite amount of time every time you set up a call through the telephone system. While a few modern switched networks are being designed for faster setup time (less than 10 seconds), most dial-in networks take as long as 30 seconds for a call setup. This time is rather long for response-time-sensitive OLTP (on-line transaction processing) applications where a 3- to 5-second response time is a required criterion.

The other factor is the cost of transferring information. This is very important, especially when business calls are charged on the basis of connect time. File transfer applications and graphic-intensive applications may not be suitable for PSTN. The user should do proper data traffic analysis in order to select a network on the basis of availability, performance, cost, software support, and other relevant factors.

6.3.2 **Performance Optimization in PSTN**[*]

End-to-end performance of remote access applications depends on the hardware and software components that are involved in an end-to-end session. The contribution of these factors is described here:

6.3.2.1 PROCESSING SPEED OF THE REMOTE COMPUTER. Remote computers are often used for stand-alone functions when they are not connected to central information processors. The capacity of the remote computer must match both stand-alone processing requirements as well as remote access requirements. As an example, Macintosh computers, using Apple Remote Access (ARA), do not use their modem's built-in data compression, but instead perform compression through the processor's software. Similarly, many new remote access software clients and Internet browsers use temporary caching techniques to avoid transmission of data elements already in the remote computer in order to improve overall performance. Therefore, a remote Macintosh computer should be based on faster 68040 or better, just as a PC should be a 166 MHz Pentium class processor or better. In the case of Java-based NCs and WIN-CE handheld computers, we may have to accept somewhat slower performance, as compared to full-function PCs. (Note that convenience and portability come at the cost of performance and function.)

6.3.2.2 TYPE OF SERIAL PORT ON THE REMOTE COMPUTER. Serial port performance has an important effect on the overall performance of a remote access system that far outweighs the attention it usu-

---

[*]*Source:* Shiva's white paper on end-to-end optimization on the Internet.

ally receives when choosing a remote PC. No amount of tuning can make a PC with an inadequate serial port interface perform well in a remote access system.

The burden placed on a PC's CPU by an interface is often inversely proportional to the maximum speed that interface can move data. Most high-speed interfaces, such as LAN cards, use sophisticated controllers that move large blocks of data with virtually no CPU intervention. Because the interfaces are so fast, sophisticated controllers are a requirement for the PC to keep up.

Serial interfaces, on the other hand, operate at relatively slow speeds, which often can be serviced directly by the PC's CPU. Until recently, most PCs were built with simple 8250 or 16450 serial controller chips that only minimally assist the CPU in sending or receiving serial data. With these chips, each time a character of data is ready to be sent or received, the CPU is interrupted from whatever it is doing, spends a few instructions figuring out why it was interrupted, and then sends or receives the next character as necessary.

By using an inexpensive serial controller chip, such as the 8250 or 16450, and relying on the CPU to service each character of data moving through the serial port, PC designers save the cost of a sophisticated serial interface controller. When serial port speeds were limited to 2400 or 9600 bps, that cost savings may have been appropriate. Improving modem technology has dramatically raised the speed that serial ports need to operate, to the point that 28,800 bps or even 115,200 bps (with compression) requirements are increasingly common. With an 8250 or 16450 serial controller, even a fast 486-equipped PC has trouble keeping up at those speeds.

When its CPU cannot keep up with its serial port, a PC experiences something called an *overrun*, when a new character arrives before the CPU has collected the previous one. Remote access data is sent in chunks called *packets*. If a character is lost from a packet because of an overrun, the entire packet is sent over again, usually not until a lengthy time-out has expired. If the CPU is much too slow, overruns occur in almost every packet, and the remote access connection simply fails. However, if the CPU speed falls just short, complex interactions between the serial port, the CPU, memory managers, and other system software may cause intermittent overruns. In that case, the remote access connection may seem to work but have intermittent problems, such as sudden drops in speed, spurious losses of connections, and so on. Such inconsistent behavior is one of the most annoying problems faced by both end users and those who support them.

In addition to the differences in performance between serial controller chips, serial performance may also vary by operating system. For example,

Microsoft Windows adds an extra layer of software for handling CPU interruptions, thus forcing the CPU to execute extra instructions each time it is interrupted to receive or transmit characters. On the other hand, DOS, as the simplest operating system, requires the least extra processing by the CPU each time it is interrupted. Therefore, a PC's maximum serial port throughput usually occurs when running DOS alone.

However, even DOS is vulnerable to unexpected serial port slowdowns. The simplicity of DOS may also lead to less consistency because it provides no context for handling CPU interruptions. Therefore, some low-level DOS software—particularly memory managers—may turn off CPU interruptions for extended periods of time, during which time characters can be lost because the CPU is not aware that it must service them.

The need to support higher and higher serial port speeds has led to the use of a more sophisticated serial controller chip, the 16550A, in some PC designs. A 16550A serial controller can receive or transmit up to 16 characters without intervention from the CPU. Thus, with a 16550A serial controller, the CPU is interrupted much less frequently and moves more data with less effort. Most 16550A-equipped PCs can easily keep up with today's high-speed modems.

For optimum remote access performance with today's high-speed modems, a remote PC must have a sophisticated serial interface controller, most likely a 16550A. Given the heavy penalty of using a PC with inadequate serial port support as a remote access client, it makes sense to ensure that most, or preferably all, of the remote PCs being used with a remote access system are equipped with 16550A serial controllers. There are several ways to add a 16550A serial controller to a PC that does not already have one.

If the PC has a socketed 16450 serial controller, it can be replaced with a 16550A. Unfortunately, socketed 16450s are very rare. The next option—if the PC has an ISA expansion bus—is a 16550A-based serial controller card. Such a card can be installed in a PC; the PC's onboard serial interface must be disabled, and then the serial controller card is configured in its place.

For a laptop with a PCMCIA expansion connector, a PC Card (PCMCIA) modem is likely to work well. Almost all PCMCIA modem cards have 16550A serial controllers on board. Most modern laptops include PC card support, and PCMCIA modems are both inexpensive and convenient. Some older laptops are either not expandable or use a proprietary expansion bus for which 16550A-based modem cards are unavailable. Such machines cannot be upgraded to a 16550A serial controller and are not well suited to use in a remote access system.

6.3.2.3 REMOTE COMPUTER OPERATING SYSTEM.

In an optimized remote access system, end users should face no restrictions on the environment from which they can establish a remote access connection. The remote access solution selected by the organization should support multiple operating systems—DOS, Windows 3.1, Windows 95, Windows NT, MacOs, Unix, and now Java. Moreover, it should be possible to establish a remote access connection in DOS, switch to Windows, and still make use of the remote access link. Client software should be able to establish and end remote access connection from either DOS or Windows; the NOS client must either be able to load and stay resident without requiring a connection or must be able to be loaded on demand in either DOS or Windows.

An optimized remote access system includes client software that can establish remote access connections from either DOS or Windows. Also, NOS clients should be capable of either loading and staying resident without requiring a connection (like the VLM version of the NetWare shell) or can be loaded on demand in either DOS or Windows (like Windows for Workgroups).

6.3.2.4 REMOTE ACCESS CLIENT SOFTWARE.

The remote access client software is an end user's primary link into the remote access system, and as such, it is the prime determinant of a remote access system's ultimate ease of use. The remote access client is also critical to compatibility with a wide range of protocols and equipment and is a major factor in the system's overall speed.

A remote access system should support any standard or de facto standard remote access client software that is available for an operating system. Macintosh users should be able to use Apple's standard remote access client, called Apple Remote Access (ARA). Unix users should be able to choose any standard serial line Internet protocol (SLIP) or point-to-point protocol (PPP) implementation. Because Windows and DOS do not yet have standard remote access clients, the remote access vendor must supply them.

A remote access client performs two major tasks: establishing and terminating the remote access connection, and sending and receiving network traffic across the network link. It is helpful to think of the remote access client as having two parts, a *dialer,* which manages the link, and a *driver* which interfaces with the network protocol stacks and the serial port to send and receive network data.

The remote access client's dialer is crucial to end-user ease of use, and also largely determines the range of data transmission equipment with which the remote access system can be used. For PC-based end users, the remote access client should offer both DOS and Windows (3.1, 95, and NT) dialers, so that users can easily establish connections in either environment; a Windows dialer is especially important for end-user ease of use. The dialer also determines the range of different modem types, telephone connections, and security systems that are usable within a remote access system. A dialer needs to have flexible modem handling that makes it easy to add support for new types of modems or other data communications equipment, such as integrated services digital network (ISDN) terminal adapters. The dialer also needs built-in support for a wide variety of existing modems, terminal adapters, and other data communications devices.

A dialer should also be able to deal with complex connection sequences. Entering credit card numbers, passing through packet networks, such as X.25, and navigating sophisticated security systems are all common roadblocks to establishing a connection. Such roadblocks require either a dialer with flexible scripting, which can be programmed to navigate a complex connection sequence automatically, or a manual mode that allows the user to navigate the connection directly, or both.

The remote access client's driver is a key factor in the number of different protocols and protocol implementations that a remote access system supports. End-user applications are more and more likely to require that a remote access system support multiple protocols. Therefore, it is critical that a remote access client work well with a wide variety of protocols.

For PC-based systems, a minimum set of supported protocols should include Novell's IPX, which is the native protocol for NetWare; TCP/IP, widely used in Unix systems and client-server databases, but also becoming a standard for many other applications; NetBEUI, used for LAN Manager and Microsoft's Windows for Workgroups; and LLC/802.2 for IBM LAN Server and host connectivity. For Macintosh computers, the protocol group is somewhat simpler—the combination of AppleTalk and TCP/IP covers almost all Macintosh applications. Unix systems are primarily covered by TCP/IP support.

Not only must the remote access client support multiple protocols, but it must also support them simultaneously. For example, a user may need to use an electronic mail package that uses the IPX protocol and a custom client/server database application that runs on TCP/IP. The remote access client should never force a user to disconnect and reconnect with a different protocol to switch between applications. In addition to supporting multiple protocols simultaneously, for full compatibility with the maximum number of protocol implementations, a remote access client must support the major PC network interface driver standards: ODI and NDIS.

Since the PC system memory is scarce, the driver should also take up as little system memory as possible.

The remote access client, in partnership with the remote access server software, plays an important part in optimizing a system for speed. Issues such as header compression and broadcast suppression, in which the client participates, are explained in a subsequent section, "Remote Access Server Software."

6.3.2.5 MODEM SPEED AND FEATURES.

Remote links have only a fraction of the bandwidth available to LANs. Networking applications are typically designed to work best at LAN bandwidths and are noticeably slower over remote connections. To minimize the difference between local and remote performance, it is critical to use the fastest modem technology available for remote connections.

Almost all modems sold today use data compression techniques to improve throughput. Claims of 4-to-1 data compression are for best-case, highly compressible data, and probably do not represent realistic, real-world performance expectations. Compression also exacts a small penalty by introducing delays as data is collected into compressible blocks rather than being sent in a continuous stream. Therefore, compression, although valuable, is not a substitute for raw, uncompressed throughput.

A V.32bis modem with a 14.4 Kbps data stream and 4-to-1, best-case compression may transmit to the remote computer's serial port at 57.6 Kbps, but with real-world data and correspondingly real-world compression, its actual throughput is likely to be about half that. An ISDN terminal adapter, on the other hand, may also transmit data to the remote computer's serial port at 57.6 Kbps, but it does not rely on compression, and thus its real-world throughput probably doubles that of the modem.

For the sake of reliability and consistency, the modem must also be able to respond gracefully when telephone conditions are not good enough to support a maximum speed connection. Modems should support fall-forward and fall-back which allow them to monitor error counts, and either drop the connection speed when error counts get too high, or increase connection speed when error counts drop back down.

For laptop use, PC Card (PCMCIA) modems are better than external modems because they have a high-speed bus connection to the system and usually contain appropriately fast serial ports with 16550A support.

The requirements for modems attached to a remote access server are a superset of the requirements for the remote PC modem. Additionally, the modems attached to the server must be able to interoperate with every kind of modem the system's users plan to dial-in with, at the maximum speed the remote modem can support. Both modems in a remote access connection must be able to perform at the desired levels, as the connection works only as well as its slower half.

The server modems for an enterprise remote access solution, which might have large numbers of ports, need to be rack-mountable, as compact as possible, and should be manageable as well. For the simplest, most manageable enterprise solution, modems should be integrated directly into the remote access server. Integrated modems save space, avoid disorderly, unreliable cabling between the server and a bank of modems, and can be managed with the same tools as the server itself.

6.3.2.6 TELEPHONE CIRCUIT CHARACTERISTICS.   Today's high-speed modems are remarkably tolerant of difficult telephone line conditions, but nonetheless, telephone connections can have all sorts of problems—for example, noise, distortion, and echoes—any one of which can drastically reduce either the reliability or speed of a modem link. Telephone connection trouble is one of the most annoying problems because it is rarely consistent and often is extremely difficult to find.

If a user is off-site and encounters an unusable telephone line, there really is not much to be done about it. Calling at a different time of day may help, because varying loads on the telephone network can cause calls between the same two endpoints to travel very different paths.

If a user has a problematic telephone line into his/her house, or some other fixed location, it may be necessary to order a data-quality telephone line, particularly in rural or high-interference areas.

In many urban areas, switched digital links, which avoid the unreliability and inconsistency of analog modem connections, are now available. ISDN and Switched56 service are fast becoming viable options for remote users who dial in from fixed locations. In addition to offering more reliable service than analog connections, switched digital links provide superior performance. An optimized remote access system should take advantage of switched digital connections wherever they are available.

6.3.2.7 REMOTE ACCESS SERVER HARDWARE.   Ideally, the hardware used for a remote access server should be designed for, tested for, and optimized for remote access. Force-fitting remote access into a general-purpose hardware platform, like a PC, is likely to result in compromises of speed, form-factor, reliability, and convenience.

The performance of the remote access server hardware is primarily determined by its ability to move data through its serial ports without much attention from the CPU and by the CPU's ability to perform the routing, filtering, and the like, that it must do, without adding undue delays as it forwards packets. The remote access server hardware should be optimized for serial port throughput and general CPU power.

The remote access server must be highly reliable and efficient. Therefore, a remote access server with few moving parts, such as floppy or hard disk drives, has a substantially reduced risk of breakdown; the hard disk, if configured, should be duplexed. Server software and configuration should reside in solid-state, nonvolatile storage, such as battery-backed-up static RAM, or flash memory, and should be upgradable via network downloading.

The remote access server should also have an appropriate form-factor for its planned use. Remote access systems that provide enterprise support—with large numbers of users and the need to fit many remote access ports in a small space—should be rack-mountable, with high port densities. A workgroup remote access system, on the other hand, is often best served by compact, tabletop server hardware.

Convenience can play a large role in the ongoing costs of managing a remote access system. The ongoing management of modems, modem cables, and telephone line interfaces can be costly. A remote access server that offers integrated, internal modems reduces the headaches that come from installing, maintaining, and managing separate banks of server and modem ports.

6.3.2.8 REMOTE ACCESS SERVER SOFTWARE. Perhaps the most important thing to look for in a remote access server is versatility, which is primarily a function of the server software. Since the remote access server makes use of scarce resources (modems and telephone lines) it should support as many uses of those modems and telephone lines as possible. This document focused on end-to-end optimization of single-user dial-in access, but for optimal use of modems and telephone lines, the remote access server should support shared dial-out and LAN-to-LAN connectivity as well.

The remote access server should provide a single point of remote access for PC, Macintosh, and Unix users. Therefore, as with the client software, it is critical that the remote access server support a wide variety of protocols, with a minimum set for PC-based systems, including IPX, TCP/IP, NetBEUI, and LLC/802.2. For Macintosh computers the combination of AppleTalk and TCP/IP covers almost all Macintosh applications, and Unix users need TCP/IP almost exclusively.

One particularly tricky issue is support by the software for an assortment of addressing schemes. Addressing schemes are different for each protocol. Some protocols, such as AppleTalk, use addresses that are completely dynamic, and require little or no administrative or user intervention. In others, such as TCP/IP, address administration is a critical issue.

Since different IP networks have different addressing requirements, a remote access server must provide flexible TCP/IP addressing options, including by-user, by-port, and client-supplied IP addresses.

By-user addressing helps optimize a system for user tracking. With by-user assignment, the server assigns the client's IP address each time a user dials in, taking that address from the user database. By-user assignment ensures that each user has the same IP address every time he/she makes a remote access connection. This assignment provides an efficient method of ensuring that users have valid IP addresses and provides tracking via the IP address of what services the user has accessed.

By-port address assignment conserves IP addresses. When a user dials into a specific port, the user is simply assigned the IP address of the port. Therefore, on an 8-port server that might be servicing 50 users, only eight IP addresses are expended.

Client-supplied addressing requires that the server allow the remote PC to choose its own address. Client-supplied addressing provides similar benefits to by-user addressing but without centralized control. It also supports TCP/IP protocol stack implementations that require preconfigured IP addresses.

The remote access server software also plays a key role in determining the speed of the overall remote access system. As mentioned before, the speed of a remote access system is influenced by the percentage of its available bandwidth that is filled with useful data. The remote access server software can use its knowledge to minimize the overhead per data packet and also to exclude unnecessary packets from the link. The remote access server software must understand both the workings of a protocol and the way it is likely to be used in order to optimize link utilization. For example, the server should support header compression, in which the server uses its knowledge that there is only a single remote PC per connection to remove addressing information and other packet overhead. Header compression is different for each protocol.

The server must also ensure that a remote access connection's bandwidth is not wasted on unnecessary broadcast traffic. Broadcast traffic on the server's network is a potential performance killer for any remote access system, particularly on large networks. Broadcast packets are received by every system on a LAN, but each broadcast packet is usually relevant to only a few of the systems on a LAN. If all broadcast packets are forwarded over the remote access connection, that connection fills up with packets that the remote system simply ignores. Meanwhile, valuable data backs up behind irrelevant broadcasts, hurting performance.

The most effective way in which a server can limit broadcast traffic is by acting as a router, which eliminates all broadcasts that are not specifically directed to the remote link. With help from the remote access client software, the remote access server may further determine which broadcast packets are truly needed by the remote system. This is particularly important in bridged environments, and for nonroutable protocols, where there may potentially be a great deal of broadcast traffic on the server's network.

A good example of broadcast traffic management by a remote access server is found in the NetBEUI protocol, which is not routable, and is particularly susceptible to performance problems caused by broadcast traffic. With NetBEUI, the server can learn all names by which the client machine is registered, respond to name lookups directly, and forward only broadcast packets that are destined for the remote machine's name.

Since all of the optimizations require that the remote access server and client software perform specific performance-enhancing tasks for each protocol, it is necessary for the software to be tuned to the inner workings of each protocol that it supports. A remote access system that is protocol blind and treats all protocols the same is unlikely to achieve optimum link utilization.

6.3.2.9 LAN EFFICIENCY. As is true with any networking solution, optimizing a remote access system requires the network administrator to monitor and gauge the efficiency of the LAN. If the LAN is heavily used, then remote access performance will be affected. In general, internetworks that are routed perform more efficiently than those that are bridged.

If a LAN segment is heavily used, further segmenting the network, via routing technology, can provide a higher level of performance, both locally and remotely. There are many network monitoring tools that provide ongoing information on LAN bandwidth use.

Summary

In this chapter, we have reviewed PSTN as a network option for remote access. This is the most common method employed by the majority of remote users. You observed that modem speeds have been going up for the past several years. However, with US Robotics' x2 technology, we may have hit the limit at 56 Kbps for downstream traffic. You also learned that modem speed is only one of the many factors that affect the performance of remote access. RAS server hardware, RAS software, serial port of the client PC, telephone line characteristics, LAN efficiency, and transport software protocol all have a bearing on end-to-end efficiency.

CHAPTER 7

Remote Access Network Options—ISDN

Since its inception in early 1980s until mid-1990s, ISDN had remained a network solution looking for a problem until remote access demands exceeded the capabilities of public switched networks. It was like a White Knight.

—Chander Dhawan

About This Chapter

Voice communication is analog; data is digital. Voice came first, and data came second; hence, the need for data to look like voice over the switched networks. This unnecessary modulation and demodulation prompted telecommunications engineers to carve out a digital network for data; they called it *integrated services digital network* (ISDN). In a telecommunications world, where there are theoretical limits on the speed of transmission that public switched telephone can support, ISDN became a technology with much greater potential. This chapter is about use of ISDN for remote access. You will learn about ISDN technology, how it provides service, various terms that you must be familiar with, and ISDN network scenarios. We also talk about configuration parameters, ISDN costs, and compare it with other options.

7.1 ISDN as a High-Speed Switched Digital Link for Remote Access*

ISDN originally emerged as a viable technology in the early '80s, but its limited coverage, high tariff structure by different regional Bell operating companies (RBOCs), lack of standards, and a scarcity of applications that demanded ISDN bandwidth dramatically stunted its growth during the '80s. It was a network service looking for a problem to solve. The joke in seminar circuits was that ISDN stood for *I Still Don't Know.* That was the scene in North America. In Europe, however, different PTTs who do not always provide compatible telecommunications services, did get together, and started offering standards-based services with interoperability across different networks. Therefore, the success of ISDN was much faster in Europe than in North America. This changed with the Internet revolution in the mid-1990s. With the increasing demand for more bandwidth, decreasing hardware adapter costs, and carriers introducing more attractive pricing, ISDN is undergoing an equally dramatic revival. Along the way, remote access and the spectacular rise in Internet traffic have together given the technology an additional shot in the arm. ISDN instal-

---

*"Understanding ISDN Solutions"—A white paper from Xyplex Corporation is a good source of introductory information on ISDN. Xyplex is a remote access system vendor. There are several books on ISDN available in the market for those who want more details.

lations doubled in 1996 (450,000 to 800,000), and are expected to reach 2,000,000 lines by the year 1999, according to some estimates.

The following is a brief description of ISDN to enable you to evaluate it as a viable option for remote access solutions.

7.2 What Is ISDN?

ISDN conveys different things to different people. However, ISDN is essentially a digital phone call—a high-speed digital service that carries simultaneous voice and data communications over existing twisted-pair phone cabling systems. In its simplest form, it is an enhancement to the telephone local loop that allows both voice and data to be carried over the same medium. It is different from PSTN described in Chap. 5, insofar as there is no modulation or demodulation of signals over the communication medium. It is a fully digital network—from one end to the other. The word *integrated* in its designation also implies multiple services (as a minimum voice, data, and image) being offered on the same network. This integration has remained only a promise so far. (See Fig. 7.1.)

Telecommunications carriers around the world have been converting their analog networks to digital switches for many years. However, the

Figure 7.1
ISDN access line as a
digital pipe.

ISDN Access Line as a *Digital Pipe*

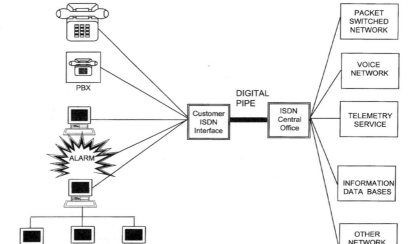

only part that they did not convert was the local loop. It is the conversion of this part that started the ISDN development.

Most companies offer ISDN service for the home or the office at attractive rates. As with regular long-distance phone service, ISDN calls are charged to the caller, and billing is based on the time connected rather than on data transmitted (as is the case with packet switching services).

7.3 Benefits of ISDN over PSTN

ISDN offers many benefits to users when compared with conventional PSTN. Some of these benefits are:

Integration of Multiple Services.   ISDN handles different types of information over the same link. It can transport voice, data, image, video, and sound.

Higher Speed.   ISDN delivers up to 128 Kbps without compression, and up to 512 Kbps with 4:1 compression.

More Cost Effective.   ISDN may be cheaper than PSTN or leased-line services when you compare cost per megabyte of data transferred, or if you consolidate voice line, data line, and fax line into a single ISDN line.

Same Wiring Infrastructure.   ISDN uses the same local loop wiring, but the telephone company installs a special network termination unit (NT1) at remote client's site for digital service.

Simultaneous Voice and Data.   ISDN service allows a voice call to take place at the same time transfer of data from a remote server or the Internet.

Up to Eight Devices Can Be Connected on a Single ISDN Circuit.   Note that the actual number of devices supported depends on the switch at the central office of the telephone company.

Use of Existing Analog Devices.   Existing devices such as telephones and fax machines can be used along with digital data transmission. In a business environment, credit card readers, and authorization devices can be used over X.25 networks.

On-Demand Pay-as-You-Use Service.   ISDN is a switched service which may be cheaper than leased-line services.

More Accuracy.   As a digital service, it offers a lower error rate for data transmission.

7.4 ISDN Applications

ISDN is a digital network option that did not find a killer application for itself until remote access to Internet, using PSTN, became a bottleneck for graphics-intensive Web applications. There are now a number of applications that are most suitable for ISDN. The applications best suited for ISDN are those that meet the following criteria:

1. Application wants to download a lot of data in a short period and is not highly interactive or chatty. An application that sends 5 MB of data in 15 minutes is more suitable for ISDN than an interactive application that sends 20 KB of data over a 6-hour period.

2. Geographical coverage of applications is within a single carrier because an ISDN interconnection between different carriers is not very reliable.

3. Bandwidth requirements are within the capability of ISDN PRI (1.536 Mbps) and ISDN BRI speeds (128K).

4. Closed user group: A majority of users do have access to ISDN. We should point out that at present (in 1997) only 1.5 to 2.0 percent of telephone circuits in North America are ISDN-ready.

5. Almost permanent continuous access: Those applications that require a constant connection are not suitable for ISDN because usage charges become very high.

The following applications are suitable for ISDN:

- Internet or Intranet access, especially for highly visual or graphics-intensive applications, such as the World Wide Web
- Large file transfer between remote clients and host server
- Video conferencing between professionals, and between field staff and the head office
- Collaborative computing (through the use of electronic whiteboards or other document-sharing methods)
- Telecommuters working at home, accessing office LAN resources
- Banking applications with videoconferencing between the customer and the bank staff, using kiosks
- Telemedicine and health care where doctors can exchange critical patient data, including X rays with a specialist, using voice and digital information.

- Telelearning at universities and other educational institutions where courses can be offered at remote or satellite learning centers
- Internet telephony

7.5 ISDN Terms You Must Know

There are a number of ISDN-specific terms that you should be familiar with when you are dealing with telephone companies.

7.5.1 SPID

SPID (service profile identifier) is a number assigned by the telephone company that looks like a telephone number, with a few additional digits appended at the end. The purpose of SPID is to identify each device attached to an ISDN line to the ISDN network. Each device must have a unique SPID on a multipoint ISDN line. The general format of SPID is shown in Fig. 7.2. SPIDs are important because you must specify that in your ISDN configuration software.

Multipoint ISDN lines can handle multiple SPIDs, and different telephone switches support up to eight SPIDs, for example, AT&T 5ESS. Custom switches can handle eight devices, and therefore, eight SPIDs. On the other hand, NT DMS-100 handles two SPIDs.

Unfortunately, there is no standardization for SPIDs in ISDN switches. As an example, some switches require one SPID for both B channels of a BRI; other switches require a separate SPID for each B channel. In fact, some switches do not require any SPID. It is expected that some sort of standardization will happen for SPIDs soon. ISDN Users Forum in North

Figure 7.2
Generic format of SPID in ISDN.

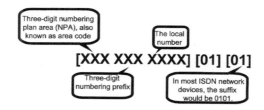

The format of the SPID is the same no matter where the line in, or who installed it.

America, along with some vendors, such as Northern Telecom, Lucent, Siemens, and Ericsson, have proposed a generic SPID for new installations. This generic SPID is shown in Fig. 7.2, cited previously.

7.5.2 ISDN Directory Numbers

A *directory number* defines the address or telephone number for the ISDN line. You may need a separate DN for each of the two B channels.

7.5.3 BONDING

BONDING (bandwidth-on-demand interoperability group) is a method for combining multiple channels into one; for example, two B channels can be combined to provide you with a single 128K channel. Note that many bandwidth-intensive applications, such as videoconferencing, do depend on bonding capabilities; you must ensure that the ISDN line configuration and ISDN adapter/software combination (CPE = customer premises equipment) support this feature.

7.6 How Does ISDN Provide Network Service?

ISDN is basically a WAN service available in two forms—basic rate interface (BRI) and primary rate interface (PRI), with speeds up to T1/E1 levels (1.5 or 2.0 Mbps, available in increments of 64 Kbps). There are two main differences between ISDN and T1/E1, however. ISDN is a switched service with pay-as-you-go rates, unlike a dedicated T1/E1 line. Also, signaling in ISDN is done on a separate channel, meaning that more bandwidth is available for data transfer.

7.6.1 Channel Types

ISDN uses time division multiplexing (TDM) to divide the available bandwidth into a number of fixed-size time slots, called channels. With ISDN, the local loop from the customer's premises comprises two basic channels, B and D, and several logical channels.

7.6.2 The B Channel

The B (for bearer) channel carries user data, as well as digitized voice and video information. It operates at a data rate of 64 Kbps, and can be used for both circuit-switched and packet-switched applications.

7.6.3 The D Channel

The primary purpose of the D channel is to provide signaling and control for each ISDN line installed. However, since the exchange of signals on the D channel rarely uses all its bandwidth, excess capacity can also be used for carrying data, although signaling always takes priority.

7.6.4 Access Interfaces—BRI and PRI

The number of logical channels is a function of the local loop's bandwidth. As mentioned, there are two standard access interface types: basic rate interface (BRI) and primary rate interface (PRI). (See Table 7.1 and Fig. 7.3.)

BRI is commonly used for remote access and Internet connections, and is available in most major metropolitan areas of the world today. BRI consists of two B channels (64 Kbps each) and one D channel (16 Kbps) for a maximum uncompressed data rate of 128 Kbps (or 144 Kbps if the D channel is used for data, as well). The BRI is usually provided through two wires with an RJ-45 jack.

Figure 7.3
ISDN basic rate interface (BRI) configuration.

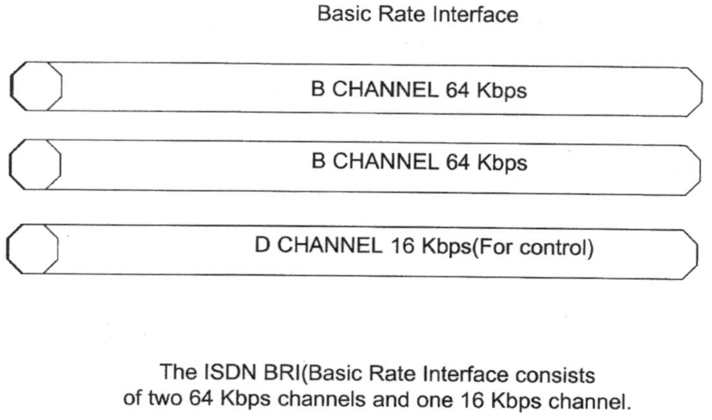

Basic Rate Interface

B CHANNEL 64 Kbps

B CHANNEL 64 Kbps

D CHANNEL 16 Kbps(For control)

The ISDN BRI(Basic Rate Interface consists
of two 64 Kbps channels and one 16 Kbps channel.

TABLE 7.1

*ISDN Access
Interface Structures*

Interface	Structure*	Total Bit Rate	User Data Rate
BRI	$2B + D_{16}$	192 Kbps	144 Kbps
PRI—USA	$23B + D_{64}$**	1.544 Mbps	1.536 Mbps
PRI—Europe	$30B + D_{64}$**	2.048 Mbps	1.984 Mbps

* The D channel operates at 16 Kbps in the BRI and at 64 Kbps in the PRI.
** This is one possible PRI configuration, and the most common today. Other configurations are also possible, such as 24B.

PRI is a higher-bandwidth interface, used at central communications server sites to aggregate BRI connections, or in applications where a native bandwidth higher than 128 Kbps is required. Generally, PRI is less widely available because of cost and bandwidth. In the United States and Canada, PRI comprises 23 B channels and a single D channel. In Europe and most of the rest of the world, it comprises 30 B channels and a single D channel. The PRI D channel operates at 64 Kbps, as do all B channels. This adds up to 1.5 Mbps in North America and Japan, and 2.0 Mbps in the rest of the world. PRI is presented as a four-wire trunk circuit connecting an ISDN port on a customer's computer communications equipment (such as a bridge/router) to the local exchange. The physical interface is DSX-1 in the United States and Canada, and 75-ohm G.703 (BNC) or 120-ohm G.703 (RJ-45) in the rest of the world.

7.6.5 ISDN Functional Devices

ISDN standards define several different devices that can be connected to a network, as well as interfaces between devices. Each interface, called a reference point, requires a communications protocol. Since ISDN was developed for telephony, common networking equipment like bridge/routers do not readily conform to ISDN standards.

7.6.6 Terminal Equipment Types

End-user devices, such as digital or analog telephones, X.25 Des, or bridge/routers are called terminal equipment. There are two types of terminal equipment. Devices that use ISDN directly and support ISDN services are known as terminal equipment type 1 (TE1). Non-ISDN devices are known as terminal equipment type 2 (TE2), and require a terminal adapter (TA) and software that enables the TA to communicate with an ISDN adapter. For example, a bridge/router with a standard synchronous serial interface is a TE2, while a bridge/router with a native BRI interface is a TE1.

7.6.7 Network Termination Types

There are two types of network terminations. Network termination type 1 (NT1) is a network terminal device, typically the local carrier's network termination unit at the customer site. This is the device to which all networking equipment is connected. The NT1 terminates the physical connection between the customer site and the local exchange, and connects the four-wire customer wiring to two-wire local loop. The NT1's functions include line performance monitoring, timing, physical signaling protocol conversion, power transfer, and multiplexing of the B and D channels.

Network termination type 2 (NT2) equipment provides customer site switching, multiplexing, and concentration of multiple ISDN lines. The NT2 is a more intelligent piece of equipment, and can include voice and data switching devices such as PBXs. The functionality of a bridge/router with a primary rate interface would fall into this category. The NT1 and NT2 devices may be combined into a single physical device called NT12. This device handles the physical, data link, and network layer functions.

7.7 ISDN Reference Points

ISDN reference points define the communication between different functional devices. Four protocol reference points are commonly defined for ISDN—called R, S, T, and U—with different protocols used at each reference point. (See Fig. 7.4*a* and *b*.)

The *R reference point* is between non-ISDN terminal equipment (TE2) and a terminal adapter (TA). There are no specific standards for the R reference point; the TA manufacturer will specify how the TE2 and TA communicate with each other. Typically, this protocol is either RS-232 or V.35.

The *S reference point* is between the ISDN user equipment (TE1 or TA) and the network termination equipment (NT1 or NT2).

The *T reference point* is between the customer site switching equipment (NT2) and the local loop termination (NT1). (In the absence of the NT2, the user-network interface is usually called the S/T reference

Figure 7.4
(a) ISDN functional devices and reference points; (b) ISDN reference points—another view. (Source: *ISDN Concepts* by G. Kessler [McGraw-Hill 1996].)

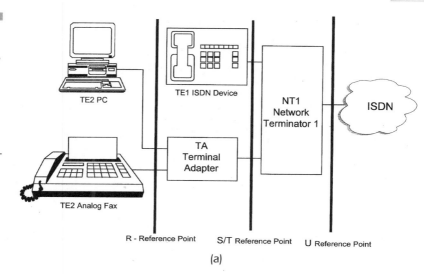

(a)

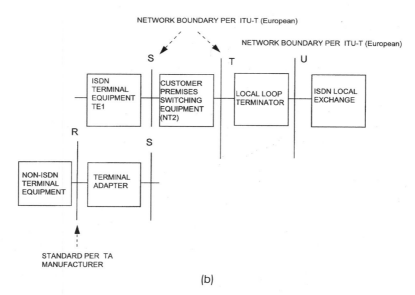

(b)

point.) The protocols used are the S and T reference points; they are specified by the ITU (CCITT).

The *U reference point* is between the NT1 and the local exchange. The protocol used at this reference point is defined by the ISDN provider.

When you place your ISDN order, find out whether your carrier supplies NT1 equipment. (Carriers will sometimes ask if you have a U or a T interface at your site. A U interface means you have an NT1; a T interface means you don't.) Some carriers will lease NT1s for a small monthly fee, but others do not offer them at all. In the United States, NT1s can also be purchased from many DSU/CSU vendors.

7.8 ISDN Bandwidth Management Options*

ISDN has the potential to be cost-effective, but has no inherent management features. Rather, it is the responsibility of the networking equipment vendor to provide these capabilities. Communications servers use sophisticated router software to manage available ISDN bandwidth automatically. Some or all of the following management features are currently available in many remote access products:

Dial-on-Demand.   Since ISDN charges are paid on a per-minute basis, this feature brings up a link only when there is data to be sent. When a router receives a packet for another network over an ISDN link, it establishes the call and transfers the data.

Bandwidth-on-Demand.   This feature in some routers (such as those from Xyplex) provides two capabilities: the ability to add bandwidth to a link and the ability to provide an alternate path when a primary link fails. Routers can be configured to add bandwidth whenever user-defined thresholds are reached.

Inverse Multiplexing (Also Called Bonding).   This is the ability to combine multiple low-speed channels across leased lines or switched services to form a single, higher-speed data path. Bridges/routers can dynamically group two or more ISDN circuits to the same partner, and dynamically balance the load over them.

Time-of-Day Circuit Control.   Network administrators can configure bridge/routers to automatically establish circuits for fixed durations at specific times during the day.

Compression.   Various kinds of compression are offered over ISDN links. Stacker LZS (TM), V.42bis, Van Jacobson header compression for TCP/IP

---

*Information based on Xyplex's white paper on ISDN. These features are present in Xyplex's internetworking software.

headers, and Compressed IPX (CIPX) for compression of IPX headers are examples. Stacker LZS compression has been known to achieve as high as four times the compression ratio. Note, however, that compression ratios achieved depend on many factors, including the type of data items and files beings transmitted.

Multilink PPP. This provides load balancing over a variety of multiple circuits (such as ISDN, X.25, and leased lines). It provides more bandwidth by using WAN services more efficiently, and quicker response times by using parallel transmissions.

Call Priority. Call priority is critical for certain client/server applications. Network administrators can set priorities for circuits or circuit groups and assign WAN services to critical data calls during the day.

Telecommuting Support. Internetworking software accommodates telecommuters by allowing PCs, workstations, and notebooks to dial-in and gain access to network resources. With PPP support across ISDN or any WAN service, telecommuters with password clearance can dial into their network through a router-based communications server, using the most economical and/or available network service.

WAN Security. With the rapidly increasing use of remote networks for financial transactions, different forms of security must be available on switched links. Three common methods are dial back, PPP authentication protocol (PAP), and WAN data scrambling. PAP requires a password and a PAP ID from the remote PPP devices before a connection can be established.

WAN Independence. As new switched services are offered, remote access server software should be able to mix any combination of WAN protocols: dial-up, ISDN, frame-relay, switched-56, and leased lines.

7.9 ISDN Hardware and Network Scenarios

A number of hardware components are involved in a remote access solution based on ISDN. These include a terminal adapter (TAs), ISDN bridges and routers, remote access servers, and ISDN PC cards.

Terminal adapters connect your PC into an ISDN line. On the PC side, it connects to the serial port, and on the network side, it connects to NT1 unit provided by the network supplier. Both internal and external versions of TAs are available from several vendors, such as Motorola, ATI, Tone Commander, US Robotics/3COM, and so forth, at an average price of $200 to $400. We would like to point out that in an arrangement where you connect your ISDN line to the serial port of a PC, and not to an ISDN adapter installed in the PC, throughput of a serial port is the gating factor in determining the effective speed of communication. Even though your ISDN link can give you 128 Kbps raw speed without compression (and up to 512 Kbps with 4:1 compression), a serial port running under RS232 protocol has a theoretical transfer data of 28.8 Kbps without compression (and 115.2 Kbps with 4:1 compression). Also, if you have multiple PCs at a remote office that must connect to a central site LAN, each of the remote PCs will need a terminal adapter. (See Fig. 7.5.)

ISDN adapters are installed in the PC. Both internal (including the ISA, PCI, and PC Card—PCMCIA) and external versions are available. These PC adapters may have integrated network terminators or may be connected to an external NT1 or NT1 plus terminal adapters. Mobile users with notebooks should have lightweight PC card ISDN adapters from vendors, such as IBM, Motorola, and 3COM.

ISDN bridges/routers are installed at the host LAN. Bridges are not capable of filtering, and are not suitable for bandwidth-on-demand requirements. ISDN routers are more appropriate devices at the LAN. They

Figure 7.5
NT1 and NT1 Plus.

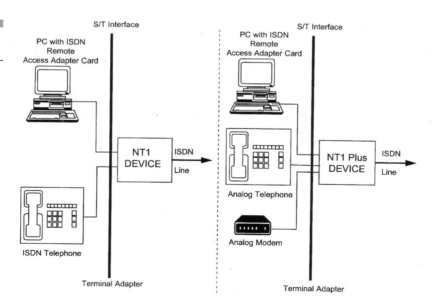

typically have an Ethernet connection, and support BRI or PRI port(s) for an ISDN connection.

For supporting remote workers, such as telecommuters, you need remote access servers. A number of vendors, such as Shiva, 3COM, Cisco, Microcom, Gandalf, and Ascend, provide such servers with ISDN support. A typical remote access server, with up to eight BRI ports, may cost you $5,000 to $6,000. It is always a good idea to acquire a flexible configuration that supports both analog and ISDN connections because not all remote offices may have ISDN coverage. Besides, some mobile workers do not need the speed provided by ISDN.

7.9.1 Communication Software Configuration Setup for ISDN Routers/Bridges

Table 7.2 gives an idea of the kind of parameters that need to be defined in ISDN bridges/routers.

7.10 ISDN Network Scenarios

There are a number of scenarios that you can consider in order to meet your remote access requirements. Figures 7.6, 7.7, and 7.8 give you an idea of six different scenarios. Network scenario #1 shows a typical remote access configuration where mobile workers, telecommuters, or small remote offices use ISDN to connect to a corporate LAN. This configuration utilizes ISDN adapters at the client end and ISDN routers at the central site. You have essentially a completely digital ISDN network from the client to the server. In network scenario #2, you are merely extending an existing LAN/WAN corporate infrastructure by the ISDN connection from remote users. The configuration in network scenario #3 shows the use of Internet as a virtual private network (VPN) to service remote users. Scenario #4, depicted in Fig. 7.7, shows the use of ISDN as a LAN/WAN infrastructure; it connects regional LANs to the corporate LAN and mainframes. Scenario #5 in Fig. 7.8 utilizes ISDN's voice phone line connectivity to integrate voice and data using LAN/WAN infrastructure. In the future, we could see more use of this type of design. Finally, network scenario #6 in Fig. 7.8 shows ISDN being used as a backup and/or supplemental network to the leased line network.

TABLE 7.2

ISDN Configuration
Setup

Configuration Parameter	User Setup	Comments
Configuration		
Switch type	DMS 100	Setup for a Northern Telecom DMS 100
ISDN type	Custom	
Callback	Off	Callback of remote user is not allowed
Line speed	64K/line	
Protocol	Compressed	Compression is turned on
Addressage time	1000	Throws out addresses older than 1000 seconds
Connection type	Auto On	Automatically calls the remote device
Packet timeout	Off	
Retry delay	30	If the call is unsuccessful, it tries calling the remote device every 30 seconds
Called number	5551212	Remote bridge phone number
Ringback number	6661111	
Security parameters		
Access status	On	Remote access of device for configuration is turned on
Client password	Exists	
Callback security	None	
Remote configuration	Protected	Device configuration is password protected
Protocol filtering		
0808 ACCEPT		Pass these Ethernet protocol types to the WAN; filter all other protocols
809b ACCEPT		Pass these Ethernet protocol types to the WAN; filter all other protocols
80f3 ACCEPT		Pass these Ethernet protocol types to the WAN; filter all other protocols
Type forwarding mode is ONLY		
Type demand mode is ANY		
Number of Ethernet addresses: 20		Bridge has learned 20 Ethernet addresses

Figure 7.6
Different ISDN scenarios: No. 1— typical remote access; No. 2—ISDN connection into the Internet.

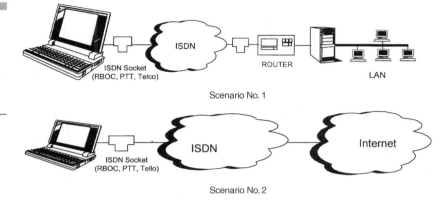

7.11 Performance Issues with ISDN and Remote Access

An ISDN network connection is more effective and faster than a PSTN dial-in connection for remote access of corporate information resources. It is also a method of obtaining a high-performance connectivity to Internet. In fact, for all those graphic-intensive applications on the Internet, ISDN connection is almost mandatory for getting a respectable response time at the present time. It could be compared to the express lanes on a

Figure 7.7
Different ISDN scenarios: No. 3— ISDN-based connection to the Internet as virtual private network; No. 4— ISDN as a part of LAN/WAN infrastructure for remote access.

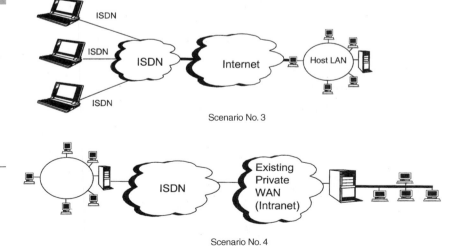

Figure 7.8
Different ISDN
scenarios: No. 5—
for voice-data
integration; No. 6—
backup or supple-
ment connection.

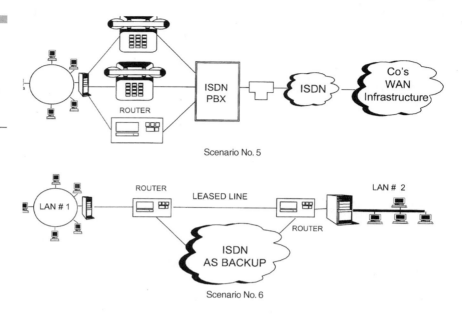

highway. However, you can increase the effective throughput of ISDN beyond its stated speed by techniques such as compression.

7.11.1 Raw Link Speed and Effective Link Speed—Magic of Compression

We made some remarks about raw link speed and effective link speed in Chap. 5. We indicated that compression is a good way of increasing the effective speed on a link. This is true with ISDN links, too. However, you should be careful when you read vendor literature while comparing different configurations and products. First of all, you rarely achieve maximum raw speed on any link. There are propagation delays, call setup time delays, and protocol overhead that reduce the achievable speed on a given link. In the case of ISDN, the maximum bandwidth you will achieve with two B channels is approximately 95 to 100 Kbps (75 percent link utilization) with optimal file transfer software and highly tuned remote access servers. You can, of course, improve this further anywhere from one to four times, depending on the type of file being transmitted. Obviously, if the file is straight ASCII (not binary), has lot of blanks, and is not already compressed by software such as PKZIP, you will achieve a much higher apparent or effective link utilization. This can be as high as 3:1.

7.11.2 Remote Access Server's Impact on Throughput

The internal hardware and software architecture of a remote access server affects the throughput that you will achieve with ISDN. Where both analog and ISDN links are supported on the same remote access server, effective speed may be further reduced because of sharing of certain components, such as the bus and upstream link. You should obtain additional performance information from the vendor with your own traffic profile.

7.12 Considerations in Selecting ISDN Routers/Remote Access Servers

Besides performance considerations, you should consider other factors in selecting the right ISDN router/server solution for your organization. These are listed here:

- Ease of use of the supported software and hardware at the client side
- Network management features—remote and host end
- Data-voice integration potential
- Shorter call setup time, as compared to PSTN (1 to 2 seconds)
- Bandwidth management features
- Availability of ISDN service where your mobile users are
- Complexity of setup at remote site
- Total cost of ISDN solution relative to other options
- Productivity gains by improved throughput
- Compatibility with existing LAN/WAN infrastructure, especially the bridges/routers and remote access servers

7.13 How Much Does ISDN Cost?

ISDN prices vary from region to region and country to country. Table 7.3 represents 1995 prices (readers should obtain current prices from the local telephone company).

TABLE 7.3

Typical ISDN Costs

	Installation	Monthly Charge	Per Minute
Nynex USA	$124	$38	$0.085
Pacific Bell, USA	$125	$20	$0.010
US West	Free	$60	Free
Bell Canada	$209 Can	$91 Can	Free, now
France	$119	$35	$0.19
Germany	$79	$42	$0.14
UK	$600	$30	$0.19

SOURCE: Xyplex white paper.

7.14 Issues in Implementing ISDN Remote Access Solutions

There are a number of issues that you must deal with in deciding to implement ISDN-based remote access solutions. These are listed here:

7.14.1 ISDN Availability, Pricing, and Support

ISDN service is standardized and available more easily in Europe. It has improved a great deal across North America (United States and Canada) during the past three years. However, it is not as universal as PSTN—your dial-in analog service. Different regional telephone companies in the United States offer varying levels of service, pricing, and support. Some make it easier to implement ISDN than others. Similarly one-time installation charges and monthly fees (fixed or per-minute usage) vary from one regional company to another, and of course, from one country to another.

7.14.2 More Complex Setup for ISDN

ISDN setup continues to be more complex than connecting a modem to a PC and loading communications software. The standardization process has improved, but configuring your ISDN equipment (adapter and software) and dealing with the telephone carrier can be a daunting exercise. Therefore, we would like to suggest that, unless you are a large organization with your own technical support staff, you should stay with a mainstream vendor's equipment. With this approach, the telephone company might be able to help you. You may even ask your carrier to recommend one or two adapters that they have the least problems with. One trade practice that is helpful is the use of ISDN ordering code that is supplied by the ISDN adapter supplier. You should provide this code to the telephone carrier who can then set up the ISDN line accordingly.

7.14.3 Incompatibility Across Applications

The vendors have been implementing standards in their equipment and software for the recent past, but there still are incompatibilities across remote access servers and ISDN applications, such as videoconferencing and Internet. You should review documentation provided with ISDN hardware and ensure that the equipment is compatible with the software applications you intend to use.

7.14.4 Client End (Remote CPE) and Server End (Host CPE) Considerations

If you are planning to use multiple devices, such as analog telephones, and employ different ISDN applications now or in future, you should consider using a stand-alone NT1 device or NT1 Plus device. A stand-alone NT1 device simply provides basic NT1 functions and line-powering functions. This device allows you to connect multiple devices to the ISDN line. On the other hand, when basic NT1 functionality is bundled within a PC adapter, you may be restricted to the kind of devices that the adapter itself supports—typically a subset of devices supported by ISDN. NT1 Plus devices include a network termination function, but they also include terminal adapter functionality allowing you to connect analog devices.

Similarly, at the server end, you should consider the number, speed, and mix of ports that your remote access solution must support. Many of the older servers allow fewer ISDN ports, as compared to PSTN ports.

7.14.5 Remote Client Node TCP/IP Address Assignment

Although assigning TCP/IP, IPX/SPX, or AppleTalk addresses to a small number of remote users is not a major issue, it does become a problem with a large network. In the case of TCP/IP, it is essentially a manual task, though DHCP (distributed host configuration protocol) allows some degree of automation. Management of this address and its specification in several different tables (e.g., in communications middleware) continues to be a problem. Hopefully, a universal TCP/IP node address management software utility will solve this problem. Meanwhile systems administrators have developed homegrown utilities to address this requirement. Address assignment and management problems extend to conveying and tracking this address information (network address and subnet mask) to remote mobile users who may be constantly growing, losing the privilege for remote access for a variety of reasons, and constantly moving around. The problem is partially being addressed by the Mobile IP group of IETF, but a versatile product-based solution has yet to come.

7.14.6 Spoofing and Filtering to Avoid Unnecessary Network Traffic

In a mobile disconnected network environment, you may lose physical connection during inactive periods of network traffic especially with switched networks. In many protocols, "stay alive" or connection acknowledgement packets are often sent between nodes, even when no user data transmission is being exchanged. *Spoofing* techniques are often employed to fool the network by acknowledging LAN packets locally, and avoiding unnecessary network connection charges. You should also review Chap. 16, in which we discuss bandwidth management techniques.

Also we should not allow unnecessary LAN traffic to wander on to the ISDN link. This is often achieved by address and protocol filtering, as shown in the configuration setup in Table 7.2.

7.15 Comparing ISDN with Other Network Services

ISDN is certainly an attractive offering for remote access where higher bandwidth is required. However, other emerging services, such as ADSL and cable modems, will give it good competition in the future. You should consider a number of other factors, besides bandwidth, before selecting a network connection for remote access. (Refer to Chap. 12 for further discussion of this topic in detail.) Table 7.4 provides a brief comparison of ISDN with other network services.

7.16 Integration of Analog and ISDN—A More Common Scenario

Whereas we have compared ISDN with other network options, the intent was not to suggest that you should use PSTN, ISDN, or ADSL. In reality, application and user requirements within an organization vary widely. Some users who use e-mail only are well served by old reliable PSTN—the cheapest and most widely available. Others who transmit multimedia files or exploit PC to PC videoconferencing should use ISDN where available. Still others—for example, telecommuters constantly logged on to the office LAN for application development—may find ADSL or cable modem appropriate. Although ADSL is not available in many areas today and not too many products support it, a vast majority of remote access servers provide support for both PSTN and ISDN in the same box, as we stated earlier. Figure 7.9 illustrates one such scenario.

7.17 More Information on ISDN

If you need more information on ISDN, there are a number of books that you can refer to. These are listed at the end of this chapter. Particularly in the context of remote access, we recommend Jeffrey Fritz's book, *Remote LAN Access*, published by Freeman Press (1996).

There is a forum of ISDN users called North American ISDN Forum. Information on the Forum's activities and a lot of introductory material on ISDN, including user experience is available on the Internet Web pages. You can also access several ISDN home pages, especially Dan Kegel's ISDN page on the Web (www.alumni.caltech.edu/~dank/isdn/) as resource material.

TABLE 7.4

Comparison of ISDN with Other Network Options

	Speed/ Bandwidth	Pricing	Advantages and Best Applications	Disadvantages
Leased line	33.6 Kbps to T3	Function of speed and distance	LAN/WAN infrastructure Cost effective for constant, high-volume use Low error rate	Expensive for occasional use
PSTN	34.4–56 Kbps	Same basis as telephone calls	Less than 2 to 3 hours per day	Limited speed, variable quality, long setup time, and poor security
ISDN	128 Kbps	Function of call duration, time of day, and distance	File transfer, Internet, videoconferencing (cheaper than leased line for infrequent use)	Not universally available, and hard to set up
ADSL	1.5–6.0 Mbps	Pricing still being determined	High downstream traffic applications	Still being piloted; not universally available
Cable modems	1–10 Mbps	Cable modem purchase, setup (less than $100 per month)	High bandwidth, cable in place in most residences	Not as widespread as telephone for business; cable infrastructure changes required for interactive applications; regulatory issues in some places
Wireless	Up to 28.8 for WAN; higher in campus	Varies with service (.06 c per MB for CDPD)	True mobile applications (e-mail, dispatch, field and sales force automation)	Very expensive; coverage and reliability are poor

SOURCE: Author's research.

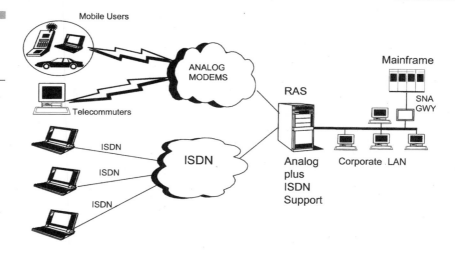

Figure 7.9
Analog and ISDN integration through RAS.

Summary

In this chapter, we reviewed ISDN as a network option for remote access. You learned about benefits of ISDN, ISDN applications, ISDN terms, and how it provides network service through various channel types. We discussed how to configure ISDN, and explained SPID—a common term that you must deal with in installing ISDN. Then we reviewed ISDN hardware and compared ISDN with other network options for remote access. Hopefully, at the end of this chapter, you have gotten a better understanding of ISDN; if you need more, refer to the source material.

Sources

ISDN for Dummies by David Angell, 1995, published by IDG Books.

Remote LAN Access by Jeffrey Fritz, 1996, published by Freeman Press.

ISDN—An Introduction by William Stallings, 1988, Macmillan Publishing Company.

ISDN—Manager's Guide by Ian Angus, 1990, TeleManagement Press.

ISDN—Concepts, Facilities, and Services by Gary Kessler and Peter Southwick, 1996, McGraw-Hill.

ISDN—How to Get To High-Speed Connection to the Internet by Charles Summers and Bryant Dunetz, 1996, published by John Wiley.

CHAPTER 8

Remote Access Network Options—ADSL and Cable

Remote access bandwidth requirements have continued to exceed network providers' capabilities. Whereas PSTN became the universal option for remote access, ISDN provided more acceptable performance on graphic-intensive applications. However, users want more—sound, image, and video. Only ADSL and cable modems may satisfy this need. Let both technologies mature, compete, survive, and blossom so that we, the users, have a choice.

—*Chander Dhawan*

About This Chapter

In this chapter, we discuss an emerging networking technology called asymmetric digital subscriber line (ADSL) that can send information at a much higher speed than PSTN or ISDN options, covered in Chaps. 6 and 7, respectively. While a number of different names and variations of this technology exist, we prefer to use ADSL as an umbrella term. We review transmission capabilities of various modem technologies on copper wire cabling infrastructure, and explain how distance affects throughput. The limitations of the local loop, and the rest of the network will also become apparent. We explain how ADSL works, describe various flavors of ADSL modulation techniques, and provide a status of trials and implementation of these technologies. Finally, we discuss cable modems.

8.1 ADSL as a High-Speed Link for Remote Access*

The popularity of Internet has significantly increased demand for faster remote access to corporate information. At the same time, it has raised demand for the integration of text, graphics, video, and even voice. The industry has been furiously looking for new remote access technologies that will deliver true multimedia capability on the Internet. ADSL and cable modems are the results of this effort. During the next several years, these two technologies will be prime candidates for handling high-speed applications over existing telephone lines and cable infrastructure.

ADSL, a new modem technology, converts our existing twisted-pair telephone lines into access paths for multimedia and high-speed data communications. Even though ADSL modems, introduced in 1997 by PairGain and other vendors, are in the 768 Kbps to 1.2 Mbps range, ADSL can transmit more than 6 Mbps to a subscriber, and as much as 640 Kbps more from the subscriber. Such transmission speeds expand existing remote access capacity by a factor of 6 to 50 (as compared to ISDN) without installing new cabling. ADSL can literally transform the existing public telecommunications networks from one limited to voice, text, and low-resolution graphics to a powerful, ubiquitous network capable of

---

*Information adopted from ADSL Forum web site (http://www.adsl.com).

bringing multimedia, including full-motion video, to everyone's home or office during the next few years.

ADSL will play a crucial role over the next ten or more years as telephone companies enter new markets for delivering information in video and multimedia formats. New broadband cabling will provide a fast alternative to urban residential clients, but will take many years to reach all prospective subscribers who are serviced by telephone lines. However, the success of these new services will depend upon the low cost of service to consumers which, in turn, will depend on reaching as many subscribers as possible during the next few years. By bringing movies, television, video catalogs, remote CD-ROMs, corporate LANs, and the Internet into homes and small businesses, ADSL will make these markets viable and profitable for telephone companies and application suppliers alike.

8.2 ADSL and Related Terminology

There are a number of terms that are used in the trade literature to denote the general class of high-speed digital technologies based on exploiting copper wire already installed by telephone companies in North America and PTTs in Europe. Terms such as ADSL, xDSL (or simply DSL), and HDSL are more common, though VDSL, VADSL, SDSL, and BDSL have also been defined and used by the industry groups. We now explain these terms with historical background and application examples. Table 8.1 lists various copper access transmission technologies.

8.3 From Voice-Grade PSTN Speeds to 1.544 Mbps and Beyond on Local Loop

In Chap. 6, we described voice-grade data modems and their limitations. Voice-grade modems can presently transmit up to 33.6 Kbps over a common telephone line in both directions, but the practical limit only 20 years ago was 1200 bps. We also described recent efforts by companies, such as US Robotics, to extend this speed limit to 56 Kbps from the network to the remote user. No one believes we can go much faster than 56 Kbps in the future, however. This is essentially so because voice-grade

TABLE 8.1

Copper Access
Transmission
Technologies

Name	Meaning	Data Rate	Mode	Applications
V.22 V.32 V.34	Voice-grade modems	1200 bps to 33,600 bps	Duplex	PSTN-based data communication
IDSL	Integrated digital subscriber line	160 Kbps[a]	Duplex[b]	ISDN service for voice and data communication
HDSL	High data rate digital subscriber line	1.544 Mbps[c] 2.044 Mbps[d]	Duplex Duplex	T1/E1 service feeder plant, WAN, LAN access, server access
SDSL	Single-line digital subscriber line	1.544 Mbps 2.044 Mbps	Duplex Duplex	Same as HDSL plus premises access for symmetric service
ADSL	Asymmetric digital subscriber line	1.5 to 9 Mbps 16 to 640 Kbps	Downstream Upstream	Internet access, video demand, LAN remote access, interactive multimedia
VDSL[e,g]	Very high rate digital subscriber line	13 to 52 Mbps 1.5 to 2.3 Mbps	Down[f] Up[f]	Same as ADSL plus HDT

SOURCE: ADSL Forum.

[a] 192 Kbps divides into two B channels (64 Kbps), one D channel (16 Kbps), and link administration.
[b] *Duplex* means data of the same rate both upstream and downstream at the same time.
[c] Requires two twisted-pair lines.
[d] Requires three twisted-pair lines.
[e] Also called BDSL, VADSL, or, at times, ADSL. VDSL is an ANSI and ETSI designation.
[f] *Down* means downstream, from the network to the subscriber. *Up* means upstream, from the subscriber to the network.
[g] Future VDSL systems may have upstream rates equal to downstream, but on much shorter lines.

bandwidth does not exceed 3.3 KHz, and Shannon's law defines the theoretical speed limit at 33.6 Kbps at a specified signal-to-noise ratio. The best part of PSTN access is that you can buy a PSTN modem for under $200. We have these faster voice-grade modems because of advances in algorithms, digital signal processing, and semiconductor technology.

Voice-grade modems operate at the subscriber premises end of voice-grade lines, and transmit signals through the core switching network without alteration; the network treats them exactly like voice signals. This has been their singular power; despite their rather slow speeds, compared to terminals today, they can be connected immediately, anywhere a telephone line exists, and there are nearly 600 million such locations in the world.

Bandwidth limitations of voice-band lines do not come from the subscriber line, however. They come from the core network. Filters at the edge of the core network limit voice-grade bandwidth to 3.3 KHz. Without filters, copper access lines can pass frequencies into MHz range, albeit with substantial attenuation. Indeed, attenuation, which increases with line length and frequency, dominates the constraints on data rate over twisted-pair wire. Practical limits on data rate *in one direction* compared to line length (of 24-gauge twisted pair) are:

DS1 (T1)	1.544 Mbps	18,000 feet
E1	2.048 Mbps	16,000 feet
DS2	6.312 Mbps	12,000 feet
E2	8.448 Mbps	9,000 feet
1/4 STS-1	12.960 Mbps	4,500 feet
1/2 STS-1	25.920 Mbps	3,000 feet
STS-1	51.840 Mbps	1,000 feet

Subscriber loop plant configurations vary tremendously around the world. In some countries, 18,000 feet covers virtually every subscriber; in others, such as the United States, 18,000 feet covers less than 80 percent of subscribers. However, the 20 percent or so remaining have lines with loading coils that cannot be used for any DSL service (including ISDN) without removing the coils. Most telephone companies have had programs underway for a number of years to shrink average loop length, largely to stretch the capacity of existing central offices. The typical technique involves the installation of access nodes remote from central offices (CO), creating so-called distribution areas with maximum subscriber loops of 6000 feet from the access node. Remote access nodes (not to be confused with communications-based remote access servers—the central theme of this book) are fed by T1/E1 lines (now using HDSL) or fiber. In suburban communities, a distribution area connects an average of 1500 premises; in urban areas, the figure is double, about 3000 premises. Of course, the number of premises served dwindles as service data rates increase. A fiber to the curb (FTTC) system offering STS-1 rates may be within reach of only 20 homes in some suburban areas. (See Figs. 8.1*a* and *b*.)

You now have enough information to be a network planner, presuming the marketing department has handed you a stable list of applications. If that list does not include digital live television or HDTV (but does include video on demand and Internet access), then a data rate of 1.5 Mbps per subscriber terminal downstream may suffice, and you can offer it to virtually everyone within 18,000 feet, the nominal range of ISDN. For sub-

Figure 8.1
Figure 8.1
(a) Line changes for
T1 and DSL. (b) Central office (CO)
arrangement for
xDSL.

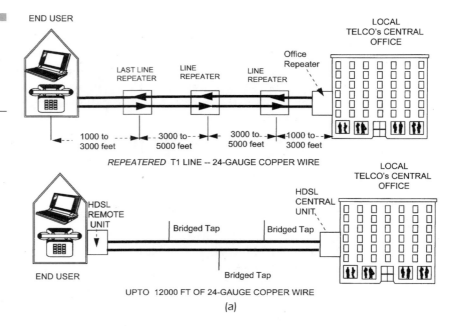

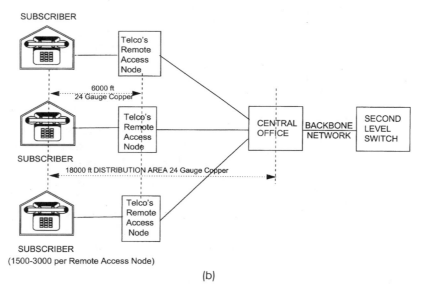

scribers with shorter lines, either to a central office or remote access node,
you can offer more than one channel to more than one premises terminal.
If digital live television is on the list, then you have to offer at least 6 Mbps,
and you may be limited to 4500 foot distances to supply more than one
channel at a time. (This fact is the heart of telephone companies' interest

in wireless broadcast digital TV.) Clearly HDTV, demanding as much as 20 Mbps, goes over only the shortest loop length.

Of course, this offering of digital services over existing twisted-pair lines requires transceivers, special modems capable of dazzling data rates, when one considers the age and the original intentions of twisted-pair wiring technology. It turns out that this effort to use twisted pair for high-speed information began many years ago.

8.3.1 DSL—Digital Subscriber Line for ISDN

The basic acronyms for all DSL arrangements came from Bellcore. In general, you could think of DSL as a modem, or a modem pair, because that is what gives the line this high-speed capability. A modem pair applied to a line creates a digital subscriber line, but when a telephone company provides DSL, or ADSL, or HDSL, it provides you with modems, quite apart from the lines, which they already own and use for voice service.

DSL itself, apart from its later siblings, is the modem used for basic rate ISDN. A DSL transmits *duplex data,* that is, data in both directions simultaneously, at 160 Kbps over copper lines up to 18,000 feet of 24-gauge wire. The multiplexing and demultiplexing of this data stream into two B channels (64 Kbps each), a D channel (16 Kbps), and some overhead takes place in the attached terminal equipment. By modern standards, DSL does not press any transmission thresholds, but its standard implementation (ANSI T1.601 or ITU I.431) employs echo cancellation to separate the transmit signal from the received signal at both ends, a novelty at the time DSL first found its way into the network.

DSL modems use twisted-pair bandwidth from 0 to about 80 KHz. (Some European systems use 120 KHz of bandwidth.) Therefore, they preclude the simultaneous provisioning of analog POTS (plain old telephone service). However, DSL modems are being used today for so-called pair-gain applications, in which DSL modems convert a single POTS line to two POTS lines, obviating the physical installation of the second line wiring. The telephone company just installs the analog/digital voice functions at the customer premises for both lines, and—presto—you get two from one. (See Fig. 8.2.)

8.3.2 T1 or E1

In the early 1960s, engineers at Bell Labs created a voice multiplexing system that first digitized a voice signal into a 64 Kbps data stream (repre-

Figure 8.2

ISDN as a digital sub-scriber line (DSL).

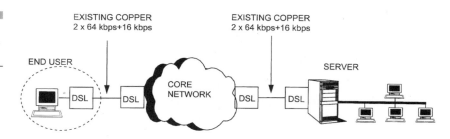

senting 8000 voltage samples a second, with each sample expressed in 8 bits). They then organized 24 of them into a framed data stream, with some conventions for figuring out which 8-bit slot went where at the receiving end. The resulting frame was 193 bits long, and created an equivalent data rate of 1.544 Mbps. The structured signal was called DS1, but it has acquired an almost colloquial synonym—T1—which also describes the raw data rate, regardless of framing or intended use. AT&T deployed DS1 in the interoffice plant starting in the late 1960s (almost all of which has since been replaced by fiber); by the mid-1970s, AT&T was using DS1 in the feeder segment of the outside loop plant.

In Europe and at CCITT (now ITU), the collection of world PTTs, other than AT&T, modified Bell Labs' original approach, and defined E1, a multiplexing system for 30 voice channels running at 2.048 Mbps. In Europe, E1 is the only designation, and stands for both the formatted version and the raw data rate.

Until recently, T1 and E1 circuits were implemented over copper wire by using crude transceivers with a self-clocking alternate mark inversion (AMI) protocol. AMI requires repeaters 3000 feet from the central office, and every 6000 feet thereafter. It takes 1.5 MHz of bandwidth, with a signal peak at 750 KHz (U.S. systems). To a transmission purist, this is profligate and ugly, but it has worked for many years, and hundreds of thousands of lines (T1 and E1) exist in the world today.

Telephone companies originally used T1/E1 circuits for transmission between offices in the core switching network. Over time, they tariffed T1/E1 services, and offered them for private networks, connecting PBXs and T1 multiplexors together over the wide area network (WAN). Today, T1/E1 circuits can be used for many other applications, such as connecting Internet routers together, bringing traffic from a cellular antenna site to a central office, or connecting multimedia servers into a central office. An increasingly important application is in the so-called feeder plant; this section of a telephone network radiates from a central office to remote access nodes that, in turn, service premises over individual copper lines.

T1/E1 circuits feed digital loop carrier (DLC) systems that concentrate 24 or 30 voice lines over two twisted-pair lines from a central office, thereby saving copper lines and reducing the distance between an access point and the final subscriber.

Note, however, that T1/E1 is not a very suitable service for connecting to individual residences. First of all, AMI is so demanding of bandwidth and corrupts the cable spectrum so much, that telephone companies cannot put more than one circuit in a single 50-pair cable, and must put none in any adjacent cables. Offering such a system to residences would be equivalent to pulling new wire to most of them. Second, until recently no application going to the home demanded such a data rate. Third, even now, as data rate requirements accelerate with the hope of movies and high-speed data for everyone, the demands are highly asymmetric—bundles downstream to the subscriber, and very little upstream in return—and many situations will require rates above T1 or E1. In general, high-speed data rate services to the home will be carried by ADSL or VDSL (or similar types of modems over CATV lines).

8.4 HDSL—High Data Rate Digital Subscriber Line

HDSL is simply a better way of transmitting T1 or E1 over twisted-pair copper lines. It uses less bandwidth and requires no repeaters. Using more advanced modulation techniques, HDSL transmits 1.544 Mbps or 2.048 Mbps in bandwidths ranging from 80 KHz to 240 KHz, depending upon the specific technique, rather than the greedy 1.5 MHz absorbed by AMI. HDSL provides such rates over lines up to 12,000 feet in length (24 gauge), the so-called carrier serving area (CSA). However, it does so by using two lines for T1 and three lines for E1 (with 2B1Q line coding, two lines with CAP), each operating at half or third speed.

Most HDSL will go into the feeder plant, which connects subscribers after a fashion, but hardly in the sense of an individual using a phone service.

Typical applications include PBX network connections, cellular antenna stations, digital loop carrier systems, interexchange POPs, Internet servers, and private data networks. As HDSL is the most mature of DSL technologies with rates above a megabit, it will be used for early-adopter premises applications for Internet and remote LAN access, but will likely give way to ADSL and SDSL in the near future.

8.4.1 SDSL—Single-Line Digital Subscriber Line

On its face, SDSL is simply a single-line version of HDSL, transmitting T1 or E1 signals over a single twisted pair, and (in most cases) operating over POTS, so a single line can support POTS and T1/E1 simultaneously. However, SDSL has the important advantage compared to HDSL that it suits the market for individual subscriber premises which are often equipped with only a single telephone line. SDSL will be desired for any application needing symmetric access (such as servers and power remote LAN users), and it therefore, complements ADSL (see following discussion). It should be noted, however, that SDSL will not reach much beyond 10,000 feet, a distance over which ADSL achieves rates above 6 Mbps.

8.5 ADSL—Asymmetric Digital Subscriber Line

ADSL followed on the heels of HDSL, but is really intended for the last leg into a customer's premises. As its name implies, ADSL transmits an asymmetric data stream, with much more data going downstream to the subscriber, and much less coming back. The reason for this has less to do with transmission technology than with the cable plant itself. Twisted-pair telephone wires are bundled together in large cables. Fifty pairs to a cable is a typical configuration toward the subscriber, but cables coming out of a central office may have hundreds or even thousands of pairs bundled together. An individual line from a CO to a subscriber is spliced together from many cable sections as they fan out from the central office (Bellcore claims that the average U.S. subscriber line has 22 splices). Alexander Bell invented twisted-pair wiring to minimize the interference of signals from one cable to another, caused by radiation or capacitive coupling, but the process is not perfect. Signals do couple, and couple more so as frequencies and the length of line increase. It turns out that if you try to send symmetric signals in many pairs within a cable, you significantly limit both the data rate and the length of line you can attain.

Fortunately, the preponderance of target applications for digital subscriber services are asymmetric. Video-on-demand, home shopping, Internet access, remote LAN access, multimedia access, and specialized PC services all feature high data rate demands downstream to the subscriber, but relatively low data rate demands upstream. MPEG movies with simulated VCR controls, for example, require 1.5 or 3.0 Mbps downstream, but can work just fine with no more than 64 Kbps (or 16 Kbps) upstream. The IP protocols for Internet or LAN access push upstream rates higher, but a 10-to-1 ratio of down- to upstream does not compromise performance, in most cases. (See Fig. 8.3.)

So ADSL has a range of downstream speeds depending on distance:

Up to 18,000 feet	1.544 Mbps (T1)
16,000 feet	2.048 Mbps (E1)
12,000 feet	6.312 Mbps (DS2)
9000 feet	8.448 Mbps

Upstream speeds range from 16 Kbps to 640 Kbps. Individual products today incorporate a variety of speed arrangements, from a minimum set of 1.544/2.048 Mbps down and 16 Kbps up to a maximum set of 9 Mbps down and 640 Kbps up. All of these arrangements operate in a frequency band above POTS, leaving POTS service independent and undisturbed, even if a premises ADSL modem fails.

As ADSL can transmit digitally compressed video, among other things, it includes error-correction capabilities intended to reduce the effect of impulse noise on video signals. The capabilities take the form of an error correcting code and interleaving; the error correcting code is always applied but the interleaving, used for greater protection, is optional because it can introduce delays of up to 20 mS. Error correction introduces about 20 msec of delay, which is far too much for LAN and IP-based data communications applications. Therefore, ADSL must know not only what kind of signals it is passing but also how much interleaving to apply (this problem exists for

Figure 8.3
ADSL connection.

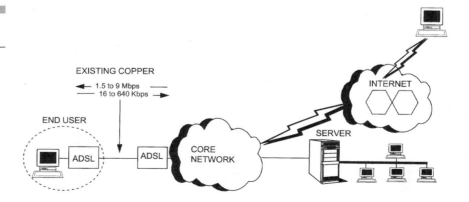

any wire-line transmission technology, over twisted-pair or coaxial cable). Furthermore, ADSL will be used for circuit switched, packet switched (such as an IP Router) and, eventually, ATM switched data. ADSL must connect to personal computers and television set top boxes at the same time. Taken together, these application conditions create a complicated protocol and installation environment for ADSL modems, moving these modems well beyond the functions of simple data transmission and reception.

8.5.1 ADSL Capabilities

An ADSL circuit connects an ADSL modem on each end of a twisted-pair telephone line, creating two information channels—a high-speed duplex channel and a POTS (Plain Old Telephone Service) channel. The POTS channel is split off from the digital modem by filters, thus guaranteeing uninterrupted PSTN or POTS, even if ADSL fails. The high-speed channel ranges from 1.5 to 6.1 Mbps in the downstream direction, and from 16 to 640 Kbps in the upstream direction. Each channel can be sub-multiplexed to form multiple, lower-rate channels which can be used for a variety of services—bit-synchronous, packet, and ATM.

ADSL modems provide data rates consistent with North American and European digital hierarchies, and can be purchased with various speed ranges and capabilities. The minimum configuration provides 1.5 or 2.0 Mbps downstream and a 16 Kbps duplex channel; others provide rates of 6.1 Mbps and 64 Kbps duplex. Products with downstream rates up to 9 Mbps, and duplex rates up to 640 Kbps will be available in the future.

Downstream data rates depend on a number of factors, including the length of the copper line, its wire gauge, presence of bridged taps, and cross-coupled interference. Line attenuation increases with line length and frequency, and decreases as wire diameter increases. Ignoring bridged taps, ADSL will perform as shown in Table 8.2.

TABLE 8.2

ADSL Speed
and Distance
Relationship

Data Rate (Mbps)	Wire Gauge (AWG)	Distance (ft)	Wire Size (mm)	Distance (km)
1.5 or 2	24	18,000	0.5	5.5
1.5 or 2	26	15,000	0.4	4.6
6.1	24	12,000	0.5	3.7
6.1	26	9,000	0.4	2.7

Whereas the measure varies from one telephone company to another, these capabilities can cover up to 95 percent of a loop plant, depending on the desired data rate. Premises beyond these distances can be reached with fiber-based digital loop carrier systems. As these DLC systems become commercially available, telephone companies can offer virtually ubiquitous access in a relatively short time.

Many applications envisioned for ADSL involve digital compressed video. As a real-time signal, digital video cannot use link- or network-level error control procedures, commonly found in data communications systems. ADSL modems, therefore, incorporate forward error correction that dramatically reduces errors caused by impulse noise. Error correction on a symbol-by-symbol basis also reduces errors caused by continuous noise coupled into a line.

Presently, ADSL models offer T1/E1 and V.35 digital interfaces for continuous bit rate (CBR) signals. Future versions will offer LAN interfaces for direct connection to a personal computer and ATM interfaces for variable bit rate signals. Over time, ADSL units will be built directly into access node concentrators and so-called premise service modules, such as set top boxes and personal computer interface cards.

8.5.2 ADSL Technology—How Does It Work?

ADSL depends not only upon advanced digital signal processing but also upon creative algorithms to squeeze so much information through twisted-pair telephone lines. In addition, many advances have been required in transformers, analog filters, and A/D converters. Long telephone lines may attenuate signals at 1 MHz (the outer edge of the band used by ADSL) by as much as 90 dB, forcing analog sections of ADSL modems to work very hard to realize large dynamic ranges, separate channels, and maintain low noise figures. On the outside, ADSL looks simple—transparent synchronous data pipes at various data rates over ordinary telephone lines. On the inside, where all the transistors work, there is a miracle of modern technology.

To create multiple channels, ADSL modems divide the available bandwidth of a telephone line in one of two ways—frequency division multiplexing (FDM) or echo cancellation. FDM assigns one band for upstream data and another band for downstream data. The downstream path is then divided by time-division multiplexing into one or more high-speed channels and one or more low-speed channels. The upstream path is also multiplexed into corresponding low-speed channels. Echo cancellation assigns the upstream band to overlap the downstream, and separates the two by means of local echo cancellation, a technique well known in V.32 and V.34 modems. Echo cancellation uses bandwidth more efficiently, but at the expense of complexity and cost. With either technique, ADSL splits off a 4 KHz region for POTS at the DC end of the band.

An ADSL modem organizes the aggregate data stream created by multiplexing downstream channels, duplex channels, and maintenance channels together into blocks, and attaches an error correction code to each block. The receiver then corrects errors that occur during transmission up to the limits implied by the code and the block length. The unit may, at the users option, also create superblocks by interleaving data within sub-blocks; this allows the receiver to correct any combination of errors within a specific span of bits. The typical ADSL modem interleaves 20 ms of data, and can thereby correct error bursts as long as 500 μsec. ADSL modems can, therefore, tolerate impulses of arbitrary magnitudes whose effect on the data stream lasts no longer than 500 μsec. Initial trials indicate that this level of correction will create effective error rates, suitable for MPEG2 and other digital video compression schemes.

8.5.3 ADSL Modulation Schemes— DMT and CAP

DMT and CAP are two different modulation systems, also called *line codes*, currently on the market for ADSL. Like any competing technologies, proponents of each technique are advocating the benefits of their scheme. The ADSL Forum is attempting to ensure compatibility across these schemes.

DMT stands for Discrete Multi-Tone; it describes a version of multicarrier modulation in which incoming data is collected, and then distributed over a large number of small individual carriers, each of which uses a form of QAM (quadrature amplitude modulation) modulation. DMT creates these channels using a digital technique, known as discrete fast-Fourier transform. DMT is the basis of ANSI Standard T1.413, issue 1. DMT works in 26 kHz to 1.1 MHz band, and divides the band between 26 kHz and 1.1 MHz into 249 4-kHz channels.

CAP, a variation of QAM, stands for carrier-less amplitude/phase modulation; it describes a version of modulation in which incoming data modulates a single carrier that is then transmitted down a telephone line. The carrier itself is suppressed before transmission (it contains no information, and can be reconstructed at the receiver), hence the adjective *car-*

rierless. CAP was developed by GlobeSpan, a former division of AT&T, and at this writing is undergoing standardization in ANSI.

Very limited comparative information between the two technologies is available at present. DMT outperformed CAP at an ANSI test event performed by Bellcore in 1973 by supporting longer distances and adapting successfully to noise on the line. On the other hand, CAP-based modems have gone further down the price decline curve, and cost substantially less than DMT modems. Of course, the pricing advantage in an emerging technology may never be permanent, and current distance shortcomings in CAP may be resolved in the future, unless there are technological barriers. In this environment, keep tuned to the latest research and comparative analysis.

8.5.4 ADSL—Conceptual Network Schematics

There are a number of ways that the ADSL service may be utilized to connect Internet to home, small business, or corporate branch offices. Figures 8.4 and 8.5 illustrate two such cases.

Figure 8.4 is a generic example of ADSL connectivity. One modem sits at the customer site and the other is in the phone company's nearest cen-

Figure 8.4
DSL extends capacity of copper telephone wiring.

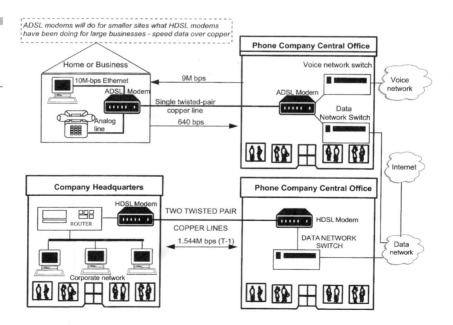

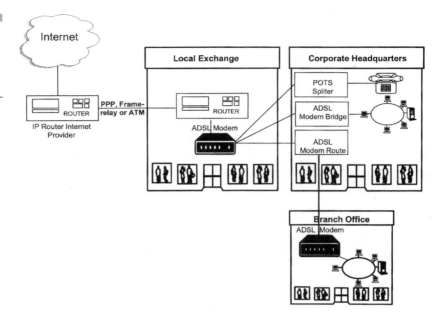

Figure 8.5
Westell's ADSL bridge
and router modems
employed for remote
access.

tral office. The phone company's CO near the remote user is connected to a CO near the corporate information server. There is another ADSL (or HDSL) modem pair between the second CO and the corporate headquarters. Therefore, there are four DSL modems involved for an end-to-end connection.

Figure 8.5 is based on Westell product implementation. Here, a telephone company provides an ADSL modem in its local exchange, which in turn connects to a corporate ADSL bridge or a router modem to connect an ISP router to a local LAN or a remote branch office LAN. This is facilitated by early products from Westell, an ADSL vendor. The Westell device, called FlexCap, provides high-speed data access over standard twisted-pair wiring, and incorporates bridging, routing, and even SNMP-compliant network management capabilities. FlexCap models support 1.544 Mbps of downstream access with a 64 Kbps upstream channel. Other models that support 6 Mbps will be available by the time the book goes to press.

8.5.5 ADSL Hardware—Early Product Implementations

A book on emerging technology can become dated easily if we include examples of first versions of products implementing a new technology.

TABLE 8.3

ADSL Product
Implementations
(circa 1997)

Vendor	Product	Feature	Price Range	Availability
Westell (www.westell.com) 1-800-941-9378	FlexCap	1.544 Kbps modem-router	$1500	Now
	FlexVision	6 Mbps modem-router	$1800	Now
	SuperVision	Mux with OC-3 (155 Mbps)	N/A	Now
PairGain (www.pairgain.com, 800-370-9670)	ADSL Modem	768 Kbps ADSL Modem	$2000 for campus environment	Now
Amati (415-903-2350)	Overture 4	4-Mbps Modem-bridge	$2500	Now
	Overture 8	6 Mbps Modem-bridge (DMT)	N/A	Now
	ADSL Access Concentrator	ADSL Concentrator	N/A	Now
Digital Link Corp. (formerly Performance Telecom) (716-654-5544)	CopperAccel ADSL	1.544-Mbps modem-bridge	$4000	Now
Aware Inc. (617-276-4000)	ADSL Internet Access Modem	7-Mbps modem	$1995	Now

SOURCE: From vendors and *CommunicationsWeek*, April 1996.

Notwithstanding this risk, we include such information on ADSL because it will be useful to early adopters of ADSL, and it will encourage these vendors. No doubt, there will be more products with ADSL features integrated with existing internetworking products. Table 8.3 provides a flavor of the first evolution of ADSL products.

RADSL (rate adaptive digital subscriber line) is another twist to ADSL technology from GlobeSpan Technologies of Red Bank, New Jersey. Their chipset allows upstream speeds to vary from 600K to 8 Mbps, and downstream from 90 Kbps to more than 1 Mbps.

8.5.6 ADSL Costs

Readers should get current information from their common carriers. Nonetheless, we provide the following rule-of-thumb numbers for preliminary evaluation of ADSL:

Product or Service	Author's Estimates
ADSL Modem	$750 for consumer to $1500 for business (expected to go down further during 1998—99)
ADSL Service (Business and consumer)	$50—200 per month going down to $40—100 by 1999 (Early introductions of ADSL at $40—60 per month are not realistic for long term profit margins. The prices may go up in future, or unlimited usage may be curtailed.)
Internet Service Provider Access	$200 to consumers to $500 to businesses (usage restrictions may apply)
Total cost	Two modems at $2000 each, router port at $1000 plus monthly service charge

8.5.7 ADSL Standards and Associations

The American National Standards Institute (ANSI), working group T1E1.4, recently approved an ADSL standard at rates up to 6.1 Mbps (ANSI Standard T1.413). The European Technical Standards Institute (ETSI) contributed an Annex to T1.413 to reflect European requirements. T1.413 currently embodies a single terminal interface at the premise end. Issue II, now under study by T1E1.4, will expand the standard to include a multiplexed interface at the premise end, protocols for configuration and network management, and other improvements.

The ATM Forum has recognized ADSL as a physical layer transmission protocol for unshielded twisted-pair media.

The ADSL Forum was formed in December of 1994 to promote the ADSL concept and facilitate development of ADSL system architectures, protocols, and interfaces for major ADSL applications. The Forum has more than 300 members representing service providers, equipment manufacturers, and semiconductor companies from throughout the world.

8.5.8 Current Status of ADSL Implementation

ADSL implementation is moving quite fast. Almost all major telephone companies either launched pilots in ADSL during 1996 to 1997 or have started offering limited trial services in certain areas. The following brief snapshot of the industry illustrates this phenomenon. However, acceptance of ADSL service as a mainstream technology for high-speed transmission will depend on its availability in major metropolitan areas, its price, the introduction of affordable modems by more than a handful of vendors, and the ease of installation. The following points present a general status of the technology.

- Several telephone companies have already conducted market trials using ADSL, principally for video-on-demand, but including such applications as personal shopping, interactive games, and educational programming. Several other telephone companies are performing ongoing trials during 1997 to 1998. Ameritech Corporation, Bell Atlantic, BellSouth, GTE, Nynex, Pacific Telesys, US West, and Bell Canada are among the early entrants in this race. More specifically, this was the status in 1997:

Ameritech	Trial in Illinois. Launch scheduled for first quarter 1998.
GTE	Trials in four states. No launch date announced as of print date.
Pacific Bell	Commercially available during fourth quarter 1997.
Bell Atlantic	Trial in Virginia. Launch scheduled for second quarter 1998.
Bell South	Testing in the lab; no launch date.
US West	Commercial ISDL/SDSL/HDSL available now. No ADSL plans.

SOURCE: *PC Week,* June 1997.

- InterAccess Co., an ISP in the Chicago area, started offering ADSL service during 1996 to 1997 at $1000 per month to businesses and $200 to consumers. Interest in personal computer applications is growing, particularly for high-speed access to Internet resources.

- Semiconductor companies have introduced transceiver chipsets that are already being used in market trials. These initial chipsets combine off-the-shelf components, programmable digital signal processors, and custom ASICS. Continued investment by these semiconductor companies will increase functionality, and reduce chip count, power consumption, and cost, enabling mass deployment of ADSL-based services in the near future.

- ADSL modems have been tested successfully by many telephone companies, and thousands of ADSL lines have been installed in various technology trials in North America and Europe.

- According to TeleChoice, a market research company, 3.5 million ADSL modems will be installed by 1999.

Our own assessment is that DSL technologies will take more than three years to be commercially available in major portions of the United States and Canada, in such a way that you can incorporate it as a serious option for the enterprise. Another reason that RBOCs will control the growth is that they do not want their revenue base of wireline services such as T1 to be eroded too fast. Meanwhile, you should use ADSL where you can get it.

8.6 VDSL—Very High Data Rate Digital Subscriber Line

VDSL began life being called VADSL, because at least in its first manifestations, VDSL will be asymmetric transceivers at data rates higher than ADSL, but over shorter lines. While no general standards exist yet for VDSL, discussion has centered around the following downstream speeds:

12.96 Mbps   (1/4 STS-1)   4500 feet of wire

25.82 Mbps   (1/2 STS-1)   3000 feet of wire

51.84 Mbps   (STS-1)   1000 feet of wire

Upstream rates fall within a suggested range from 1.6 Mbps to 2.3 Mbps. The principal reason T1E1.4 decided against VADSL was the implication that VDSL would never be symmetric. Some providers and suppliers hope for fully symmetric VDSL someday, recognizing that line length will be compromised. (See Fig. 8.6.)

Figure 8.6
Distance limitations for ADSL and VDSL.

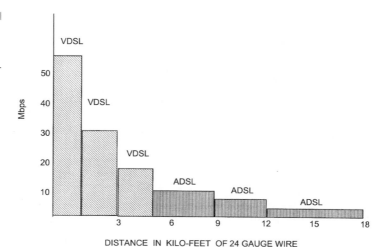

DISTANCE IN KILO-FEET OF 24 GAUGE WIRE

In many ways, VDSL is simpler than ADSL. Shorter lines impose far fewer transmission constraints; so, the basic transceiver technology is much less complex, even though it is 10 times faster. VDSL targets only ATM network architectures, obviating channelization and packet handling requirements imposed on ADSL. VDSL also admits passive network terminations, enabling more than one VDSL modem to be connected to the same line at a customer's premises, in much the same way as extension phones connect to home wiring for POTS.

However, the picture clouds under closer inspection. VDSL must still provide error correction, the most demanding of the nontransceiver functions asked of ADSL. As a public switched network, ATM has not begun widespread deployment yet; although it will take many years to become ubiquitous, VDSL is a candidate for transmitting conventional circuit and packet switched traffic. Furthermore, passive network terminations have a host of problems, some technical, and some regulatory, that will surely lead to a version of VDSL that looks identical to ADSL (with inherent active termination), except for its capability for higher data rates.

VDSL may operate over POTS and ISDN, with both separated from VDSL signals by passive filtering.

8.6.1 Other Terms—VADSL, BDSL for VDSL

VDSL had been called VADSL or BDSL, or even ADSL prior to June, 1995, when T1E1.4 chose VDSL as the official title. The other terms still appear both in technical documents created before that time and in media presentations unaware of the convergence. ETSI TM3, the European counterpart to T1E1.4, has also adopted VDSL, but temporarily appends a lower case *e* to indicate that, until the dust settles, the European version of VDSL may be slightly different from the U.S. version. This is the case with both HDSL and ADSL, although there is no convention for reflecting the differences in the name. The differences are sufficiently small (mostly surrounding data rates) that silicon technology accommodates both.

8.7 Cable Modem as an Alternative*

The telephone network is not the only path to the Internet. The cable television industry has wired 95 million homes in North America with

---

*Information source—www.boardwatch.com—*Fable of Contents* by Neal Schnog.

very high bandwidth coaxial cable. Cable modems are now available, and they are very fast. In the same way that telephone companies want to exploit their copper wiring, the cable industry is trying to use its coaxial infrastructure to transmit high-speed data. In fact, the cable industry has entered into direct competition with ADSL, by introducing cable modems and offering 1.2 Mbps or even higher speeds at monthly prices, which are, in fact, lower than ISDN prices.

CyberSURFR, Motorola's first generation of cable modem, offers throughput speeds of up to 10 Mbps in the downstream path, and 786 Kbps upstream. Zenith's current cable modem, the Homeworks Universal provides 4 Mbps, both down- and upstream; but, according to Michael Scott, Zenith's Data Communications regional manager, "The next generation of Zenith's cable modems will be capable of 27 Mbps downstream and 10 Mbps upstream." There are several other cable modem manufacturers, including Toshiba, Hewlett-Packard, and Lan City; but the one common thread is that all the products are ready for market. The modems work and are matched only in throughput speeds by direct Ethernet connections which on average run at 10 Mbps. These new high-speed modems will mean big changes in the remote access marketplace.

8.7.1 Cable Modem Speed and Costs

Data throughput with cable modems varies from 1.2 to 10 Mbps. Cable is capable of higher speeds up to 27 Mbps, but unlike ADSL, this bandwidth is shared among subscribers. Like ISDN, monthly service charges vary from city to city, but are generally in the $40 to $80 per month range. Cable modems cost approximately $400 to $500 per unit.

8.7.2 Cable Modems' Advantages

Besides being fast, cable modems will deliver other performance enhancements. First (and most notable) is the ability to use the modem and watch television at the same time. Unlike phone lines, which become busy once in use, cable modems use only a small fraction of the available bandwidth of cable systems. What this means for consumers is that they can surf the Internet, watch television, and use the telephone all at the same time.

The bandwidth advantage also gives cable modems instant access to the Internet. Unlike phone systems, no dial-up and login process is needed; the service is always available. Cable systems will essentially be configured as wide area networks (WANs), allowing users to send and receive data at will. In addition, a computer can be left on the network 24 hours a day without using any network or system resources, except when data transfers are under way. This means that e-mail would no longer be static, but instead be configured much like e-mail on today's local area networks (LANs).

8.7.3 Implementation Timetable for Cable Modems

The cable television industry is betting on the potential of cable modems for remote access to Internet. By mid-1997, cable modem manufacturers had announced firm commitments to deliver over a million units to the U.S. domestic market. The buyers are cable television companies, including the three largest MSOs (multiple system operators) in the United States. Forrester Research of Cambridge, Massachusetts, estimates that 7 million cable modems will be installed by the year 2000.

TCI (Telecommunications Inc.), the largest cable operator in the United States, plans to roll out its first cable modem services this spring in Silicon Valley's Sunnyvale, California, with the start-up of @Home service, a high-speed network. It will provide real-time multimedia news, entertainment and advertising content, access to the Internet, e-mail, and other services to consumers, via cable systems and their personal computers.

During 1996, over two dozen U.S. cable systems started cable modem trials. In 1997, many U.S. and Canadian cable companies started offering production-level services in metropolitan areas at an attractive price of $40 to 50 per month with unlimited Internet access, multiple e-mail addresses (up to five), and a home page storage. These services are clearly aimed at computer-literate, middle- to higher-income families.

The cable industry is offering a bundled service at present. They may change this strategy later by asking the customers to buy modems. This will happen only when compatibility issues are resolved. Portability of the modems is dependent on compatibility between cable systems and two-way capability of the majority of cable systems. The compatibility issue is already being addressed through a consortium of cable operators called Cable Labs. Cable Labs has been testing modems for some time, and has been key in promoting a single standard for the entire industry. Many cable systems, however, do not have two-way capability.

According to the National Cable Television Association (NCTA), approximately 25 percent of the cable systems in the United States have two-way capability, and 95 million homes have cable television available.

Simple calculations would translate to an immediate market potential for cable modems of only 24 million households, but this number could increase substantially over the next five years. According to Michael Luftman, vice president of Corporate Communications for Time Warner Cable, "Time Warner has been aggressively rebuilding cable systems for quite some time and now has about one third of its systems capable of two-way operation." In addition, TCI and most of the other major MSOs continue to install fiber-optic cable and upgrade their older systems. Even as cable operators become capable of delivering two-way service, they will still have to convince customers that their new systems, unlike current systems, will be the most reliable way to access the Internet.

Finally, there is one factor out of the control of cable operators—the speed of the Internet backbone network. Most Internet connections are made through T1s and T3s. While these connection points used to be considered the fastest available, they may soon be the cause of data bottlenecks. For this reason, it may be some time before consumers can take full advantage of the 10 Mbps speeds of cable modems.

8.7.4 Asymmetric Nature of Cable Network

The reader should note that cable network topology is different from that of a telephone network. There are two inherent problems. First, telephone networks were designed with switching capability in all their components. Since the switching of signals was not a requirement in the cable industry, networks were designed for unidirectional broadcast of signals. Therefore, cable is generally noisy in the reverse direction. As such, it is less suitable for two-way interactive traffic. Second, the bandwidth is shared among users, just as it is on a single segment of a LAN. Without switching implemented inside the cable network, the first cable modem user on a segment of a cable network may get very good response. However, every subsequent user diminishes the performance of the network, just as with an Ethernet LAN. Research by CableLabs indicates that response time by individual users decreases significantly when more than 100 users are accessing Internet for heavy graphics-intensive or video. Most CATV systems have 300 customers per neighborhood hub. As far as two-way transmission capability is concerned, only 25 percent of cable plant in North America has this inherent capability. Cable operators are working hard to resolve this problem, as described in the previous section.

The switching problem in the cable industry is more difficult to tackle than in the telephone industry, where conversations are unique and consume relatively small bandwidth. In the cable industry, usage patterns are not known, some bandwidth is shared, and each transfer may be multimegabit, especially with video signals. The design of switches with this multimegabit transfer, in an extremely dynamic fashion, is a challenge, if not an impossibility, to switch designers. The second problem of shared access on cable is partially addressed by proposed bandwidth reservation protocol by Microsoft, whereby users could request, "I need 1.2 Mbps bandwidth for a video transfer application between 3:00 and 4:00 P.M. for a customer presentation."

Also, the cable modem industry is suffering from a lack of standards, though some work is going on with IEEE group P802.14. This problem may be offset by the fact that many cable companies are offering a bundled service with cable modem included in the per month price.

8.8 Integration of ATM with ADSL and Cable Modems

ADSL and cable modem technology do not obviate the need for switching at appropriate points in a LAN-WAN internetworking solution. In fact, that is the challenge for you as a network designer—where to use what technology. Several vendors are developing products that will integrate ATM switching with cable modems and/or ADSL. General Instrument, Fore Systems, and Com21 are extending ATM switching technology into the cable network. General Instrument and Fore are testing a product called Surfboard that will allow users to lease cable modem-based service (for around $50), and then integrate it with a standard desktop ATM card (for about $250). Similar solutions will be employed by telephone carriers. Thus, ATM backbones can be integrated with cable or ADSL desktop solutions for corporate users.

8.9 Various Remote Access Network Services Compared

We believe that each of the network options presented in this chapter has its role in network design. Some of these options will fall off the list of popular and widespread implementations. However, one solution does not fit all. As a network designer, you should compare the features and strengths of a specific network service, and match them against application requirements. Also, do not lock yourself into a single technology, such as ISDN or cable. More flexible solutions—that match user needs against network options and can be upgraded, as solution rationalization takes its due course—are most likely to survive the test of continuous changes in our industry.

You should be aware of the following noteworthy points:

- Both ADSL and cable modem network technologies are asymmetric. Therefore, they are best suited at the edges of a network where a lot of information is moving from servers to remote desktops or notebooks. They should not use the backbone portion of a network for server-to-server or peer-to-peer services, which have significant traffic in both directions.

- Unlike PSTN and ADSL, cable modems are not switched services, but provide permanent connection between a central office and the customer's location. In many applications, this connection bypasses the voice switch.

- A remote access connection using ADSL or cable modem service is only one element of an end-to-end connection. You should check with telephone carrier or cable modem service providers as to how many remote users they have on each network node comprising a virtual LAN at their site. Just as a single LAN cannot support hundreds of local users doing graphics-intensive applications, a cable modem network cannot support too many users. Readers should refer to Chap. 6 for a better understanding of end-to-end performance.

Table 8.4 summarizes various services.

Summary

In this chapter, we reviewed ADSL, ADSL-related, and cable modem network options for remote access. We looked at basic electrical characteristics associated with a telephone carrier's copper wiring infrastructure, and explained how telephone companies can extend the speed from PSTN (56 Kbps) and ISDN class (128 Kbps) to multimegabit per second. We explained various flavors of symmetric and asymmetric digital subscriber lines—ADSL, SDSL, VDSL, and HDSL. These services will certainly extend the

TABLE 8.4

Comparison of
Various Network
Options

Service	Speed—Downstream/Upstream	Availability Now	Best Application	Pros and Cons
PSTN	1200 to 56 KBPS symmetric	Yes (700 million lines)	Remote access from consumers	Low speed, but widest availability
ISDN	128 Kbps symmetric	Good in most areas of North America, Europe, and Japan	Power Internet user and videoconferencing	High speed, low error rate; relative costs high
HDSL	1.54 Mbps symmetric	Limited	Private corporate network	Limited availability; pricing is high
ADSL	1.54 to 6 Mbps upstream; 64 to 640 Kbps downstream	Limited trials and some areas only	Internet access	No new cabling required
Cable modems	10 Mbps to 40 Mbps upstream; 28 Kbps to 15 Mbps downstream	Technology and market trials now accessing Internet	Consumers and telecommuters infrastructure; and corporate LANs	Limited cable not suitable for two-way video conferencing
Direct broadcast satellite	400 Kbps to 30 Mbps upstream; 28 Kbps downstream	Available now	Video broadcast	Limited interactivity

SOURCE: Author.

capabilities of high-speed remote access to telecommuters, consumers, and small businesses. We described how ADSL works, what the current state of implementation is, and illustrated, through network schematics, how you could use this option in your network design. Finally, we reviewed the cable industry's answer to the bandwidth problem. Certainly these options will become prevalent in the years to come.

CHAPTER 9

Remote Access Network Options—The Internet

Internet has affected every facet of computing and communications. It has extended the concept of remote access to users inside and outside the organization. It may become the universal interface for remote access.

—Chander Dhawan

About This Chapter

In the previous chapters of this book, we have discussed various network options that utilize private network solutions. Whether you use PSTN, ISDN, or higher-speed ADSL, you set up a private network. In many cases, these solutions require a significant amount of investment of time and money toward planning, design, and installation. Now, we want to consider an option that creates a virtual private network for remote access. Internet is being increasingly used for remote access especially when remote users are spread throughout the country and across the globe. In this chapter, we discuss the benefits of using the Internet, describe common technical terms, and look at various scenarios of using a component of your enterprise network.

9.1 The Internet—Increasingly Popular Option for Remote Access

It is difficult to describe the Internet. Perhaps we need not define it. The Internet is a phenomenon that is bigger than many technological revolutions. To say that it has taken over computing as we knew it in the 1980s and early 1990s is not too much of an exaggeration. Internet has a role in all aspects of computing. Therefore, it is not much of a surprise to say that it has a significant role in remote access and has become an increasingly popular option for businesses. What has become obvious is that rather than dialing directly into the LAN, many organizations are using Internet as a way of letting their telecommuters, mobile users, and remote offices connect to the corporate network.

Much has been written about Internet—on browsers, the World Wide Web, search engines, homepage development, e-mail, Java, Web servers, and more. Although we do not discuss these topics in this book, instead we concentrate our focus on the following topics related to accessing information from corporate information servers residing inside and outside the Internet, by using Internet as a connectivity vehicle:

- Internet as a remote access network option for mobile users
- Internet as a virtual private network (VPN)
- Internet as the total solution for remote access
- Intranet

- Internet as a hybrid solution (private network plus Internet)
- Considerations in choosing network options for Internet access

9.2 Benefits of Using Internet for Remote Access

Before we discuss remote access to the Internet, we should understand the benefits of using Internet as a connectivity vehicle. We list some of these benefits here:

Universal User Interface.   Internet browsers (Netscape or Microsoft Internet Explorer) are becoming the user interfaces of choice for remote communications, especially for e-mail, file transfer, LAN applications, and accessing public information from the Internet and Intranets. Users find it convenient to use a universal and consistent user interface for their business and consumer needs.

A Single Communications Interface.   Individual users, inside or outside the organization, do not have to worry about installing proprietary client communications software, whether it is from Shiva, Cisco, IBM, Microsoft, 3COM, or others.

One Network for Internal and External Communication.   It is cheaper to have a single remote connection for both internal and external communications.

Remote Speeds from 56 Kbps to 10 Mbps.   Most ISPs support PSTN, ISDN, and T1 today. Several ISPs are supporting ADSL and cable modems, albeit at relatively higher prices per connection. In the future, the price curve will come down even further; $50 to $75 per month for unlimited use connection at 1.2 Mbps will become commonplace.

Automatic Network Concentration.   From a network manager's point of view, at any medium or large organization, the Internet provides automatic concentration of network traffic from hundreds or thousands of users.

Cost.   Internet access is by far the cheapest way of accessing remote information. You simply dial into your local ISP, and then through Internet, into your corporate server. This way, you do not incur any long-distance charges, especially if the server is in another long-distance area code. Instead of having hundreds of low-speed ports (dial-in and ISDN) at your corporate remote access server, you have one thick pipe from the ISP to the corporate location. With dial-in access through voice-grade lines starting at $9.95 per month, you can rarely beat Internet for per port cost as compared to private remote access servers. This is because ISPs have more leverage with higher-capacity servers that have cheaper per port costs. Also, it is worth noting that queuing mathematics is such that you can support more than twice as many users with twice as many ports on a server. If you want to understand this phenomenon more, you should read a telecommunications network design book.

Instant Network Installation Across the World.   With Internet connection becoming a commodity and extremely painless to install, you can implement a worldwide global network in days rather than months.

Automatically Outsourced Remote Access.   With Internet-based implementation of remote access, the network manager is not burdened with the problems of operating a remote access network, tariff management, problem management, and related issues. There are a number of tasks that centralized staff has to perform for providing remote access services:

- Acquire, install, and configure remote access servers and modem pools
- Keep up with client software
- Monitor network traffic on remote ports and reconfigure, if necessary
- User access identification management
- Pay for business lines for dial-in ports and long-distance charges
- Continuously upgrade remote access servers to support new access methods, such as ISDN and ADSL
- Technical support for users having problems with accessing network

With outsourced service, the network manager is concerned only with managing a single or a few thick lines from the Internet to the internal LAN or front-end processor. This reduces the amount of work in managing your mobile users and telecommuters—quite often more than the cost of the outsourced service.

Bandwidth on Demand.   With Internet, you can essentially acquire variable amounts of bandwidth, depending on your needs. With cable modems and ADSL, bandwidth will become a commodity in the future, and T1 lines will become much cheaper. Your needs for access speed may vary from time to time and application to application. You may not need cable modem or ADSL speed today, but you may need it tomorrow. Some users may be happy with dial-in speed; other power users may benefit from cable modem speed. With future software protocols, you will be able to reserve bandwidth from your client workstation. The Internet gives you this flexibility and bandwidth upgrade potential.

9.2.1 Concerns with Using Internet for Remote Access

While there are a number of benefits of using the Internet for remote access, there are also a number of concerns with Internet-based design. Security, lack of transparency with LAN or mainframe applications, and increased response time (in some cases) to LAN database applications are some of these concerns. We discuss these and other considerations in a later section of this chapter.

9.3 Internet Applications Suitable for Remote Access

There are a number of applications that are suitable for remote access through the Internet. The following applications are particularly suitable:

- *E-mail to and from internal and external users.*   Internet is especially more attractive if the users deal with external organizations, suppliers, vendors, and customers. When your e-mail group is a closed group or when you want to control external frivolous chat, Internet is not a good solution.

- *Accessing information from the Internet.*   Many professionals find Internet to be a rich reservoir of information for evaluating products and services. Market research people find a tremendous amount of information on competitive strategies.

- *Providing product and pricing information to sales force and customers on the Web.*   Easy to maintain and update in a single database.

- *Technical support.*   As a bulletin board for problems, fixes, and tips to consumers.

- *Internet as a place to do business.*   More and more organizations provide a place to buy merchandise from the Internet.

- *Remote faxing and paging services.*

- *Internet phone.*   Long-distance communication of the future.

9.4 Internet Terms and Issues You Must Know

You should be familiar with Internet-specific terminology when you are dealing with ISPs for your remote access needs:

HTML (HyperText Markup Language).   A markup language that defines attributes and links that are used in a Web document. These attributes may consist of tags that indicate how to display a piece of text or graphic in a document, or it may direct the browser to another file or document. Please note that any LAN-based application that you want to access remotely through Internet may have to be *Web-enabled,* that is, forms may have to be changed to HTML format, and client application modified with appropriate back-end changes for accessing LAN or mainframe databases.

HTTP (HyperText Transfer Protocol).   The protocol that negotiates delivery of text and other elements from a Web server to a Web browser.

FTP (File Transfer Protocol).   A specific protocol used for file transfer on the Internet. In the early implementations, primitive Unix commands were used for file transfer from a remote computer. However, modern FTP implementations use graphical (Windows-type) interface that ask you to sign on to the remote file system by entering an appropriate password before transferring files. Many information delivery or software update sites allow anonymous as a password, and then let you see the directory listing of the remote system. Then you can select the file and with a click of a mouse, transfer the file. (See Fig. 9.1.)

PPP (Point-to-Point Protocol).   A protocol that allows a remote computer to connect to the Internet by using a dial-in modem and a standard POTS (plain old telephone service). This is a de facto standard for connecting to the Internet for remote clients.

Internet Dialer.   A piece of software that is used to set up the configuration (telephone number, password, speed, modem initialization strings, etc.) for telephone, ISDN, or any other connection to the Internet. It may employ PPP, SLIP, or other protocols. (See Fig. 9.2.)

Intranet.   A term employed to describe implementation of Internet protocols and technology (hardware and software) inside an organization's own network infrastructure. This is becoming a popular implementa-

Figure 9.1
FTP screen
(QUALCOMM).

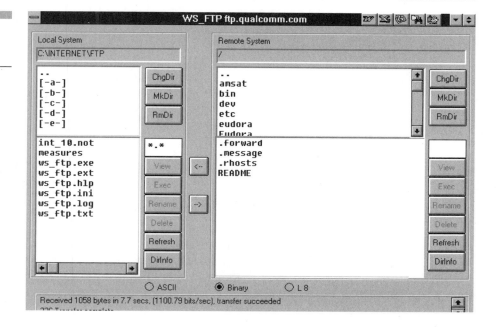

Figure 9.2
IBM internet dialer.

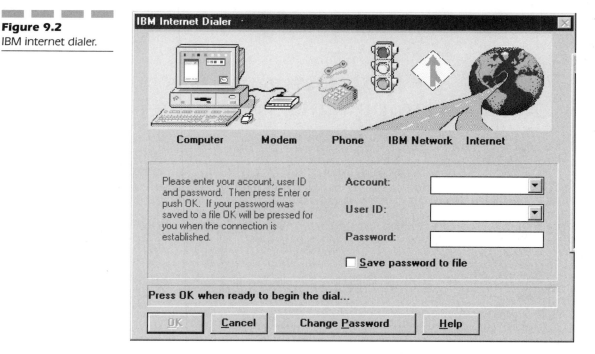

tion because it gives you the flexibility of moving information servers from inside the organization's private network into a public shared network.

Extranet.   A term that denotes the Intranet concept being employed to share information with your business partners, suppliers, and customers.

IMAP4 (Internet Message Access Protocol Version 4).   An evolving protocol that allows a client to access and manipulate e-mail messages on a server. IMAP4 is designed for disconnected e-mail users, just like the paper-based mail. It performs functions such as remote management of mail folders, viewing just subject lines, and selectively downloading messages and attachments based on criteria that you can define (e.g., source, size, etc.).

LDAP (Lightweight Directory Access Protocol).   An emerging directory service protocol that uses a subset of the X.500 directory standard to provide a common way to identify user and group information.

SMTP (Simple Mail Transfer Protocol).   A standard protocol that defines how e-mail messages are transferred between servers. SMTP defines ASCII text content only, and requires other standards, such as MIME (Multipurpose Internet Mail Extensions) for file attachments.

9.5 Understanding Internet Components, Including Its Backbone

Let us explain what the Internet is, and what its components are, especially the backbone. This will help you understand the maze of computers and connections that you go through in order to access a piece of information, or send an e-mail message to somebody in another part of the country or continent.

9.5.1 What Is the Internet?

The Internet is a global network of networks. The Internet began as a project of the U.S. Department of Defense (Defense Advanced Research

Project Agency or DARPA, to be specific) in the late 1960s. It was intended to link scientists working on defense and research projects around the country. It was an experiment in packet switching, shared circuits, and decentralized control. In the 1970s, DARPA supported the development of internetworking protocols—TCP/IP. During the 1980s, the National Science Foundation took over the responsibility for the project, and extended the network to include the major universities and research sites in the United States. Links were then established to similar emerging networks in other countries. Today, the majority of countries in the world are linked in some way to the Internet. More connections are being made every day.

It is important to understand that the Internet is not a single network entity. In fact, it is made up of many different networks which are able to communicate with each other. Each separate network is managed by its own network administration staff. The Internet works because all the different networks comply with Internet standards. There is no single body that is in charge of all the networks connected to the Internet. The Internet is an amazing example of cooperation and collaboration.

9.5.2 Who Owns the Internet?

Nobody *owns* the Internet—everybody owns an individual piece of the Internet—although the Internet backbone within the United States is owned by the National Science Foundation (NSF). It is currently run under contract by Merit, a joint venture between IBM and MCI.

Registration services are provided by AT&T under contract with the NSF. The leased lines that hook most of the network together are owned by the regional Bell operating companies and the long-distance carriers, such as MCI, AT&T, Sprint, and WilTel.

9.5.3 The Internet Backbone

The backbone of the Internet is the long-haul communications links that provide the main connections between its connected networks—much like the interstate highway system feeds a network of local roads. There is one difference with the highway system: there are computer nodes, like traffic routers, at the end of any link. The backbone originally ran at 56 Kbps, considered sufficient for the traffic in 1970s. In the 1980s, the backbone was upgraded to 1.544 Mbps. In the early 1990s, it was upgraded to 45 Mbps per second. In 1995, some portions of the network, especially the MCI portion, were upgraded to 155 Mbps, and then in 1996 to 622 Mbps. We should see 2 gigabits per second soon. So the speed keeps on increasing, but demand, as measured by the number of users, type of applications, and usage keeps on outstripping the increased speed.

Figure 9.3 shows a conceptual and logical schematic of the Internet. The physical schematic changes so frequently that it is difficult to keep up for any trade press reporter, much less an author of a book. Essentially, your client workstation connects to a regional ISP's router, which in turn is connected to a larger and national ISP, often called the National Service Providers (NSP). It is these NSPs who connect to the backbone. T1 is a common speed of the first-level regional ISP link to the bigger NSPs. From NSPs to the backbone, you may connect with a T3, or an even faster link. However, most ISPs rely on multiple links for redundancy and hardware configuration reasons. Many ISPs started off with smaller configurations, which they could not always upgrade to faster links, thus leading to multiple routers rather than fewer larger routers. NSPs have designated Internet traffic exchange points, where packets are routed to appropriate links and destinations; for example, there are NSPs called AM West, Pacific Bell NSP, located in San Francisco, AM East in Cornell University, and so forth. These are the locations where NSPs like MCI, BBN, and UUNet exchange traffic destined for their respective sites.

Figure 9.3
(a) Logical schematic
of Internet backbone.

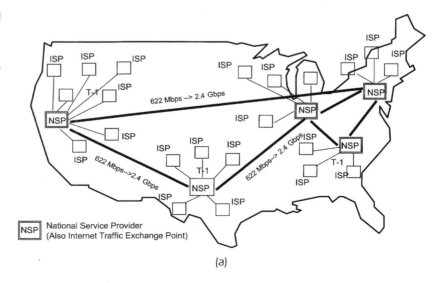

(a)

Figure 9.3
Continued (b) Internet in the United States. Canada, and Europe interconnected; (c) UUNET's Internet backbone—1997 (622 Mbps OC-12 metro rings and trunks).

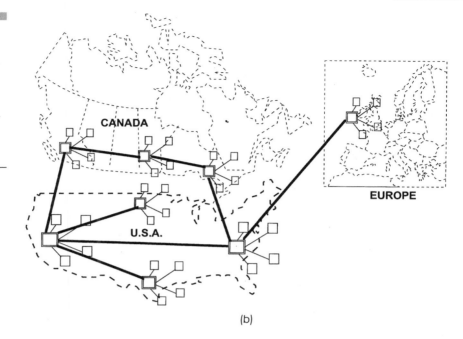

(b)

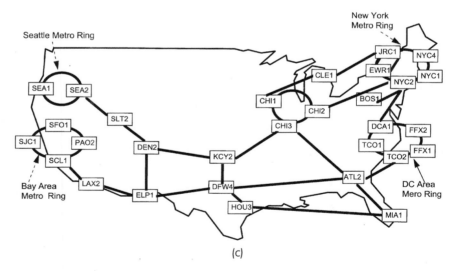

(c)

9.5.4 Internet Router—Vital Component of the Internet

A router, like the Shiva Integrator 200, or a PC workstation configured to act as a router with appropriate software, identifies packets that are destined for the Internet and routes them between your Ethernet and the Internet. The router formats and transmits your packets, provides support for protocols, such as IP and IPX, does packet retries, and error recovery. Other router functions include: dialing the phone—if you access the Internet over a modem—and packet filtering to restrict access to your network to specific hosts, networks, or perhaps specific users (acting as a firewall).

Optionally, if your router has the capability, it can also:

- Act as a mail gateway for your network
- Receive, store, and serve news articles to your network
- Route packets between your internal networks
- Directly execute Internet applications, such as Mosaic, Gopher, Archie, FTP, and Telnet
- Act as an FTP server to make specific files accessible to your customers, suppliers, and so forth
- Act as a gopher server to provide information about your company to potential customers and so forth

9.5.5 Internet Example—How Does My Mail Get Delivered?

Let's say you're sending mail to *cdhawan@netsurf.net*. When you've finished composing your e-mail, your mailer, a piece of software performs a *gethostbyname* system call to look up the IP address of the remote host *netsurf.net*.

Normally the host would be looked up from */etc/hosts*, but since you're on the Internet, your system queries an Internet *name server* to find the address of netsurf.net. Your mailer opens a virtual circuit over the Internet to *his* mailer. It communicates with the remote mailer by speaking the *Simple Mail Transfer Protocol (SMTP)*. If all goes well, the mail is usually delivered within a few seconds, even if the recipient's machine is on the other side of the world. If the remote site is down, your mailer will keep trying every hour or so, usually for a couple of days. If the site still can't be reached, the mail will be returned to you as *undeliverable*.

9.5.6 TCP/IP—Universal Internet Transmission Language

Standard computer protocols are a set of rules which enable communication between different kinds of computers. The Internet relies on TCP/IP protocols to exchange information. Those providing information on the Internet must decide what protocol is most appropriate to convey their information to the intended audience. The major Internet protocols are:

- World Wide Web (HTTP) for multimedia information
- Gopher (GOPHER) for hierarchical menu display
- File Transfer Protocol (FTP) for downloading files
- Remote login (TELNET) to access existing databases
- Usenet newsgroups (NNTP) for public discussions
- Electronic mail (SMTP) for personal or mass mail correspondence

IP uses a 4-byte addressing scheme which is usually written as 249.101.203.157 to uniquely identify each node in the Internet. Domain names, such as *www.Mobileinfo.com*, are translated into numerical addresses (described previously) by the name server.

9.6 Other Internet Components Related to Remote Access

Components you may have to consider when implementing private remote access, Internet-based remote access, and Intranet-based remote access are outlined as follows:

Web Browser. The Web browser is a desktop viewer that allows the user to access information provided on the Web server. It provides the user-friendly interface that is critically important to Internet technology acceptance. The browser runs on top of the IP stack, and communicates with the server using the industry standard point-to-point protocol (PPP). One of the popular browsers (e.g., Netscape, Microsoft Internet Explorer, or others) should be installed on every desktop and remote computer.

Web Server.   There are many different Web servers on the market today. A Web server is a Unix- or Windows-based application that runs on a Sun Sparc Workstation (Unix), Windows NT, or Silicon Graphics personal computer system to name a few. Web server software ranges in functionality from simply displaying information to providing secure, encryption-based transactions.

Client Dial-In Software.   In order to establish seamless connections, remote dial-in client software should be installed on remote user PCs. While most operating systems, such as Windows 95/NT and OS/2, provide dial-in software with basic functions, remote access hardware suppliers, such as Shiva, provide client software as well (e.g., ShivaRemote client software is available for distribution with Shiva LanRovers). Since many end users operate heterogeneous computing environments, the remote access client should be available for different operating systems. ShivaRemote is available for Windows, Mac, and OS/2 environments. The client software initiates the call, negotiates the connection, and terminates the connection when the remote session is over.

ISDN or Leased-Line Connection.   There are several ways to connect a corporation to the Internet. Any ISP that provides service to the corporate market can arrange an ISDN or leased-line connection through the local telephone company. Leased-line connections are available in the following speeds: 56 Kbps, fractional T1 or a full T1 at 1.544 Mbps, or E1 at 2.048 Mbps. The choice of bandwidth depends on the amount of activity between the corporation and the Internet Service Provider, also known as a point of presence. It is easy to upgrade dedicated service as your access needs grow.

CSU/DSU.   CSU/DSU connects to the leased-line router and determines the speed and type of connection. For instance, depending on the CSU/DSU, a connection could be 56 Kbps, fractional T1, or full T1/E1.

Firewall Security.   Internet connectivity creates new security needs. A firewall is typically Unix- or Windows NT—based software that filters IP addresses at the application level, screening out unwanted users. There are many firewalls on the market today. We discuss this subject briefly at the end of this chapter, and then again in greater detail in Chap. 14.

9.7 Internet Network Scenarios

There are a number of ways you can use Internet infrastructure for your remote access requirements. We review some of these scenarios here.

9.7.1 Internet as a Remote Access Server

This is, by far, the most common scenario whereby your mobile users and telecommuters do not dial into your remote access server on the corporate LAN, but into the nearest ISP of their choice with the connection of their choice, that is, PSTN, ISDN, ADSL, or cable modem. The corporation may impose certain standards of browsers and client application software in order to ensure compatibility with LAN applications. As long as you are exchanging e-mail messages, doing file transfer, and accessing HTML documents, there may not be any problem. However, if you are trying to access LAN DBMS applications, you have to ensure that a compatible browser is used on the client machine. Similarly, if you are using new generation of Java applets, you may require compatible Internet software on your router on the LAN. Figure 9.4 shows a schematic of this scenario. It allows your remote mobile users, telecommuters, your remote offices, and even customers to access information on your corporate LAN. These users can also access legacy applications and data through an SNA server on the LAN.

9.7.2 Allowing LAN Users Access to the Internet

Allowing any user on a LAN to access the Internet directly through a serial port and a modem is an easy task, but allowing hundreds of users on a

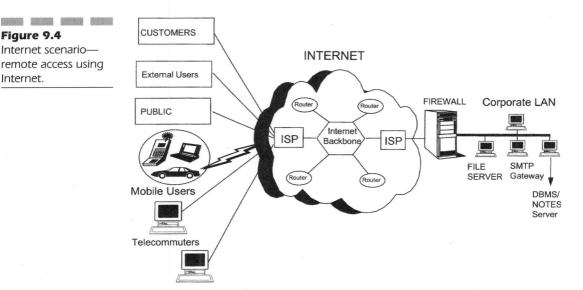

Figure 9.4
Internet scenario—
remote access using
Internet.

LAN in an organization the same type of access using individual modems can be very expensive. Also, there is no control regarding the personal or business use of Internet, no organizational security control, and no firewall. This can lead to uncontrolled access, and high telephone charges, without knowing whether this use is legitimate or frivolous. There is a cheaper and more controlled way of allowing access to the Internet by sharing several connections through an Internet router installed on the LAN. The number of users that can be supported on a single connection to the Internet depends on the average time spent by each user and the speed of the connection. Ask your ISP for advice that may be based on rule-of-thumb estimates or a more accurate mathematical model. (See Fig. 9.5.)

9.7.3 Internet Virtual Private Network— LAN-WAN Integration

The Internet is a dynamic network model that connects computers throughout the world. The Internet community of ISPs has demonstrated extreme responsiveness to the needs of its customers. They have been increasing the capacity of the backbone by an order of magnitude almost every six months. Some of the Internet vendors may lack network management capabilities today, but that is improving, too. The telephone companies are either becoming ISPs themselves, buying ISPs, or aligning themselves with specific ISPs. These trends are creating an industry of

Figure 9.5
Internet scenario—
accessing Internet
from a LAN.

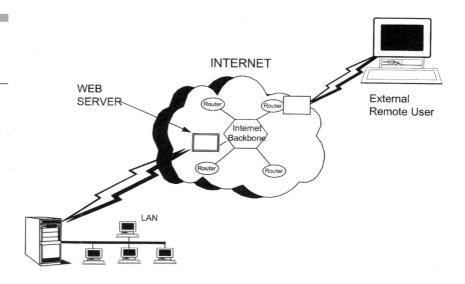

powerful network service providers (NSPs) with significant bandwidth capacity, rather than pure Internet Service Providers (ISPs) who deal with Internet users. With this combination of traditional telecommunications network infrastructure, and now the Internet services, these NSPs can easily replace your private networks with Internet-based virtual private networks (VPNs), as shown in Fig. 9.6. They can offer not only a cheaper network solution, but effectively offer an outsourced remote access solution. Once these NSPs also start providing content, we shall have an information network utility industry.

9.7.4 Evolving to a More Advanced Internet Model—Java Applets and Web Servers

So far, we have not talked about making any changes to the software applications in the client workstations (desktops, notebooks, and PDAs) or in the back-end servers. Figure 9.7 shows a network model where we show more advanced implementation in which existing Windows applications get replaced by Web-enabled native Java applications, and information databases are moved inside the Internet (either logically or physically). These information servers in the Internet may be either replicated copies of the master databases or the master copies themselves, depending on the security design, type of data, or database architecture.

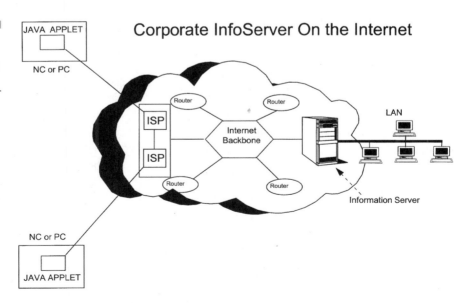

Figure 9.6
Internet scenario—
information server on
the Net.

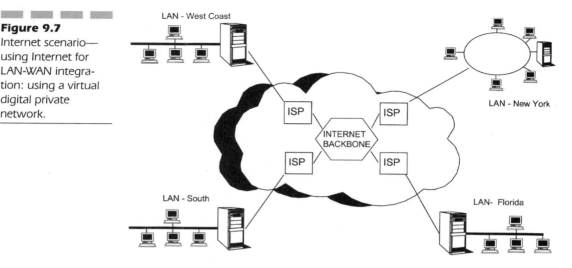

Figure 9.7
Internet scenario—
using Internet for
LAN-WAN integra-
tion: using a virtual
digital private
network.

9.8 Internet Remote Access Server Product Examples

There are a number of products that are available in the marketplace to implement Internet-based remote access solutions. In fact, almost all the major vendors provide a Web server capability. In this section, we discuss a couple of examples for illustrative purposes.

9.8.1 Shiva's WebRover

Shiva offers a product called WebRover that allows remote employees to access information on the corporate LAN, private Web server, and the Internet by dialing one number. Therefore, this product offers both a traditional remote access server functionality, as well as Internet connectivity. WebRover is typically connected to an ISP through a leased-line connection. This is Shiva's version of the schematic described earlier and illustrated in Fig. 9.8. Any client PC with Shiva's dial-in client software can log into the corporate LAN where a dedicated Web server could display special product information, catalogues, and other promotional material.

WebRover consists of Shiva's LanRover remote access server and the Shiva Integrator 200 (a router) that supports up to a full T1/E1 dedicated connection to the Internet. Shiva Integrator 200 routes both IP and IPX traffic. WebRover can be configured with 8, 16, or 24 analog ports, or a 20-port configuration that supports a mix of analog and ISDN lines.

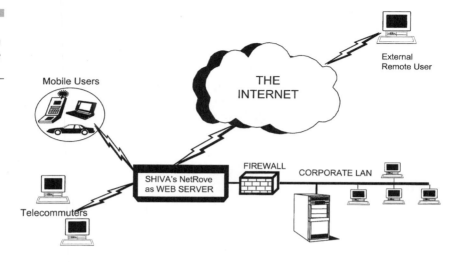

Figure 9.8
WebRover—between remote users and the Internet.

9.8.2 Novell's Web Server

Novell's NetWare Web Server is a software implementation (as a NetWare Load Module—NLM) that converts a NetWare 4.1 server into an Internet server that connects to the World Wide Web and thus establishes an Internet presence. In this way, your LAN users can connect to the Internet through a common server.

NetWare Web Server supports common features of the World Wide Web—forms, Java applets, the remote common gateway interface (R-CGI), local common gateway interface (L-CGI), access controls and logging, plus Basic and PERL script interpreters. Since Novell NLMs can run on an Intel platform, in either a shared or a dedicated server environment, this solution offers an inexpensive entry point solution. While NT implementation of Web access servers is becoming popular, NetWare Web Server is a viable option in the case of extensive Novell infrastructure in the organization.

9.8.3 Microsoft's Internet Information Server for NT

This is Microsoft's answer to implementing a Web server; it is based on an NT™ operating system. All you need to install is Microsoft's Internet Information Server software and connect to an ISP. Your ISP will provide your server's IP address, subnet mask, and default gateway's IP address. The default gateway is the computer through which your computer will route all Internet traffic.

9.8.4 Cisco Internet Routers

Cisco is the leading supplier of Internet access servers for both private (Intranet) and public (Internet) access by remote users. Cisco offers a range of models to suit the needs of different user organizations.

The Cisco 2500 Series Access Servers provide both dial-up service and single-user access to the corporate LAN from a single platform, with up to 16 asynchronous lines that support V.34 speeds, simultaneously, and two synchronous serial ports for remote access routing.

The Cisco AS5100 Access Server provides asynchronous modem dial-up services as a stand-alone system of channel service units/data service units (CSUs/DSUs), channel banks, 48 managed modems, and up to 48 access server ports, all in a single chassis. This high degree of component integration in one chassis eliminates the incompatibility problems that are common with multibox, multivendor installations, and increases the overall system reliability.

The AS5200 is a universal access server to deliver hybrid asynchronous serial and ISDN line service to accommodate both mobile users and high-bandwidth dedicated telecommuters. By terminating both analog modem and ISDN calls on the same chassis from the same trunk line, the AS5200 enables ISPs and enterprise network managers to meet traditional analog dial access needs, while supporting the demand for high-speed ISDN access.

All Cisco access servers offer network management and multiple protocol support—IPX/SPX and TCP/IP. Dynamic port assignment supports all protocols and services on all asynchronous ports. This allows a single modem pool to support users dialing in via various protocols, as well as asynchronous routing and reverse Telnet connections for dialing out from the network.

9.9 Factors in Evaluating Internet as Remote Access Server

There are a number of factors that must be considered in evaluating Internet options for remote access. You should analyze the following factors before you decide that Internet is the technically, economically, and operationally feasible solution for you:

■ Number of users
■ Average time per user

- Which types of links—PSTN, ISDN, ADSL, or cable modems
- Whether remote or LAN users (mobile users and telecommuters) need to access Internet services
- Ease of use
- Network operating systems and protocols supported
- Performance
- Costs—monthly connect charges, modem costs, and cost of private servers
- Network management
- Security management

Now we discuss each of these factors.

9.9.1 Number of Users and Average Time Spent by Each User

These two factors determine the capacity of the server and the number of ports that you or your ISP has to provide to support your users. The design objective should be that remote users should not find a busy signal as they dial in for 90 percent (or a similar design objective) of the time. While more accurate mathematical models are available for calculating the number of ports, a rough rule of thumb is a 10:1 ratio between the number of remote users and ports. This rule assumes an average use of less than one hour per day spread at different times of the day. Please keep in mind that telecommuters who work out of home may occupy the port for a much longer period of time. The propensity of heavy users will decrease this ratio, and lighter users will increase it.

9.9.2 Which Types of Links—PSTN, ISDN, ADSL, or Cable Modems

The Internet supports a number of options; you should offer as many options as your users can justify from a business case perspective. Many technophyte users will always want the latest, fastest, and functionally richest connection, whether they can justify it or not. While programmers and telecommuters may justify ADSL and cable modems, sales professionals doing an occasional file transfer can be easily served by PSTN.

As far as connecting LAN to the Internet is concerned, that link is best served by a high-speed leased connection, the actual speed depending on the aggregate data transfer requirements over that link.

9.9.3 Whether Remote or LAN Users (Mobile Users and Telecommuters) Need to Access Internet Services

Many of the remote users need to access corporate information. Whether they need to or must access Internet is a different matter. It depends on the job description. Access to the Internet also leads to wasted time for browsing the Internet for personal entertainment; this does not lead to improved productivity. There should be a corporate policy on who and which group should have access to the Internet. In fact, many organizations have started monitoring the use of the Internet by its employees.

9.9.4 Ease of Use

One major reason for providing service over the Internet is the ease of access by remote users, telecommuters, and mobile professionals alike. The Internet provides a universal or a consumer-oriented interface that is graphical, intuitive, and requires no training. In fact, that is one of the biggest reasons for the Internet revolution. The Internet makes it so convenient and easy to get information from different locations—hyperlink is all that is required. It is easy for organizations to provide these hyperlinks. At the same time, the task of making this information available has been drastically reduced in complexity. Publishing (a new phrase whose meaning is with respect to information on the Internet) requires installing a Web server with information (text documents, graphics, images, sound, or video) loaded on to this server. What used to take months on old infrastructure takes weeks on the Internet, and what took weeks takes days. Of course, your network administrators, database administrators, and your ISP technical support staff may have to do a lot of configuration setup and customization. Your application programming staff may have to develop new applications with languages such as Java. The greatest excitement of Java as a new development language is its platform independence, ease of development, and object-oriented design.

9.9.5 Network Operating Systems and Protocols Supported

The network operating system on your LAN is certainly one of the considerations that you have to think about while developing your Internet strategy. While there may be many different network operating systems out there in customer organizations, a big majority belongs to Novell's NetWare and Microsoft's NT implementation. As for transport protocol, TCP/IP is the universal protocol for the Internet, and most organizations are moving toward that. Novell's IPX can be integrated with the Internet by tools provided by Novell. One desirable attribute of a good network is that it has a very small number, possibly one or two protocols only. However, remember that one important factor in any design is the inherent tradeoff between efficiency of performance and simplicity. In many situations, you may find that TCP/IP is not as efficient a protocol as IPX. On the other hand, IPX was not designed for unreliable and lower-speed WAN connections.

9.9.6 Performance

Performance of Internet connections, as measured by the number of bytes transferred per second, is dependent on a number of factors, including, but not limited to, the type of network connection (PSTN, ISDN, ADSL, or cable modem), remote link speed, the type of modem employed by the remote user, the type of modem or CSU/DSU employed by the ISP, the communications server/router hardware platform, the average number of users supported per port, ISP's connection to the second-tier NSPs, and most important, the speed of retrieving information from the information server. Typically, the weakest link, in many cases, is the inability of the information server to spew out information to a large number of remote requests. The number of hops in the physical connection, and the overstressed connection between the ISP and NSP are other important reasons for delays in getting your information from the Web server or legacy database connected through the Internet.

There are not that many tools available to monitor end-to-end performance, but we would like to mention one. Service Management Architecture (SMA), from Jyra Research Inc., is a Java-based monitoring tool that samples response time and lets network managers identify specific problems and use that information in order to enforce service-level agreements. Figure 9.9 gives a glimpse of how this tool works. There are two components of SMA—a service level monitor (SLM) and a midlevel manager (MLM). These modules collect performance measurements and report the data back to the network administrators. SLM runs in a workgroup server and handles response time measurements, which it feeds to a local MLM. The local MLM also collects SNMP information about network device traffic. The network manager can specify which application SLM should monitor, and at what intervals.

Two methods of monitoring are used. One is *traceroute*, which is part of ICMP (Internet Control Message Protocol); it pings devices between the workstation and application host to determine location, while measuring round-trip delays. This provides a fairly comprehensive picture of delays between each IP address, helping network managers identify the cause of sluggish application performance and target the equipment or circuits in need of upgrades.

Please note that traceroute methodology does not always work because some ISPs deactivate ICMP, and firewalls can be configured to dump pings. Even with these limitations, some conclusions can be made with respect to those parts of the network which do support ICMP.

The second method of monitoring is called *synthetic transaction*. It adds a simple task to the stream of application traffic coming out of a remote workstation. The results of the task are sent to the host where delays can be calculated. Synthetic transactions give end-to-end response time, not component response time. However, they work over any network even

Figure 9.9
Service-level monitoring for Internet bottlenecks.

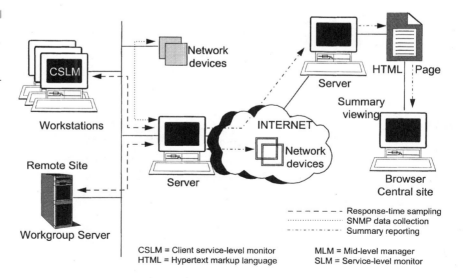

CSLM = Client service-level monitor
HTML = Hypertext markup language

MLM = Mid-level manager
SLM = Service-level monitor

when ICMP is turned off. A good network performance analyst can use both methods, and come to definite conclusions about bottlenecks.

We encourage the reader to refer to Chap. 17 for a more detailed discussion of this topic because many of the factors that affect performance are the same in various network options.

9.9.7 Costs

Although we discuss the total cost of ownership (TCO) in Chap. 15, here we shall concern ourselves with network connection charges only. With Internet connection, it is relatively easy to estimate costs because there are four major elements that can be easily estimated. These elements are monthly connection charges paid to the ISP, the remote user's modem costs, CSU/DSU costs, and the monthly cost of the high-speed link between the LAN and the ISP at the other end. There are more elements involved in a private remote access solution—all these elements must be carefully estimated after getting proposals from network suppliers or integrators. The biggest cost factor is the ongoing support costs for two solutions, one based on the Internet and the other based on a private remote access solution. In the case of the Internet model, much of the support costs are shared by ISPs and the remote users directly.

9.9.8 Network Management

The Internet-based remote access solution is much easier to manage because this task gets outsourced to ISPs from the perspective of central information publishing organization. The only portion of the network that you must still manage is the back-end connection, user ID management, and password management for operational applications and data. It is easier to integrate this portion of the back-end network with enterprise network management systems from IBM, Computer Associates, or Hewlett-Packard.

9.9.9 Security Management

Security on the Internet is one of the biggest concerns of organizations for a number of reasons. Not only do you want to protect your operational and critical data from being modified or accessed by hackers, information spies, and unwanted guests, but you also want to control who is accessing the Internet, and for what purpose. If you are planning to do electronic commerce, security assumes an entirely different magnitude. The spread of viruses from the Internet is still another reason that you would like to control access and scan for any unwanted virus traveling on to your network.

Security policy, practice, and policing must be consistent with the organization's business needs to utilize the Internet as a tool for information delivery and access, and keeping its operational data protected. The requirements for security vary from one organization to another. You can neither expect to have a completely sealed environment like Fort Knox, nor can you keep your doors ajar so that anybody can walk in or out. Since the Internet revolution began and organizations saw the potential for electronic commerce, tremendous technical advancements have taken place in this area. Packet filters, firewalls, the encryption of data, digital certificates, and public/private key schemes are only a subset of the many different options that you must evaluate.

Packet filtering is a capability implemented in many routers that allows traffic to pass through, based on authorized TCP/IP addresses. For example, if you allow e-mail access only, then you should use packet filters to allow SMTP traffic only. You may also allow UP traffic from known addresses only. Ask your ISP to apply filters so that nobody can talk directly to your router. It is not always a good idea to allow your ISP to control your router as a part of their domain, so that you have to share passwords with them. It will be a better policy to have a second router that is completely controlled by you.

A *firewall* is a specialized piece of software that sits in between the corporate network and the Internet. It is essentially a routing software with an extremely tightly configured set of packet and address filters. Apart from the packet filtering described earlier, connection filtering is another common service offered by firewalls. With connection filtering, TCP connections are allowed only if they originate in a certain specified domain or direction. As an example, a Telnet connection may be allowed only if it originated within the internal corporate network. It will be denied if somebody from outside tried accessing corporate information using a Telnet connection from the Internet. Some firewalls, such as BorderWare, do allow external Telnet connection, but only after satisfying a user-specific challenge. After entering the PIN number on the electronic card, the user verifies the challenge, and is given the response. The next Telnet attempt would require a different response. (See Fig. 9.10.)

Some firewall servers provide the functionality of a bastion host and packet filtering. A *bastion host* is an exposed gateway machine that pro-

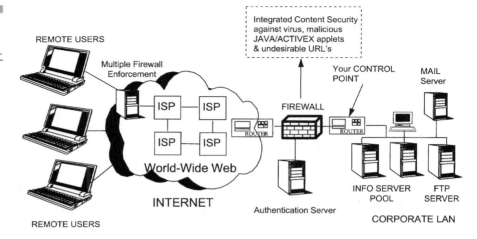

Figure 9.10
Security firewall in an Internet environment.

vides services between an organization and the Internet. A *proxy* is an agent or application that makes a connection from one side of a firewall to the other, through the firewall. The BorderWare Firewall Server's proxies create an illusion of transparency to the client application.

9.10 Virtual Dial-Up Private Networks (VDPNs)—Internet-Based Remote Access*

In the first part of the book, we talked about modem-based private dial-up links that connect remote users to centralized remote access servers. In this chapter, we discussed Internet as a vehicle for remote access. Users from anywhere in the world can now connect to corporate information through the Internet. However, this type of connection does not always support IPX/SPX, NetBEUI, or other network protocols in a transparent fashion. What is emerging now is the concept of outsourced remote access that allows any remote user to set up a point-to-point protocol (PPP) connection to an ISP, and then create a private connection over the Internet to your private network. (See Fig. 9.11.)

A virtual dial-up private network (VDPN) lets users set up an end-to-end connection by using the two WAN components of the Internet, that

---

*For more information, see "Internet-Based Multiprotocol Remote Access" by Mike Fratto, *Network Computing,* April 1997.

Figure 9.11
(a) The virtual digital
private network
(VDPN) connection.
(b) Client-established
VDPN with NT server.

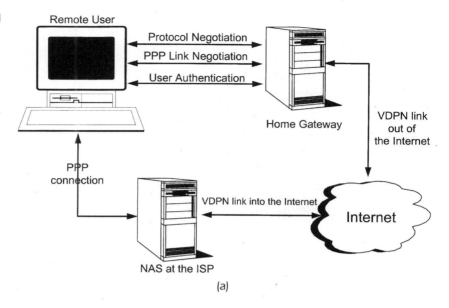

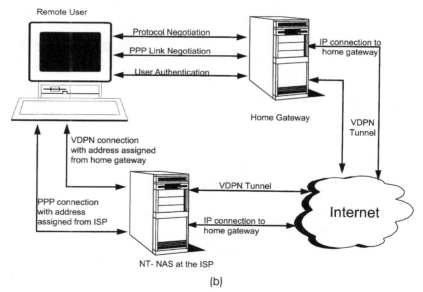

is, the dial-up link as one portion and the Internet as the other. By using VPDN, you can extend a PPP session between a remote client device and a remote access server, such as Shiva's WebServer to a home gateway on the network. The home gateway serves as the other end of a PPP session and also acts as a remote access server, including the security function of user authentication and protocol negotiation.

There are two technologies used for this purpose—PPTP (Point-to-Point Tunneling Protocol) and Layer 2 Forwarding (L2F). The first is an Internet Engineering Task Force (IETF) draft standard that is promoted by several companies, such as Ascend, 3COM/US Robotics, Microsoft, and ECI/Telematics. The Cisco-promoted L2F is also an IETF draft standard.

In VDPNs, you employ two multiprotocol servers—one at the ISP site and the other at your corporate information LAN site. Now the link between the two servers becomes a tunnel for carrying multiprotocol traffic. A remote client dials an ISP and sets up a PPP session with the NAS at the ISP site. The NAS answers the call and sets up one end of the tunnel. Then the NAS tells the home gateway (the other end of the tunnel) that a VPDN session has been requested. The NAS receives the name and user password from the client, and forwards it to the home gateway. After authentification, the NAS and the home gateway establish a tunnel, and assign a session ID that specifically identifies the user and the assigned tunnel. Once the user has been authenticated and the tunnel established, the client and home gateway negotiate the PPP session, setting up the protocol and allocating network addresses to the client.

You should take care of security—such as encryption or token authorization—in the usual manner, and not depend on VDPN to provide this functionality. You can employ Password Authentication Protocol (PAP) or Challenge Handshake Authentication Protocol (CHAP), RSA's RC4, and DES technologies. (Refer to discussion of security in Chap. 14.)

L2F is a technology being promoted by Cisco. It offers a similar set of services, but with a bit more flexibility. L2F does not make any assumptions about the underlying transport layer carrying the PPP packets. Instead, it defines a generic encapsulation protocol that transports OSI layer 2 PPP frames. This scheme was originally developed for non-IP protocols.

9.10.1   Product Implementations of VDPN

Many vendors have implemented VDPN functionality in their RAS products. The following list represents a subset of products with this capability:

- Cisco's AS5200 Universal Access Server
- Shiva's LanRover Access Switch
- Microcom's Access Integrator
- US Robotics' Total Control Enterprise Network Hub
- 3Com's AccessBuilder Enterprise LAN/WAN Switch

Some of these products—for example, Microcom's Access Integrator and Shiva's LanRover—have implemented both PPTP and L2F. This is not surprising, because IETF (Internet Engineering Task Force) has recommended merging of two datalink protocols into one standard—layer 2 tunneling protocol. Other products have implemented only one of the two methods, so far. You should find out if they intend to support other methods, when the IETF standard is finalized.

9.10.2 VDPN Features to Be Considered While Evaluating Products

The following features are important in evaluating VDPN capabilities of a product:

- LAN protocols supported, for example, IP, IPX, NetBEUI/NETBIOS, and AppleTalk
- Support for PPTP, L2F, or IP direct
- Call-routing features, such as subaddressing, outbound dialing, and so forth
- Support of security features, such as RADIUS, TACACS/TACACS+
- Maximum number of tunnels supported
- Multiple tunnels per user (see chapter 7)
- Support for multilink PPP (see chapter 7)
- Multilink support across devices
- Reporting features—user log files, VDPN logging, error logging
- Overall performance of VDPN in terms of effective Kbps on a specific interface

You should map your current and future requirements against the product features before you select any of these products. Your ISP may help you in this selection process so long as it supports multiple products.

9.10.3 VDPN Leading to Outsourcing of Remote Access Networks

You can easily see the potential of VDPN from three key points of view. First, it can provide high bandwidth on demand between the remote user and corporate LAN. Second, the economics of a VPDN are definitely in its favor when compared with a private remote access network. Third, the technical support and headaches associated with it can be outsourced to the ISP. Therefore, it is not surprising that both ISPs and Telcos have started offering VDPNs to large corporations, either on their own or through strategic relationships. In fact, several Telcos have already started buying equity stakes in large ISPs to position themselves for the future.

Purely on the basis of cost and technical support requirements, VDPNs are less expensive than commercially provided private networks, such as frame-relay. However, the more expensive frame-relay networks may provide consistent and higher performance, better security, and service level guarantees.

We would like to point out that you have less direct control over the performance of your VDPN because the ISP may oversell its capacity to lots of customers. The reverse is equally true—you may get the benefit of new technology because ISPs have to continuously upgrade their infrastructure to meet demands from other customers. It is always a trade-off.

9.11 Comparing Private Networks with Internet-based VDPNs— An Application Perspective

There is no doubt that the Internet-based VDPNs will have a profound impact on future corporate networks. However, the reality is that for the time being, the most cost-effective and efficient approach is one based on a hybrid design—a combination of traditional data links and the Internet. Initially, remote access and less-sensitive nonoperational applications will move over to the VDPNs first. Table 9.1 looks at this issue from an application perspective.

Summary

In this chapter, we discussed the benefits of using the Internet for remote access, issues involved in using the Internet, and applications that are more suitable for remote access through the Internet. We explained the structure of the Internet backbone, its components, and the impact this may

TABLE 9.1

Private Networks versus Internet-based VDPNs

Application	Current State of Private Network Solution		Current Internet Solution		Future Internet Solution
	Pros	Cons	Pros	Cons	
E-mail—How	*Proprietary e-mail systems from major software vendors and Network providers*		*SMTP-based Internet mail*		
Pros and Cons	Reliable and secure	More expensive Disparate systems Message conversion	Low cost Internal and external	Limited security	Real-time delivery Improved security Standardized forms and attachments
File transfer—How	*Dial-up, dedicated, and virtual links*		*File attachment in e-mail, FTP, Gopher, or WWW*		
Pros and Cons	Timely, secure delivery of corporate data	Remote users must pay for private network	Excellent for file transfer	Limited security Unpredictable transfer times	Faster transfer rates More links to corporate data Higher security
Database access—How	*Dial-up or dedicated lines*		*Use VDPN or Internet dial-up*		
Pros and Cons	Secure, reliable, high performance	Expensive and limited to internal users only	Suitable for infrequent access by mobile and external users	Poor security, unpredictable performance Intranet better	Performance guarantees Better security than current

248

Application	Private network		Internet		
	Done through frame-relay or ISDN		*Use dial-up or ISDN to Internet*		
Workgroup/document sharing	Predictable performance and secure	Requires additional hardware and bandwidth	Low cost, performance meets requirements	Poor security for confidential documents	Default network for such applications with better security
	Over BRI or PRI ISDN		*Usable over ISDN connection to Internet*		
Videoconferencing	Guaranteed performance and security	Expensive	Cheaper than private network	Poor performance, quality unpredictable	Guaranteed performance Superior security
	EDI		*Web-based payment and electronic malls*		
Electronic commerce	Highly secure, standards-based, supported by major value-added networks	Systems closed to closed group of trading partners	Low cost, mass consumer-oriented, convenience	Poor security, lack of consumer confidence	Standard-based security Interoperability Integration with existing legacy back-end systems

SOURCE: *CommunicationsWeek*, Feb. 1996.

have on your solution. We reviewed various network scenarios for exploiting the Internet, and gave some example of products from the marketplace. Finally, we reviewed how virtual private networks based on the Internet use PPTP and L2F technologies. The Internet is not necessarily the optimal solution for all remote access applications. Therefore, we showed through a table where the Internet is best suited in the present context. Tomorrow, it might be a different story.

CHAPTER **10**

Remote Access Network Options—Wireless Networks*

Wireline networks are multilane highways that are inexpensive, fast, and wide. But you must be on the ground to travel on them. If you want to fly like a bird anywhere, anytime, you'd better use the wireless airways—they are expensive, slow, and narrow. But look at the freedom that the birds enjoy.

—*Chander Dhawan*

---

*Most of the material for this chapter has been drawn from the author's book entitled *Mobile Computing—Systems Integrator's Handbook*, published by McGraw-Hill in November 1996.

About This Chapter

Remote access has become one of the most sought-after requirements in networking during the last few years. With the increasing movement toward telecommuting, mobile computing, and virtual office, remote access enables workers and professionals to obtain information and stay in logical touch with their physical offices.

So far, the emphasis on remote access solutions has been on using wireline networks. Therefore, these solutions were limited to places where there was either a LAN, a WAN, or an Internet connection available. Since early 1994, remote network access solutions from companies such as Ascend, Shiva, Cisco, and 3COM have extended the reach of LAN applications to remote users who are close to a telephone jack. This connectivity, although quite universal, still depends on a tethered connection. However, wireless network—based connectivity frees you from this dependency on a tether. All of a sudden, true mobility becomes possible. This is the focus for this chapter. We explore unique business requirements, understand wireless networks, explore wireless computing architectures, review wireless switches, and discuss implementation issues with providing ubiquitous *anywhere, anytime* access to remote applications on a LAN, minicomputer, or mainframe. In particular, we discuss the following topics:

- Why explore wireless networks for remote access?
- Business imperatives and financial justification for wireless computing applications
- How do wireless network—based solutions work?
- Understanding the basic nature of wireless networks
- Brief discussion of different wireless networks
- Architectural issues with wireless networks
- Product examples for wireless-based remote access

10.1 Why Explore Wireless Option for Remote Access?

The primary motivation for including this chapter in this book is based on the following phenomena in the marketplace:

- Mobile computing evolution is progressing slowly, but steadily. Competitive pressures are dictating that we reengineer our business processes using remote access technologies, based on both wireline and wireless networks.

- There is an increasing demand for universal and ubiquitous access to information. Remote access solutions based on switched wired services (PSTN, ISDN, ADSL, and cable modems) are today's solution to our ultimate requirement for anytime, anywhere connectivity that can be provided only through wireless networks.

- Remote access vendors, such as Cisco and 3COM, are going in one direction with their remote access plans. Wireless solutions remain a niche requirement. However, we strongly believe that this need not be the case.

- There are many different wireless networks with their unique characteristics and cost structures. Devices supported, software interfaces supported, speeds, and coverage available vary significantly with each of these networks. Considerable analysis and planning are warranted for developing an optimal solution.

- Presently, wireless networks do not support—at a reasonable cost—high-bandwidth remote access applications, such as bulk file transfer, multimedia, and videoconferencing. However, for many mission-critical applications and low volume of remote information exchange, wireless networks do offer a solution that is affordable today. Tomorrow, it will get better from both cost and coverage points of view.

10.2 Business Imperatives and Financial Justification for Wireless Network—Based Applications

In order to provide a high level of customer service, mobile workers and sales professionals must stay in touch with their customers and home offices at all times. Many categories of workers rarely return to the office on a regular and frequent basis. Locating a spare telephone jack for a dial-in connection from a remote site is not always easy. Calling into the home office for messages is a time-consuming process when secretarial staff is rare in today's downsized organizations. Improved levels of productivity can be achieved if remote access solutions provide anytime, anywhere connectivity that is possible only through wireless networks.

The level of remote access to e-mail and operational database applications on the LAN or legacy mainframe systems varies in different organizations. Most organizations provide in-house LAN-based messaging for in-house exchange of information. Some organizations provide remote access through dial-in, ISDN, or Internet to those employees who deal with the outside world; a few of the more advanced organizations also provide wireless network—based access to a selected group of users. Irrespective of the current status, many organizations are currently evaluating the value and cost of universal access to messaging through wireless networks.

The Internet has propelled electronic e-mail into the number one application demanded by mobile workers. Besides e-mail, there are a number of other remote access applications, such as operational or traditional OLTP (on-line transaction processing) database inquiry and update, decision-assist queries, groupware applications, document exchange, or collaborative applications where you need not be permanently connected through a private network. Not all of these applications can be justified on wireless networks today. However, in several vertical industries, wireless networks are either mandatory or can be easily justified on an economic payback basis. For public safety (police, fire, and ambulance), computer-aided dispatch for field force automation, courier, and route distribution companies, the use of wireless networks is not a choice, but a necessity for tracking where field staff is at any given time. Even in sales force automation applications, you can achieve improved productivity of sales staff by giving them access to the most current product and inventory information. You cannot always expect these mobile workers to locate a telephone jack and make a dial-in connection.

IDC estimates that the payback period on mobile computing applications can be as low as one and a half to two years. It has been shown by research that benefits increase significantly with anywhere, anytime access through wireless networks. So do the costs, of course. The impact of improved payback from universal access to corporate information is shown in Fig. 10.1.

10.3 How Does Wireless Network Change Remote Access Solutions?

Figures 10.2 and 10.3 illustrate remote access solutions based on wireless networks. The first illustration, Fig. 10.2, shows a relatively simple messaging application on a wireless network. Our schematic shows the mini-

Figure 10.1

Relationship of mobility, business value, and cost.

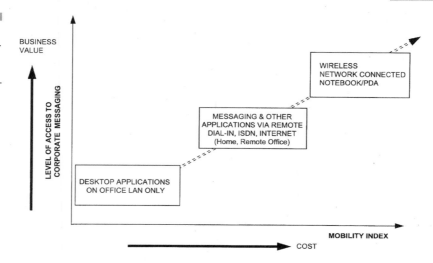

mum number of changes required to use messaging on a packet-switched wireless network, such as ARDIS or Mobitex-based RAM Mobile Data. As a minimum, a network-specific modem and corresponding network-specific software driver are required in the notebook, PDA, or any other device used, including the soon-to-be-launched network PC. Once the connection is established, it works like any remote client. As compared to this simplistic illustration for messaging, the number of changes required to provide general-purpose wireless-based access from LAN applications depends on a number of factors, such as the number of users, the type of client devices, the type of wireless network employed, the application platform, and so forth.

E-mail messages can be retrieved by mobile users anywhere the wireless network has adequate coverage—you do not need a telephone jack, ISDN termination point, ADSL jack, or a cable modem. Users see the same interface as they would see in the office. In view of the slower speed, security issues, and premium on endorser's time, many messaging vendors provide a different interface for remote wireless clients than what they offer for a wired LAN client. Both Lotus cc:Mail and Microsoft Mail offer mobile versions of their mail client software.

Figure 10.3 is a picture of the hybrid configuration, showing both wireline and wireless network—based remote access. It may be observed that we have shown one mobile communications server switch (MCSS) serving both dial-in and wireless network users. In practice, there are very few communications servers that meet the needs of both wireline remote access and wireless access in the same box—we comment on this need later in the chapter.

Figure 10.2
Ubiquitous e-mail
access on a wireless
network.

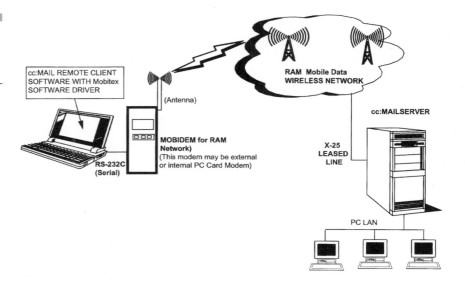

Figure 10.3
Hybrid remote
access—wireline and
wireless.

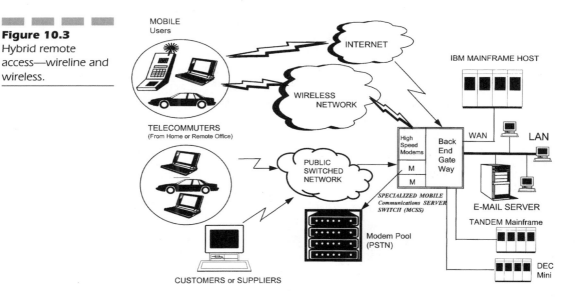

10.4 Understanding the Components of a Wireless Computing Solution

Figure 10.4 provides a schematic of a wireless network in overall architecture of a wireless computing solution. We briefly describe some of the components of such a mobile solution:

10.4.1 End-User Devices—Notebooks and Radio Modem

You should carefully evaluate the devices that the wireless WAN supports. On the front end, you have client devices, such as notebooks, Windows-CE organizers, PDAs, two-way pagers, and in the future, digital PCS telephones (such as Nikon 1900 PCS) that can be used to access Internet for e-mail or simple low-information-content applications, such as stock quotes—while playing golf! You should ensure—from the software driver and modem availability points of view—that the wireless network that you are considering does support these devices. The power of the radio and the base station controller determines whether the radio must be installed inside a vehicle because of its bulky size or it can be carried around like a cellular telephone. Low-power radios in PC cards require more powerful base station controllers. The radio's integration into the end-user device is an important consideration for many mobile applications. If the only radio modem that the network supports is large and bulky and won't fit in a PC card slot, then this network is no good for those who must carry their notebooks around.

10.4.2 Specialized Mobile Communications Server Switch (MCSS)

On the other side of the network, you have an MCSS that acts as a traffic cop between the remote end-user device and the back-end information servers. MCSS is an important piece of the overall configuration. It performs a number of important functions, such as the identification of mobile devices through their radio network addresses, mapping of these addresses to TCP/IP addresses, network protocol handling, compression, and network security. MCSS may have limited capability to support dial-in clients as well.

10.4.3 Components of a Wireless Network

Figure 10.4 shows the components of a basic wireless network. Almost all wireless networks utilize a base station controller (BSC) or a cell that has certain coverage.

10.4.4 Frequency Reuse Concept

The *frequency reuse concept,* illustrated in Fig. 10.5, allows a radio network designer to reuse a frequency outside the coverage area (typically 5 to 7 miles), and therefore, increase the coverage area as the demand increases. This concept is utilized heavily to increase the number of subscribers in an area. The other concept used in wireless network design is called *multiple frequency broadcast.* Under this design, multiple frequencies are used in the same area.

Figure 10.4
Components of a wireless network–based remote access. (Source: Mobile Computing Handbook by C. Dhawan [McGraw-Hill].)

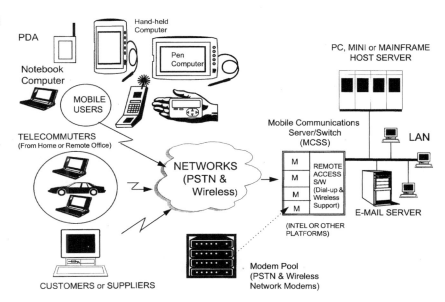

Figure 10.5
Frequency reuse plan.
(Source: Bishop Train-
ing Company.)

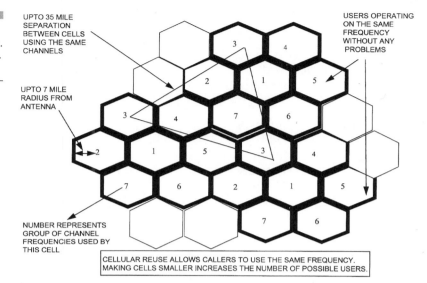

UPTO 35 MILE SEPARATION BETWEEN CELLS USING THE SAME CHANNELS

UPTO 7 MILE RADIUS FROM ANTENNA

USERS OPERATING ON THE SAME FREQUENCY WITHOUT ANY PROBLEMS

NUMBER REPRESENTS GROUP OF CHANNEL FREQUENCIES USED BY THIS CELL

CELLULAR REUSE ALLOWS CALLERS TO USE THE SAME FREQUENCY. MAKING CELLS SMALLER INCREASES THE NUMBER OF POSSIBLE USERS.

10.5 Understanding Basic Nature of Wireless Networks

Having discussed how mobile computing applications work in a wireless network context, we are ready to try to explain the unique nature of wireless networks. The following characteristics are worth considering in evaluating wireless network solutions:

10.5.1 Wireless Spectrum Is Finite and Wireless Networks Have Lower Transmission Speed

Wireless networks utilize a specific radio frequency in a given geographical area. The FCC in the United States and similar bodies in other countries assign (through auction in some cases) these frequencies for specific uses. A network provider, an equipment manufacturer, or an end user must acquire a license to use this frequency or to sell equipment operating on a given frequency. Radio spectrum is a scarce resource. As such, there is a limit to the amount of bandwidth available on wireless networks. Also, the effective transmission speed that you can achieve on wireless networks is almost always lower than what you can achieve on copper-based wireline connections. Whereas LANs operate in the 10 to 100 Mbps range, and wireline WAN connections operate from 56 Kbps to T3 speeds, wireless wide area networks typically operate at 4800 to 19,200 bps range, with some exceptions using spread spectrum technology at 28,800 bps or higher. While remote dial-in and ISDN access to Internet is available at 28.8 Kbps and 128 Kbps, respectively, wireless links can transmit at only 4800 bps on ARDIS (19,200 bps in selected areas), 8000 bps on RAM, and 19,200 bps on CDPD (cellular digital packet data). ISDN level speeds on wireless networks are not expected before two to three years when broadband PCS networks may be in production in metropolitan areas. Therefore, message transfer times are much higher on wireless networks than on LANs and wireline WANs.

10.5.2 Wireless Networks Have Variable Coverage and Lower Reliability than Wireline Networks

Wireless networks require expensive infrastructure (base stations, antennas, and wireline connections to Telco's switching centers). The amount of coverage varies with different networks, but is almost always less than wireline networks. Because of the medium used (radio frequencies in air), the wireless networks are affected by atmospheric disturbances, geographical terrain, buildings, tunnels, and many other factors much more than wireline. Most communications software packages are not robust enough to withstand the coverage problems or high error rates encountered on wireless networks.

10.5.3 Many Proprietary Wireless Network Technologies

There are a number of wireless networks that utilize different frequencies, employ distinct modulation techniques, and have specific network design characteristics. More important, they require different radio modems, software drivers, and communications software interfaces. We describe these networks in detail in Sec. 10.6.

10.5.4 Wireless Networks Are Generally More Expensive than Wireline Networks

Wireless networks cost more than wireline networks, in most cases, for carrying the same amount of data. One exception is for wireless bridging in the downtown core or campus environments where you may use wireless bridges to connect messaging servers. Table 10.1 gives representative figures.

10.5.5 Lack of Standards in Wireless Networks

There are few standards in the wireless network industry. Most wireless networks use proprietary modulation techniques, modems, and communications interfaces. Many wireless networks do support TCP/IP at the transport layer, but common APIs are rare. This means that applications written to a specific network interface may not work with another wireless network. Wireless middleware may solve this problem as long as appli-

TABLE 10.1

Representative Costs for Wireless Messaging

Type of Network	Cost of Shipping a 600-Character E-mail Message	Cost of Sending One Page of Fax
Circuit-switched cellular	$0.40 to $0.70 (30 sec to dial; 15 sec handshake; 1 sec to transmit)	$0.40 to $0.70
CDPD	$0.15 to $0.20	Not available, requires session connection
Paging	$2.40 to $6.00	Not suitable for fax
ARDIS	$0.05 to $0.39	Not available; requires session connection
RAM Mobile Data	$0.15	Not available; requires session connection
ESMR	National pricing not available, expected to be comparable to packet networks	Pricing not set
Satellite	$2.00 to $3.00 (1.2 min)	$2.00 to $3.00 (1.2 min)

SOURCE: Author's research.

cation developers standardize on that middleware. No middleware standard, de facto or de jure, has emerged so far. Accordingly, none of the major messaging vendors (Lotus, Microsoft, or Novell) use this middleware, which can be relied upon for building custom messaging or operational OLTP applications.

10.6 Exploring Wireless Network Options

We describe various wireless networks in this section. We have covered short messaging service based on two-way paging on narrowband PCS in greater detail, because we expect this service to become more pervasive due to lower costs. Also, the paging extension to e-mail is becoming a very common method of contacting remote workers, who can then access partial contents of important messages through a new breed of electronic organizers, such as Pilot, HP200LX, and Sharp's Zaurus.

10.6.1 SMR/PMR

In 1974, the FCC in the United States and similar bodies in other countries created the specialized or private mobile radio (SMR/PMR) service. Today, SMR systems provide services in the United States to more than a million radio users through over 7000 SMR systems nationwide. According to the FCC, the total SMR market was well over $1 billion in 1995, is expected to become a $4 to $5 billion industry by 2000.

SMR operators provide services in the 800 MHz and 900 MHz bands on a commercial basis to bodies eligible to be licensed under this category. SMR systems have been primarily used for dispatch applications in public safety (police, fire, and ambulance) organizations, the taxi industry, the trucking industry, and in many maintenance services—oriented industries. These networks provide two-way radio (voice) and data capabilities.

As a result of current economics, there is a general trend to move away from private SMR systems to public shared wireless networks, except in those cases where you can cover the entire set of users by a small number of base stations. (See ESMR in a later section.)

10.6.2 Paging Networks

Paging is by far the most popular form of mobile communication. The reasons are obvious. Pagers are small, affordable, simple to use, ubiquitous, and reliable. About 25 million pagers are in use today in the United States; almost 60 million people around the world carry them. The industry continues to grow and is forecasting substantial increases with the widespread adoption of two-way paging. One study by McLaughlin & Associates forecasts that the market for pagers in the year 2000 will be worth $21.5 billion worldwide. Whether pagers will eventually be replaced by a new breed of PDAs, personal communicators, or wireless-enabled electronic organizers, or whether they will simply become functionally ever richer, providing more functions is a moot point—the basic concept behind paging will always be in demand.

Most of the current paging networks provide one-way paging, that is, they allow one-way transmission of brief messages (numeric or alphanumeric) to a low-power receiver. Receivers can have their own display or they can direct the display output to a laptop, PDA, or personal communicator.

10.6.2.1 ELECTRONIC MESSAGING VIA TWO-WAY PAGING ON NARROWBAND PCS—A PROMISING NEW TECHNOLOGY. In 1995, several companies (SkyTel being the first) started offering, on the new narrowband PCS frequencies, a two-way messaging service that allowed users to acknowledge receipt of a message. The author believes that two-way paging will become an increasingly effective way of wireless communication for field-dispatching and other applications where cost is an important factor. A corresponding SMS (short messaging service) based on Europe's GSM (global system for mobile) standard has been available there since 1993. (See Fig. 10.6.)

Research in Motion, based in Waterloo, Ontario, Canada and Motorola of Schaumberg, Illinois, introduced Inter@ctive Pager and PageWriter devices, respectively, with a full qwerty keyboard. This new breed of two-way pagers acts as transceivers—transmitters, as well as receivers. Although only two types of two-way pagers have been developed as of now, the potential is much greater. The first type returns basic acknowledgments such as *yes, no,* or *page received.* Brief return messages can be composed and sent back with the second type. (See Fig. 10.7.)

With two-way paging, delivery is guaranteed, and the response is immediate. Customer service response times will improve, and users will

Figure 10.6
Skytel's two-way
paging network.
(Source: SkyTel Tech-
nology Backgrounder
Paper.)

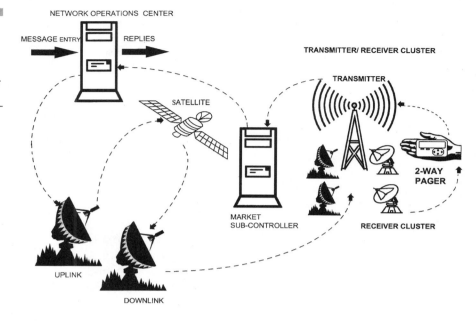

benefit from savings in time and money, as well as increased personal
effectiveness and responsiveness.

In the current implementation, users have four ways of responding to
messages:

Figure 10.7
Research in
Motorola's
Inter@ctive Pager.
(Courtesy of RIM,
Waterloo, Ontario.)

Inter@ctive™
P A G E R

Your customers,
your office and the
Internet are now
in the palm of
your hand.

1. *Automatic message confirmation.*

2. *Multiple choice.* All messages can include a set of optional responses that recipients can use in reply.

3. *Predefined responses.* SkyTel pagers come preprogrammed with a standard set of responses. These can be customized to meet the needs of individual applications.

4. *Freeform responses.* Users can create freeform responses or initiate wireless messages, by using a short cable to connect SkyTel two-way pagers to palmtop or laptop devices, such as the HP-LX.

Under the second implementation, messages can be sent to users from an e-mail application. In this scenario, if a user is not logged on to the office LAN and a message is urgent, a mail server can send a page. Even if the recipient does not belong to an in-house workgroup, a page can still be sent. Most modern help desk applications are able to automatically alert a service vendor or a specific support person in case of a problem. With the advent of workgroup applications, such as Lotus NOTES and Microsoft Message Exchange, the initiation of a paging function has been greatly simplified, and can be invoked on the basis of a predefined filter (essentially a codification of paging criteria, such as an urgent message or an emergency). This type of pager gateway in an e-mail server is illustrated in Fig. 10.8.

10.6.2.2 WIRELESS MESSAGING AND INTERNET. Internet is being increasingly used as a backbone transport for mobile computing.

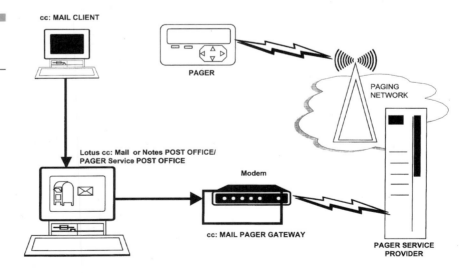

Figure 10.8
cc:Mail to pager
application gateway.

Several vendors are now offering wireless network connection to Internet. InfoWave from GDT in Vancouver, British Columbia, is one example of a messaging application with wireless network interface into Internet. Figure 10.9 illustrates this application.

IBM offers WebExpress, a wireless Internet access product that allows you to develop wireless applications based on the HTML model. WebExpress may be used with IBM's ARTour mobile communications server switch that can act as a gateway into Lotus NOTES—based messaging, Microsoft Exchange, or legacy databases. ARTour may be run on an Intel server or RS/6000 under AIX.

10.6.3 Circuit-Switched Cellular

Since there is now extensive cellular network coverage throughout the United States and the rest of the world, opportunity exists for mobile workers to use cellular networks for data transmission. Once a notebook computer is equipped with a PC Card modem and a cable to connect it to your cellular phone, you have a physical connection that is quite similar to a public-switched telephone network. Many sales professionals and other mobile workers are using circuit-switched cellular networks already—but so far with only mixed results, because of problems of coverage, noise, poor quality of signals, temporary disconnection due to handoffs, and the

Figure 10.9
Wireless interface into Internet mail.

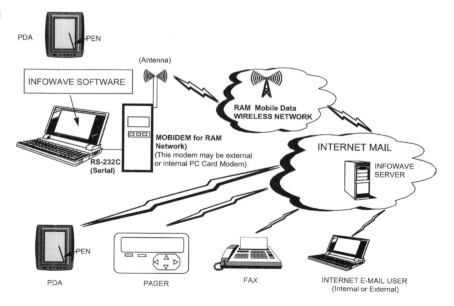

like. Specialized modems, such as ETC from AT&T and AirTrue protocol, do help, but only partially, with the result that you get an effective through-put of only 14,400 bps. (See Fig. 10.10.)

10.6.4 CDPD (Cellular Digital Packet Data)

Cellular digital packet data (CDPD) is a wireless packet-switched WAN that uses idle spots in cellular channels not used by voice at any given time. The CDPD technique, originally developed by IBM, allows data to hop from one idle channel to another idle channel through frequency hopping. The CDPD network is being deployed by several cellular carriers, such as GTE MobileNet, providing coverage to most of the nation, through association with other network providers. (See Fig. 10.11.)

CDPD in its most basic form can be considered as a wireless extension of an existing TCP/IP network. It allows workstations to talk to host computers to retrieve sales data, inventory, billing information, and so forth, in a similar manner as packet radio networks, such as ARDIS and RAM Mobile Data.

CDPD differs significantly from a traditional circuit-switched cellular connection. Using circuit-switched cellular technology, once a connection has been established, it owns that link and the users are paying for it, whether or not data is actually flowing. Furthermore, any line noise,

Figure 10.10
How data works over cellular circuit switched network. (Source: Adopted from Byte Magazine.)

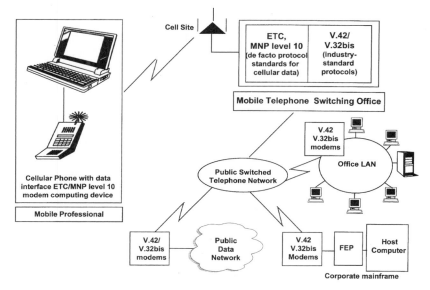

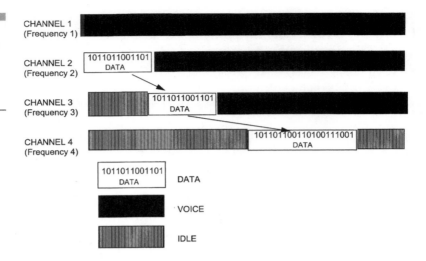

Figure 10.11
Empty spots in cellular voice channels and frequency hopping employed in CDPD.

dropped connections, or other impairments to which modems are susceptible are visible to the modem. Often connections have to be reestablished at an additional cost. CDPD avoids these problems. (See Fig. 10.12.)

CDPD network modems operate at 19,200 bps, and are available from several different vendors. The main concern in selecting a CDPD network is whether it is available in the areas where your users are located, and the possible adverse impact of heavy voice traffic on the response time of mission-critical applications during peak periods. However, CDPD service providers are desperately looking for paying customers, and would be willing to beef up their network for any sizable customer. Some CDPD providers, such as AT&T Wireless, have decided to offer dedicated channels for data applications—a concern that channel hopping creates in areas with too many cellular voice users.

Like ARDIS and RAM Mobile Data packet-switched networks, CDPD tariffs are based on packet charges.

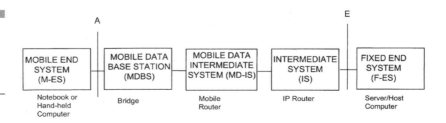

Figure 10.12
Components of a CDPD network. (Source: GTE MobileNet.)

10.6.5 ARDIS—A Packet-Switched Wireless Network

The essential characteristic of a packet-switched wireless network, such as ARDIS, is that you share a single physical connection (or a data pipe) between different users and send information in packets. The packet-addressing scheme identifies the source and the destination. There is an elaborate protocol to bidding and releasing the physical connection.

ARDIS was the first and continues to be the largest wireless data network provider in the United States. ARDIS was formed in 1990 as a joint venture between IBM and Motorola. The network was designed and built by Motorola for IBM as a specialized private wireless network for IBM's customer services application. It was used primarily to dispatch service technicians, to query databases for parts availability, and to determine call status. In 1995, IBM sold its interest in the network to Motorola.

The ARDIS network infrastructure consists of 1400 base stations throughout the United States connected to local switching nodes, which in turn are connected to a single national switching node through a high-performance, nationwide telecommunications backbone. (See Fig. 10.13.)

ARDIS provides national coverage with seamless roaming, which allows users to travel anywhere there is coverage, and be registered as soon as they power up their modems. ARDIS covers the top 400 metropolitan

Figure 10.13
ARDIS network schematic. (Source: Bishop Training Company.)

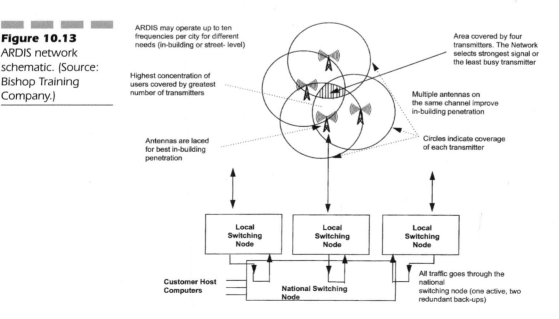

areas (10,700 cities and towns) of the United States—which includes more than 90 percent of all businesses, and 80 percent of the total population. ARDIS had 60,000 users connected in 1997 and expects to increase its penetration further as a single wireless network with national coverage. ARDIS is aggressively promoting its two-way messaging service that has been available nationally before two-way NB-PCS service was offered. Now it is promoting it with RIM's Inter@ctive pager or Motorola's PageWriter.

ARDIS is optimized for in-building penetration and provides good in-building coverage, ensuring continued wireless connectivity inside tall buildings. The ARDIS network runs at 4800 to 19,200 bps, depending on the geographical area.

10.6.6 RAM Mobile Data (Mobitex)

RAM Mobile Data (Mobitex) is part of a worldwide mobile packet radio data network that provides national coverage for remote access to data and two-way messaging for wireless computing applications. The technology on which Mobitex networks are based is similar to that used in ARDIS and cellular telephone systems. In 1995, there were public Mobitex networks in 14 countries on four continents. Mobitex is a worldwide standard, introduced and controlled by Ericsson of Sweden, and administered by the Mobitex Operators Association (MOA). RAM Mobile Data in the United States and Cantel in Canada are two such Mobitex operators.

RAM operates at 8000 bps—faster than ARDIS's 4800 bps speed in most places but slower than ARDIS's 19,200 bps speed in large metropolitan areas. However, RAM can optimize the network for a specific application by having its individual channels utilized to a lower level. Like ARDIS, it is suitable for short messaging and on-line transaction processing applications. RAM Mobile Data has recently announced its intention to jump headlong into two-way messaging service in direct competition with two-way paging service based on narrowband PCS. (See Fig. 10.14.)

10.6.7 ESMR—Nextel

The SMR networks described earlier employ older technology, and consequently do not use the frequency band efficiently. Enhanced specialized mobile radio (ESMR) networks are essentially SMR networks that operate in the same frequency band, but with enhanced capabilities that address the shortcomings of SMRs. ESMRs differ from SMRs in the following ways:

- ESMRs use more advanced digital technology, providing data and voice.
- ESMRs use cellular techniques, such as channel reuse.
- Wider in geographical coverage, ESMRs are regional and even national.

Enhanced SMRs are only now emerging in the United States as a result of a recently adopted FCC policy to encourage consolidation, and the establishment of regional and even national SMRs that can use the scarce spectrums in the 800 MHz and 900 MHz bands more efficiently. Telecommunications engineers know that more than twice as many users can be supported on a single 10-channel system than on two 5-channel systems.

Under FCC's new policy, FleetCall, a start-up company, aggregated the assets of several fleet dispatchers (taxi companies, construction companies, and others) and has implemented a new digital wireless network called Nextel.

The Nextel ESMR network (see Fig. 10.15) offers the following services:

- Dispatching with two-way radio
- Two-way alphanumeric paging, with acknowledgment and other data applications
- Cellular telephone capabilities
- Session or packet-data transmission without a separate modem

Figure 10.14
RAM Mobile Data.

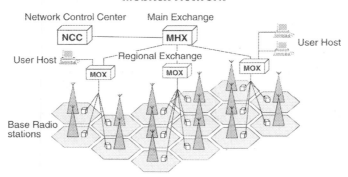

Figure 10.15
Nextel's ESMR net-
work. (Source: Bishop
Training Company.)

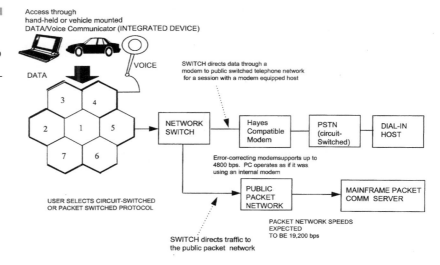

10.6.8 Metricom Ricochet—WAN or MAN?

Metricom Ricochet is a specialized campus area wireless network that extends wireless spread-spectrum technology of a MAN (metropolitan area network) into a limited-distance WAN environment. It utilizes low-cost wireless controllers installed on the tops of buildings that do not require the expense of installing towers. These low-power cells pick up signals from computers equipped with Ricochet modems. The cells hand off the data from one cell to another until it connects with the receiving computer. Ricochet modems operate at 28,800 bps. (See Fig. 10.16.)

Metricom offers unlimited wireless access to the Internet for $29.95 per month in the San Francisco Bay area, Seattle, and Washington at present. However, Metricom recently signed a deal with KeySpan Energy to expand its services to 16 states in the Midwest and Northeast during the next two years.

10.6.9 PCS and Digital Cellular

Personal communication services (PCS) or personal communication network (PCN) is a brand-new technology that promises to completely revolutionize the way we use telephones and wireless communications. Although the fundamental concept behind PCS from a voice communications point of view is the assignment of a personal, unique, cradle-to-

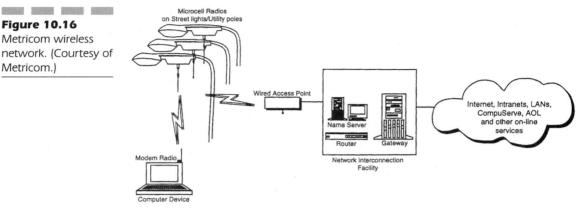

Figure 10.16
Metricom wireless network. (Courtesy of Metricom.)

grave worldwide communications number, the digital nature of the network and high bandwidth make PCS the most attractive from the wireless computing perspective. The first set of PCS pilots are for enhanced voice communication services, though data will be introduced soon thereafter—in our estimate, in two to three years from now. The equivalent network standard in Europe is GSM.

AMPS-based cellular networks of today will be upgraded to digital cellular—CDMA and TDMA during the next five years. Without getting involved in the debate of which system is better, we can say that both technologies will be implemented, though CDMA has a higher (20 times) frequency reuse gain factor than enhanced TDMA (10 to 15 times). However, you should not wait for these new networks to become operational across the nation for several more years. Remember, it is not just the network, but modems, middleware software, and MCSS support that determine whether you can use an emerging network technology.

Europe started implementing corresponding wireless network technology called GSM for voice and SMS (short messaging service) for data two to three years ago. Actually, GSM is an overall digital technology and SMS is a subset of that for data applications.

10.6.10 Satellite Networks

Several satellite networks, such as Iridium, Inmarsat, and MSAT are available for messaging applications over a much wider geographical area. (See Fig. 10.17.) The Iridium network is being designed and implemented by a consortium, with Motorola acting as a major coordinator. It is expected to go into production service in late 1997 (or more likely in mid-1998). It will provide a full suite of messaging and Internet applications with global coverage. If you subscribe to Iridium service in one country, you will be able to send a voice, fax, or e-mail message to someone in another country who subscribes to the same service in his/her country. Directory services will be coordinated by the Iridium consortium. In some respects, this capability is not dissimilar to what is increasingly being offered on Internet. The major difference is that the Iridium network will have an inherent wireless front-end and back-end connection to sender and receiver, respectively. Moreover, Iridium will be a closely coordinated and managed service. However, costs are expected to be higher than on a wireless interface on Internet.

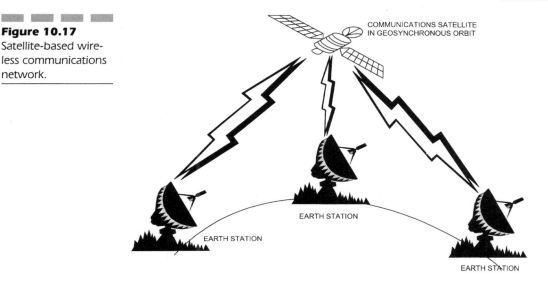

Figure 10.17
Satellite-based wireless communications network.

10.6.10.1 THE RACE TO CONTROL THE WIRELESS SKY. There are several satellite-based worldwide infrastructure networks that are already operational, are being implemented, or are in the planning stages. Over a dozen suppliers are competing for frequency allocations to offer voice, data, paging, and radio-based position-sensing services.

A significant player in satellite networks is Inmarsat, which is already in wide use for voice and data purposes. Inmarsat could use its 11-GEOS (geosynchronous; 22,300-mile or 35,887-kilometer altitude) satellite network to offer services similar to Iridium.

The American Mobile Satellite Corporation (AMSC) has been providing services similar to QUALCOMM's OmniTRACS since 1992. AMSC's Skycell, whose satellite footprint covers the United States, Canada, Cuba, Puerto Rico, and the Virgin Islands, is a fill-in service, providing voice and data communications services in areas where no terrestrial cellular service is available.

Telesat Canada, a satellite services company based in Ottawa, Ontario, started offering mobile data services in Canada from the ANIK E satellite network in 1995.

The following proposed satellite systems should also be mentioned:

A consortium called the TRW Space and Electronic Group expects to have launched by 1999 into an MEO (mid-earth orbit; 1800-mile or 2897-kilometer altitude), a 12-satellite network called Odyssey. TRW believes that an MEO is more economical than an LEO (low-earth orbit) because it requires fewer satellites, and the satellites stay in orbit longer. Odyssey will cover the entire earth with only two satellites, each providing 2300 channels for voice, data, radio-location, and messaging services.

Loral QUALCOMM Satellite Services, Inc., a joint venture of Loral Corporation and QUALCOMM, has proposed Globalstar for 1999, a network of 56 600-pound (272-kilogram) satellites—including eight spares—in eight 750-nautical-mile (1389-kilometer) LEO orbital planes.

Two other proposed LEOs are Constellation Communications Inc.'s Aries system and Ellipsat corporation's 12-satellite Ellipso system.

10.6.11 Comparison of Various Wireless Networks

Table 10.2 gives a summarized comparison of various networks.

10.7 Future Trends in Wireless Networks

Wireless networks will change radically during the next few years. You should ensure that your wireless computing solution can be upgraded to future wireless networks. You should expect the following changes in the wireless network area during the next few years:

- More infrastructure coming onstream, giving you more capacity and bandwidth

- Continuing reduction in network prices (15 to 25 percent annually) as additional capacity comes into the market

- CDPD availability and coverage will get better, with more competitive prices than ARDIS and RAM Mobile Data; consortium of CDPD network providers will provide national coverage

- Increasing use of narrowband PCS for two-way paging computing applications

- Wideband PCS for data will become a reality in three years, but national infrastructure with mainstream software support may take up to five years

10.8 Architectural Issues with Remote Access and Wireless Networks

In order to appreciate architectural issues in a large enterprise that is evaluating wireless network options for their remote access applications, you should review a general-purpose mobile computing technology architecture as proposed in Chap. 5. This architecture is based on a mobile communications server switch (MCSS) from a communications perspective, and utilizes a client-agent-server paradigm from an application partitioning perspective. You should also consider the mobile-aware design concepts described in Chap. 12 for modifying your existing applications or for developing new applications.

Essential attributes of this architecture are briefly listed here:

- Client-agent-server model from application partitioning point of view

- Multiplatform and open design based on standards (de facto or de jure, where they exist)

- Support for multiple applications (messaging is one of them)

- Applications written to standard APIs that recognize mobile-aware design

- Support of wireless networks by these APIs or vice versa

- Mobile-aware middleware, such as Oracle's Mobile Agents, Racotek's KeyWare, or IBM's WebExpress on ARTour

- Multiple wireless network support to allow migration to higher performance, cheaper, and more reliable wireless networks of the future

- Mobile TCP/IP as a transport

While no single product suite matches this description completely, we are seeing a splintering of products that follow these concepts. We are suggesting that serious professionals should evaluate to what extent these concepts apply to their own environment, and how they can influence vendors in modifying products based on the proposed architecture.

TABLE 10.2

Comparison of Different Networks

Criteria	SMR	Paging	Cellular (Circuit Switch)	ARDIS	RAM Mobile Data	CDPD	ESMR	PCS	GSM	Satellite
Frequency band (MHz)	800 and 900 MHz	930—931	800—900	806—821, 851—866	896—901, 935—940	800—900	Mostly 900	901—930, 1850—1990	849—890, 935—960 1.7—1.8	Varies (IRIDIUM: 100061—1000.625)
Private/ Public	Private shared (mostly)	Public	Public	Public shared	Public shared	Public shared	Public shared	Public shared	Public shared	Public shared
Voice/Data	Yes	Very few	Both	Data	Data	Data	Data	Both	Voice now, data later	Both (including video)
1-way/ 2-way messaging	Mostly 1-way	Both	Yes	Yes	Yes	Yes	1-way now 2-way later	Yes	Yes	1-way mostly
Availability (now/ future)	Yes, since 1980	Available	Yes	Yes, since 1990	Yes, since 1992	Limited in 1995	Limited in 1995	Trials in 1995	In Europe, but not in the US.	Some now, more in the future
Coverage	Excellent (95%)	Excellent (95%)	Very Good (90%)	Good, better than RAM	Good	Limited, but, expected to grow	Limited	None for Data in 1996	Limited, but now growing	100% in US and Canada
Building penetration	Fair	Excellent	Not optimized, but improving	Very good	Good	Same as cellular	Fair	Not known	Not known	Limited

Speed		Very low (512 bps)	2400—9600 bps rarely 14.4 Kbps	4800—19,200 bps	8000—16,000 bps[1]	19,200 bps	4800 bps	>19,200 bps	9600 bps for SMS	2400 bps
Data volume	Low	Very Low	High	Medium	Medium	Medium	Low-medium	Not known	Low	Low
Complexity of use		Very simple	Simple	Medium	Medium	Simple-medium	Medium	Not known	Not known	Simple (Trucks)
Hardware cost per user	$800—$3000 per Mobile	Low ($100—) $250	PCMCIA modem only ($350—$500)	Medium ($500—$700)	Medium ($500—$700)	Higher now ($1000)	Not available	Not known	Not known	$4500 (Omni-TRACS)
Fees	$15—$20 dispatch, $45 (inter-connect)	$10—$25 per month	25—50 cents per minute Low for volume	See item 8.6.1	See item 8.6.2	Lower than ARDIS/ RAM ($0.20/KB)	Not available	Not known	$1.20—$1.40 per minute	$0.05 per message plus $0.02/char.
Reliability	High	Good	OK for batch data; poor for interactive	Very high (redundancy)	Very high (redundancy)	No field data Better than cellular	No field data	Better than cellular	No fielda data	Excellent
Unique features	For private network	Cheap for short messages	Good for long file transfer type applications	Good for in-building OLTP	Good for short OLTP applications	Cheaper, low data priority	Good for dispatch applications	Voice 1st data later (U.S. only)	Voice 1st, data later (Europe)	Good for trucking Ubiquitous

NOTE: Chart shows author's estimates for future network services.

10.9 Implementation Issues with Wireless Networks

Remote access networking integration based on wireline and wireless networks is a complex discipline, dealing with many proprietary, noninteroperable, and emerging technologies. Wireline remote access is often the starting point for users—ultimately they would want wireless network—based implementation. Here we would like to briefly mention some of the implementation issues and perhaps offer some advice on managing these issues.

10.9.1 Wireless E-mail as a Starting Wireless Application

Even if wireless e-mail is not the only or major application for the organization, there are distinct advantages in implementing it as the first one:

- It comes bundled as a fully supported application from network service providers.
- It allows the organization to gain early experience in mobile computing.
- Costs are easily identified and controlled with individual user billing.

10.9.2 Selecting the Right E-mail Package

It is recommended that the same package that is used inside the office be used for wireless e-mail. However, if the chosen package does not have a mobile-aware or a network-optimized version of the e-mail, you may consider switching even at the cost of training. Also, evaluate server platforms supported, performance, scalability, and software integration issues for operational OLTP applications.

10.9.3 Who Gets Wireless-Based Privilege to Applications?

Proper business analysis and financial justification is warranted for deciding who gets wireless application privileges. Remember wireless messaging is costly, especially if you start sending files over the wireless network. You need some control mechanism.

10.9.4 Internal versus External Messaging Service

If the organization uses an internal messaging standard that the mobile user community is familiar with, it should be considered as a preferred messaging platform. If it is appropriate, an external messaging service should be considered.

10.9.5 Selecting the Right Network and Service Provider

Careful analysis is required for selecting the right network and service provider. Network coverage, reliability, communications software support, and the vendor's ability to do end-to-end systems integration are factors that must be evaluated. Remember that it is far more difficult to migrate at a later stage, even if the logical analysis of network planners strongly favors such a migration. Messaging is likely only one—albeit, the first—application that professional and sales users may eventually implement. Decisions as to what wireless network to use for messaging should be made in the context of other applications that will also be offered to mobile users in the future.

10.9.6 Selecting the Right Technology and Network Architecture

Careful analysis is required for selecting the right network and service provider. There are a number of half-baked products in the market that address a portion of the mobile computing puzzle. Remember that it is far more important to do it right the first time around.

Enough attention should be paid to the end-user devices, messaging packages, communications software, wireless modems, and communications servers. At a time when the industry is evolving, stay with the mainstream products and vendors.

10.9.7 Tactical versus Strategic Decisions

Wireless technology is just emerging. New network options are appearing in the marketplace every six months. The technology scene is confusing and full of proprietary solutions. Any short-term tactical decision should be taken in light of technology trends in such a way that these solutions can be thrown away in the future.

10.9.8 Security and Network Management

These are two critical issues in wireless networks. Investigate if your corporate standards can be extended to wireless applications rather than the other way around.

10.9.9 Pilot Before Rollout

It is a good idea to pilot wireless-based messaging applications.

10.9.10 Business Process Reengineering as an Integral Part of Mobile Computing and Messaging

Those mobile computing projects that have reengineered business processes have been considered far more successful by executive management than those which replaced a single process with an automated one. Far greater payback is possible if messaging is considered as a starting application with interfaces into source order entry, inventory updating, and other operational applications. Federal Express, UPS, Merrill Lynch, New York Stock Exchange, and many others who have invested hundreds of millions of dollars on their mobile-computing projects are standing examples of this issue.

10.9.11 Sociological Issues

If messaging-based applications become comprehensive and thereby change the business culture of the organization, pay enough attention to factors that affect the personal and social lives of organization members. Telecommuting and mobile computing are blurring the line when work finishes and personal life starts. Control, supervision, personal productivity, and methods for compensation need to be reevaluated.

Summary

In this chapter, we covered the topic of wireless computing from a remote access perspective. First we discussed the business imperatives and justification for wireless networks. We explained how wireless networks are different from wireline network options described in previous chapters. Then we reviewed different wireless networks. We compared features of these networks. We looked at future trends in wireless networks, and speculated on the merging of wireless and wireline remote access solutions. Finally, we briefly described some of the architectural and implementation issues.

Remote Access in Enterprise Network Context

- Switching
- Comparison of Network Options

11

Remote Access and Switching— In Enterprise Network Context*

If you place a component without considering overall architecture of its surroundings, it may appear as a misfit. Remote access is no exception to this principle.

—Chander Dhawan

---

*The material for this chapter was provided by Greg Ma, Principal, Networking Consulting Practice, IBM Canada. The author has made minor editorial changes to match the style and context of the book.

About This Chapter

Many remote access solutions are being designed and implemented in isolation, without understanding the impact of such decisions on the enterprise backbone network or Intranet design. Remote access is only one of the services that organizational users want. This chapter addresses the relationships between remote access network solutions and the enterprise network architecture. We also consider the impact of ATM switching technology on remote access network.

11.1 The Remote Access Network— An Integral Component of the Enterprise Network

As established in previous chapters in this book, the ability for authorized users to remotely access enterprise networking resources is becoming an essential element for corporations to maintain high employee productivity and gain a competitive advantage. The Gartner Group Research Reports have also consistently predicted enterprise remote access as one of the fastest growth areas in the I/T environment. For example, in its "Enterprise Remote Access: Defining the Workplace of the Future" (July 31, 1996) and "Trends in Remote Access" (June 1997) reports, the Gartner Group predicted that "in 1997, more than 2.3 million business users will engage in remote access (0.9 probability). This strong growth usage will continue unabated, reaching more than 55 million business users by year-end 2000 (0.8 probability)."

In order to successfully implement a remote network access solution that meets user requirements and expectations, is cost-efficient, and minimizes security exposures, the enterprise remote access must be incorporated and included as part of the enterprise network strategy, architecture, and design. This chapter outlines the evolving enterprise networking environment and offers some suggestions on how to effectively incorporate the remote access network into the enterprise networking infrastructure.

11.2 The Enterprise Network— A Hierarchical Approach

An effective way to build an enterprise network is to adopt a hierarchical approach by segregating the network into a core backbone network which provides connectivity to a number of access networks or services. Figures 11.1 and 11.2 are high-level examples of such hierarchical network design. Following this hierarchical approach, it is easy to focus on the functional requirements of the respective networks.

11.3 The Evolving Enterprise Network

The enterprise network is evolving at an unprecedented rate to keep pace with new business and user I/T processing requirements, such as network-centric computing, client/server processing, data mining, multimedia and image processing, desktop videoconferencing, and so forth. There is a constant demand for higher network bandwidth.

In the campus and local area network environment, the direction is from shared media (e.g., from 10 Mbps Ethernet or Token Ring technologies) to dedicated or switched media (e.g., using LAN switching or ATM—asynchronous transfer mode—technologies), or high-speed LAN technologies (for example, using Fast or 100 MBPS Ethernet technology). In the wide area backbone network environment, frame-relay is commonly used. Some organizations are beginning to migrate to public and private ATM services.

Figure 11.1

Example of hierarchical enterprise network design.

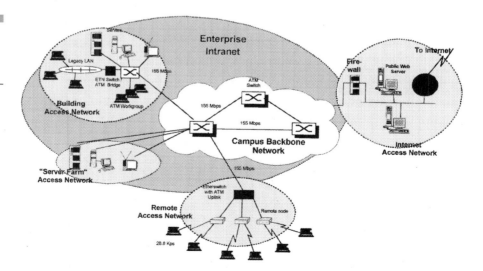

Figure 11.2
High-level example of
enterprise network.

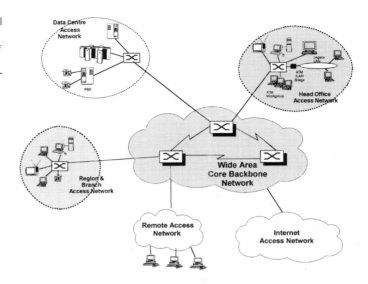

11.3.1 Shared versus Switched Media LANs

Figure 11.3 graphically portrays the differences between shared and switched media LAN architectures. Traditional LAN architectures employ a media access control (MAC) protocol to enable workstations to share use of the local area network media to transmit and receive data. This can be via a carrier sensing, multiple access with collision detection (CSMA/CD) mechanism, as in the case of Ethernet and Fast Ethernet (100Base-T), or via a token passing mechanism, as in the case of Token Ring and FDDI, or using a polling protocol as with 100Base-VG. However

Figure 11.3
Shared and switched
media LANs.

Shared Media LANs

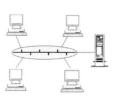

Switched Media LANs

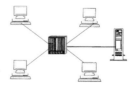

▫ **Examples**
  – Ethernet
  – Token-Ring
  – FDDI
  – 100 VG
  – 100 Base-T

▫ **Examples**
  – Switched Ethernet
  – Switched Token-Ring
  – ATM

as network traffic and demand for bandwidth increase, there will be more contention among workstations attached to the same LAN to compete for use of the network media, resulting in lower workstation throughput. A way to maintain a high level of network performance to the end user is to reduce the overall traffic flow, and hence competition, within the LAN. This can be achieved by reducing the number of workstations attached to a LAN segment. With less workstations competing for network resources, the network response time should improve. In the ultimate case, if a workstation has exclusive use of a LAN segment, then there will be no contention. The workstation will be able to send information as it pleases. The only limitations will be the speed of the network media and the interface adapter. The challenge, of course, is that if you have a micro LAN segment of one workstation, who are you communicating with, and why do you need to be connected to a network.

11.3.2 LAN Switches

The premise of a switched media network design is to provide a dedicated network segment to each workstation and employ a high-speed electronic switch to establish the communications path and direct network traffic from one network segment to another. The workstation on one network segment can then communicate with a workstation on another network segment. There are two types of switching technologies: packet switching and cell switching.

Packet switching is typically the underlying technology used in Ethernet or Token Ring switches. The packet switch actually functions like a multiport transparent bridge. Upon receiving a frame of data, the packet switch will move the data to an output port based on the destination address in the MAC frame header. By attaching a workstation to a port on the LAN switch, the workstation is provided instant access. If this is an Ethernet station, it will be like having its own dedicated collision-free network segment; it does not have to listen before transmission. If this is a Token Ring station, it will not have to wait for a token to transmit. Some LAN switches can support full duplex data transfer, meaning that it can simultaneously receive and transmit data. This can effectively double the usable bandwidth and is ideal for server applications because they can be receiving requests from one user, while simultaneously sending responses to another.

LAN switches are widely implemented in local and campus area networks today. They can provide better network performance by dedicating bandwidth to critical resources, such as server farms and power work-groups. Typically, servers are removed from existing LAN segments and connected to a LAN switch, using dual port or full duplex connection per server, if required. Existing LAN segments (minus the servers) can be attached to the LAN. This will greatly enhance LAN performance because workstations will no longer be contending for bandwidth with the application, file, or print servers. The servers can be servicing workstations on multiple segments simultaneously. This can preserve corporate investments in traditional shared media. Figure 11.4 is a typical campus network implementation with LAN switches.

11.3.3 Cell Switching

Cell switching is developed on the principle that all communications are bursty. (See Fig. 11.5.) If information (voice, image, video, data) can be encoded into very small packets or cells, and if there is a high-performance switch that can handle redirection and multiplexing of cells efficiently and consistently with low delay, it will be possible to carry all communications in one network.

ATM (asynchronous transfer mode) is the new communications technology based on cell switching. (See Fig. 11.6.) Under ATM, all information (voice, image, video, data) is transported through the network in small fixed size cells. Each cell has a 5-byte header and a 48-byte payload. Information entering an ATM network is segmented into a stream of 48-byte

Figure 11.4

Sample campus area network with LAN switches.

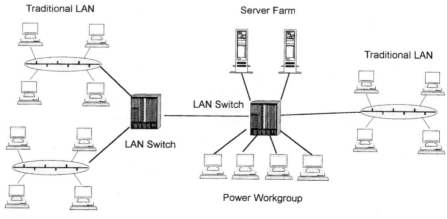

Figure 11.5
All communications
are bursty.

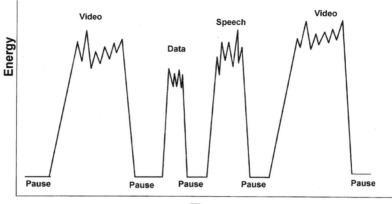

Figure 11.5
All communications
are bursty.

cells. A 5-byte header is added so that the cells can be switched to the correct destination. At the destination end, the cells are reassembled back to the original format.

ATM networks are beginning to be implemented in the campus and wide area backbone network environments. Figures 11.1 and 11.2 are illustrations of ATM implementations. The advantages of ATM are:

- Like LAN switches, ATM can provide dedicated full duplex connections for data transfer.

- Unlike LAN switches, ATM can support all information types (voice, images, video, and data) with guaranteed quality of service.

Figure 11.6
How ATM works.

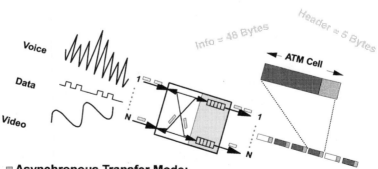

Asynchronous Transfer Mode:
A Transport Protocol, Implemented in Hardware & Software
- Integrated Voice, Video and Data
- Scaleable in Speed
- Consistent Protocol across LAN and WAN
- Standard for High Speed Networking

- The bandwidth is scalable. ATM networks can operate from 25 million to giga bits per second transmission rates.
- ATM supports seamless LAN-WAN integration.
- Existing applications are supported and can coexist with ATM applications via LAN emulation.

With ATM, communicating workstations are connected via virtual circuits, which is made up of two unidirectional virtual channel connections (VCCs). The ATM switch(es) are responsible for setting up the virtual circuit connection across the ATM network. Traditional LANs use a different media access control, framing, and addressing protocol from ATM. They can still participate in an ATM environment by using LAN emulation (LANE), "Classical IP" (RFC 1577), or MPOA (multi-protocol over ATM) standard provisions. In a typical server farm implementation, each server has a dedicated ATM connection to the switch, usually 155 Mbps. Traditional LANs are attached to LAN switches that have ATM uplink connections to an ATM switch. Through LAN emulation, workstations attached to traditional LANs have high-speed access to the critical server resources.

11.3.4 Protocol Switching

The router is an important building block of enterprise networks. Routing provides key capability in local, campus, and wide area network environments, including segmentation, filtering out unnecessary traffic, communication between virtual LANs, MAC-layer translation, and traffic control. However, as the enterprise network evolves to a switching fabric, the latency inherent in the traditional routing process becomes a bottleneck. Router vendors are seeking to improve performance by introducing "Layer 3" or protocol switching. In the traditional routing process, each data packet received is stripped of its MAC header and the type of protocol data unit is determined and checked against the filter; then, based on the protocol address and the routing table entry, new MAC frames are created and transmitted to the destination. In protocol switching, only the first packet has to go through the traditional process. A state table entry is then created containing information about the connection. Subsequent packets of the connection are sent directly through the "pipeline." This method greatly reduces the amount of router processing, and is especially beneficial for client/server, file transfer, and graphically oriented sessions that have long sessions and/or involve larger volumes of data transfer.

11.4 Remote Access Network Infrastructure Considerations

As mentioned earlier in this chapter, the remote access network is becoming an essential component of the enterprise network. The main objectives of the remote access network are to provide facilities to enable mobile users, telecommuters, business partners, and other authorized users to access LAN resources, log on to host applications, or use the consolidated gateway to access the Internet. This section discusses some design and planning considerations for the remote access network, as the enterprise network evolves to a switching fabric.

11.4.1 Remote Access Network

Remote access to the enterprise network can be provided through an access network connected to the backbone network, as illustrated in Fig. 11.7.

Figure 11.7
Remote access
network.

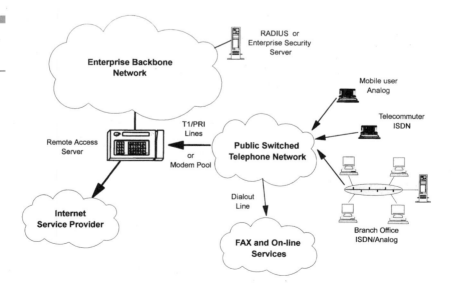

11.4.2 Bandwidth and Performance Constraints

A design objective of the remote access network is to provide end-user functionality as close to local access as possible. A big challenge is the bandwidth disparity between LAN and WAN.

There are two broad classes of remote access servers in the marketplace today: fixed-port and concentrators. Fixed-port access servers are typically 8- or 16-port chassis with individual analog phone lines or ISDN BRI (basic rate interface) connections at the server side. Concentrator access servers use T1 or ISDN PRI (primary rate interface) trunks to connect at the server side. Individual remote calls are multiplexed or concentrated onto the trunk lines by the common carriers. Today's remote access concentrator can handle anywhere from 100 to 600 concurrent connections. At the users' end, they can be accessing the server using analog modems operating from 14.4 Kbps to 56 Kbps, or using ISDN BRI at 64 or 128 Kbps. Within the remote access server, there is a routing function to channel the calls to the LAN. Figure 11.8 illustrates some of the functional blocks within a concentrator remote access server.

11.4.3 Network Optimization Considerations

A good remote access network design involves trade-offs from many considerations: performance, cost, security, ease of use, network management, tariff management, and so forth. Optimization should be planned from an end-to-end perspective. A complete remote access network includes remote computers, application and operating system software, modems,

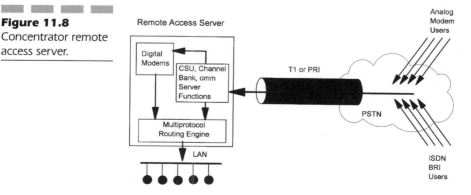

Figure 11.8
Concentrator remote access server.

telephone lines, communication protocols, remote access servers (remote node or remote control), LAN connection, security system, network management system, and so forth. As an integral component of the enterprise network, the remote access network should be planned, architected, and designed based on the same vision, strategy, and architecture principles. Alignment with business goals and objectives, plus understanding and meeting user requirements and expectations, are keys to success.

Optimization discussions are covered in more detail in other chapters of this book. From a network architecture viewpoint, since the remote access server typically provides logical connections to enterprise network resources to multiple remote clients concurrently, the LAN connection between the remote access server and the Enterprise backbone network should be designed to operate as efficiently as possible. As the enterprise network infrastructure evolves to switching (such as LAN switch or ATM) and fast LAN (e.g., 100Base-T) technologies, a first priority should be to upgrade the remote access server LAN connection to a LAN switch (full duplex, if possible) or high-speed port (e.g., ATM). If the remote access server is connected via routers to the enterprise backbone network, then protocol switching should be used in the routers.

Summary

In this chapter, we have discussed remote access in the context of overall network design for the entire enterprise. We also introduced ATM switching concepts and explained how remote access solution can be used along with ATM switching in the back-end network infrastructure.

12

Remote Access— Comparing Network Options

There will come a day when the only thing we need to worry about will be how we access the network infrastructure. The network utility will simply ask us how thick a pipe we want. Until that day arrives, we, as network professionals must worry about which type of pipe to use, how much water flow we need, and what control mechanism to employ.

—Chander Dhawan

About This Chapter

In Chaps. 6 to 10 of this book, we have surveyed the following five network options for remote access:

1. Conventional analog switched network is by far the most pervasive option for remote access. This dial-in network is also called public switched telephone network (PSTN) in more formal circles, and POTS in less formal ones.

2. Following on the heels of PSTN is the somewhat faster, and truly digital network called ISDN. It is gradually assuming the position of the second most popular network, especially in metropolitan areas where it is available.

3. For those who want megabit speed, there are two choices, if you are lucky enough to reside in an area where these options are available: ADSL and cable modems.

4. Then we have the wireless networks that work in their own proprietary, and at times peculiar, fashion. These networks are much slower, but provide the ubiquitous access when none of the previous options are available.

5. Finally, the Internet offers a method of using a universal backbone for connecting a remote user to the corporate information server. While the Internet is not a network option on the same basis as the first four, it does, nonetheless, consolidate remote users into a single pipe at the front end. In fact, you can connect to the Internet by any one of the first four methods.

While discussing these networks, we have described the special attributes of these networks in the context of remote access. While we have compared some of these networks in a micro sense in individual chapters, we want to compare them in a global sense, now that we understand their capabilities and unique applications.

12.1 Factors Affecting Network Selection

Telecommunications professionals studying remote access solutions face a number of network options. We have listed these alternative technologies in a previous paragraph. Evaluating these network technologies is not a straightforward exercise. There are a number of factors that you must consider in order to make this choice. Even after a careful and structured analysis, there may not be a single right answer that is optimal for all your application needs. You may end up choosing multiple networks. A group of engineers in an aircraft manufacturing organization may choose an ADSL or cable modem solution to connect to another engineering group in a different location because they transfer data-intensive drawings and documents. Telecommuting application programmers in the information systems department may use ISDN from their homes, whereas those in sales may use dial-in access to the Internet to exchange e-mail with external users. Finally, customer service engineers may use wireless networks to access critical parts information from the corporate parts inventory database server. In fact, it is unlikely that the same shoe will fit everybody. Therefore, an organization may have different groups of users employ different networks for their unique needs. The following list represents a fairly complete set of factors that network professionals must consider for this purpose:

- Application suitability
- Batch or interactive nature of the application
- Bandwidth and speed of access requirements
- Response time requirements
- Availability and coverage issues
- Ubiquitous (anywhere, anytime) access
- Telecommunications network usage costs
- Compatibility with organization's network architecture
- Protocol and communication software integration issues
- Security and network management issues
- Reliability, stability, and maturity of network offering
- Future outlook for the network

Let us review each of these factors, and understand how various networks fare against these criteria.

12.1.1 Application Suitability

We believe that application requirements should guide the network choice more than any other factor. Therefore, this factor should be given the highest priority in your evaluation. While a file-transfer-intensive application that is not sensitive to real-time response requirements is best served by the Internet, a mission-critical on-line transaction processing application may not be suitable on the Internet. Although the Intranet model is being increasingly employed by large organizations for publishing information to outside users, it is because of a common communications architectural model rather than the kind of connection used. Here, we are talking about the connection method. Therefore, you should map out the application's data flow characteristics and determine how these requirements can be met by different networks.

12.1.2 Batch or Interactive Nature of the Application

The way information is accessed and how often it is accessed during the day will determine if remote access technology is suitable. Remote access networks are not suitable if there is a need to be permanently connected and the application is highly interactive. In such cases, ports in the remote access server are not available to other users.

12.1.3 Bandwidth and Speed of Access Requirements

This is the most talked-about measuring stick for comparing networks—the faster, the better. One of the key reasons for this is that applications are becoming more bandwidth-hungry, and users want to use multimedia capabilities with multiple types of information—graphics, color, sound, and video. However, we should not allow users to run amok with their demands for transferring huge amounts of data unless they can justify it on a business-need basis. Although network providers are talking about ADSL and cable modems that can deliver megabit speeds to access the Internet, keep in mind that there is a cost implication at the server end, too. A group of 1000 users accessing multimedia applications from a legacy warehouse at ADSL speed could impose significant work for your front-end communications processors. You should always study the impact of your user requirements on all the components, including remote access server and the database server.

On the other hand, wireless networks just cannot deliver high bandwidth at a reasonable cost. Therefore, you should seriously question your user's need for multimedia applications delivered through wireless networks. If it becomes a necessity that is cost-justified, you should employ special compression techniques for such applications.

12.1.4 Response Time Requirements

Response time requirements of your application, and the amount of data that you must transport will dictate the minimum speed of access that your network must deliver. As we have seen in previous chapters, speed varies considerably with PSTN, ISDN, ADSL, and wireless networks. Besides data transfer time, you should also keep in mind call setup time, especially for PSTN. Here again, it is not just the link speed from the remote user to your organization's remote access server or ISP that determines end-to-end response time but also the number of hops and speed of the backbone network. (You should review Chap. 13 for an understanding of end-to-end response time in remote access networks.)

12.1.5 Availability and Coverage Issues

PSTN is available anywhere there is a telephone jack. ISDN is available in most parts of the United States, Canada, and Europe in large urban centers. ADSL and cable modem technology is just emerging, and it will take many years (two to four) before it is available across the nation. Coverage is a big issue with wireless networks. You should obtain coverage maps of wireless networks from your network supplier.

12.1.6 Ubiquitous (Anywhere, Anytime) Access

If your users are not always close to a telephone jack, and you want ubiquitous access to information, you have no choice other than wireless networks. This is especially true for mobile users like the truckers, law enforcement officers, and service personnel.

12.1.7 Telecommunications Network Usage Costs

Perhaps, the second most important factor is the operational cost of using a particular network. While there exists tremendous competition in the telecommunications marketplace, the cost of using a network is dependent on a number of factors. One of the biggest unknowns is how much information the users will transfer on a daily basis. In fact, a common surprise that a network manager faces after a couple of months is the telephone bill, especially if the usage is based on per minute charges. Most estimates look at the obvious only, and incorrectly guess the traffic pattern.

12.1.8 Compatibility with Organization's Network Architecture

Networks should be transparent to applications; however, they are not always so. Remote access networks should integrate seamlessly with the overall network architecture of the organization. Quite often, they do not do so. The remote access component of the network stands alone in isolation in many organizations because RAS vendors provide richness of features that many internetworking vendors do not. It is, therefore, important to validate remote access architecture against corporate network architecture.

12.1.9 Protocol and Communication Software Compatibility Issues

TCP/IP is the most prevalent standard for remote access, but there may be other communications software considerations that you must evaluate. As an example, if you want to access IBM mainframe legacy applications, you may have to implement an SNA gateway on your LAN.

12.1.10 Security and Network Management Issues

Security is an important issue in remote access. Security support for different remote access network options is not always consistent. Your favorite security software solution may not be supported on the selected network, especially if the network technology is relatively new. You should also not assume that the remote access server that supports ISDN will support ADSL.

12.1.11 Reliability, Stability, and Maturity of Network Offering

Technology matures with age. If you use an emerging network technology, you should be prepared for paying a price for more extensive problem resolution. During the early days of the introduction of a network technology, there are very few standards. You may have to use proprietary hardware. Modems or corresponding termination units may not be available from too many third-party suppliers.

12.1.12 Future Outlook for the Network

Finally, you should find out how a particular network will evolve in terms of faster speed and support by major remote access server hardware suppliers, such as Cisco, 3COM/US Robotics, and Ascend.

12.2 Evaluation of Network Options

Figure 12.1 outlines a methodology that you could follow in evaluating various network options for your applications.

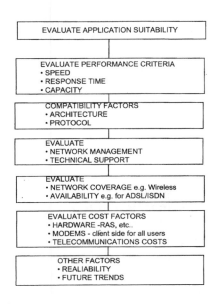

Figure 12.1
Methodology for comparing remote access network options.

12.3 Comparison of Network Options

First, we shall compare the five network options on the basis of the criteria previously listed. For this comparison, we use a qualitative measure indicating how well these networks stack up against each other. Where appropriate (e.g., while comparing speed and cost), absolute measures will be provided. Then we look at characteristics for common remote access applications and suggest networks that are most suitable for those applications.

Table 12.1 shows different characteristics for different networks. Table 12.2 matches networks with applications.

12.4 Changing Scenario of Networks for Remote Access

Remote network access started with dial-up connections using public switched telephone (PSTN) networks. However, 28,800 bps speed on PSTN did not match the 10 Mbps LAN speed, even though the former link was dedicated to a single user. Waiting in the wings was ISDN, a technology invented more than a decade ago. This provided a faster and more affordable link when it involved faster data transfer. ISDN is still gaining ground, and more intensive users—such as telecommuters—are switching from PSTN to ISDN. Market forecasters are predicting ISDN usage to grow to 2 million lines by year 1999. There is even movement toward higher ISDN speeds: PRI, T1, FT1, and so on.

Already, we have seen the emergence of another interesting telecommunications link technology—called DSL, HDSL, or ADSL, depending on its implementation. (We described ADSL in Chap. 8.) When it is available in most areas at reasonable cost, ADSL will allow ordinary twisted copper wire medium strung throughout our homes and businesses to be used for 1 to 6 Mbps speed. With this speed, many more bandwidth-intensive applications, such as videoconferencing and multimedia will become available to the remote mobile workers. However, this will take time.

Another technology competing with ADSL is cable—facing similar timetable and implementation challenges. We believe that cable has similar potential as ADSL, in terms of speed for users. While both ADSL and cable have the technical potential of providing increased bandwidth,

TABLE 12.1

Comparison of Different Network Options

	PSTN (Analog Modem)	ISDN (Digital)	ADSL (Digital)	Cable (Digital)	Wireless (Analog)
Application suitability	E-mail, file transfer, Internet with text browser	Telecommuting, file transfer, limited videoconferencing	Large file transfer, graphics-intensive browsing on Internet, videoconferencing	Internet browsing, telecommuting for application developers	Low volume data transfer, mission-critical, dispatch
Speed	14.4 to 56 Kbps	64 to 128 Kbps	1.5 to 6.0 Mbps	1.2 to 10 Mbps	4.8 to 28.8 Kbps
Traffic flow	Symmetric	Symmetric	Asymmetric	Both"	Symmetric
Availability and coverage	Available for a long time where there is a telephone jack	Available in most urban areas; not as universal as PSTN	Limited availability now; mostly pilots in 1997	Limited availability now; mostly pilots in 1997; 25% areas have potential	Available since 1992; spotty coverage outside urban areas
Widespread availability	Available now	1997–1999	Two years (1998 and 1999)	2 to 4 years	More capacity in 3 years with broadband PCS
Performance	Initial setup time is high; acceptable for text or database access	Good for graphics-based browsing	Ideal for video and multimedia	Much faster than PSTN and ISDN	Much slower; not suitable for Internet browsing
Compatibility with RAS products	Strongest support	Most products support ISDN	Limited support	Transparent	Limited; specialized MCSS required
Security	Uses public network; need encryption and authentication	Shunted over public network; need encryption and authentication	Over point-to-point connection; less risk	Cable shared among subscribers; use encryption	Security weak over public airwaves; use encryption
Typical cost/ user modem/ usage*	$100 to $250+, $15 to $25/month	$300 to $500; $40 to $60/month	$1500+; $30 to $50/month	$700 to $1000; $20 to $40/month	$29.95 /month at Metricom (loss leader)
Issues	56 Kbps limit Connection set up time is high Slow for multimedia	Pricing is high relative to ADSL and cable	Asymmetric nature; prices could rise—ask for long-term price guarantees	As number of users sharing cable increase, performance would decrease Support may not match business requirements	Slow, expensive, limited coverage Only for critical applications

SOURCE: Adopted by the author from a special Report on "Internet Access Options," published in Telecommunications magazine in the Feb. 1997 issue.

* Usage costs are for initial introduction of consumer services for cable and ADSL. Business prices are higher. These prices could rise in the future.

" See Chap. 8 for additional comments on the symmetric/asymmetric nature of cable.

TABLE 12.2

Which Network
Matches Applica-
tion Requirements

	Application Characteristics	Best Choice of Network
E-mail	Short 150 to 500 character messages, received in disconnected mode and picked up at user's convenience	All network options are (acceptable file attachments not realistic with wireless networks)
File transfer	Straight transfer of various types of files	Small files—PSTN acceptable Large files—ISDN, ADSL, and cable client-server—PSTN acceptable; ISDN better
Database access or on-line transaction processing	In client-server or mainframe terminal mode, short messages may be transferred In LAN-based database applications, a lot of data may be exchanged—use remote control mode	
Multimedia (sales force automation and others)	A lot of data transmitted, especially for video	ISDN, ADSL, and cable (where available)
Telecommuting modem tomorrow,	Working at home, need same environment as office LAN	ISDN today; ADSL or cable where available
Mobile worker (sales force automation)	Several connections during the day Need to pick up mail and other information in one session Always on the move	PSTN and wireless network combination (with client-agent-server implementation à la Xcellenet's RemoteWare product)
Field force dispatch applications	Short messages requiring immediate response (within a few seconds)	Wireless network
Videoconferencing	A lot of data per frame (over 50 to 100K) depending on resolution and compression techniques	ISDN today; ADSL or cable modem tomorrow
Distance learning (based on videoconferencing)	A lot of data per frame (over 50 to 100K) depending on resolution and compression techniques	ISDN today; ADSL or cable modem tomorrow
Telemedicine and remote diagnosis	High resolution X rays generate a lot of data	ISDN (PRI) today and ADSL tomorrow
Ubiquitous, anytime, anywhere	Only essential data transmitted	Wireless

SOURCE: Author.

their success at getting large enough market share will depend more on which industry will be able to put up a multibillion dollar investment in the infrastructure, market its services, and support the end users to their satisfaction. Telephone companies certainly are more cash-rich and have deeper pockets. In competition are market visionaries, like Bill Gates, who have invested huge sums in the cable industry to make it successful.

Irrespective of how the market forces play themselves out, we believe that both ADSL and cable will succeed. There is enough room for both of them. The users deserve the choice. Meanwhile, ask hard questions from ADSL and cable vendors. Be wary about coverage and inherent risks with any new technology. Remember that if it sounds too good to be true, it probably is.

Figure 12.2 illustrates this progression in the dominant network option for remote access over the next five years.

12.5 Impact of Internet over Network Selection

The Internet is not *a* network option; it is *the* network. It will increasingly replace private remote access solutions. As we explained in Chap. 9, more

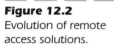

Figure 12.2
Evolution of remote access solutions.

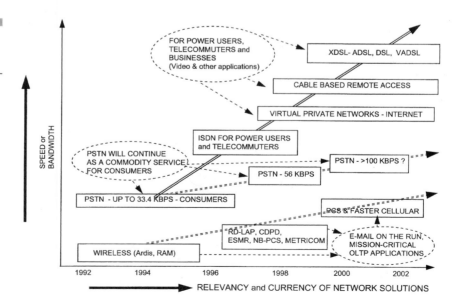

and more virtual private networks are being implemented by using the Internet as an intermediate connectivity vehicle between remote users and corporate back-end information servers. (The reader should refer to Sec. 9.9 for this.)

12.6 Need for Hybrid Solutions— Multiple Networks Through One Server

We are always striving to find simple answers to complex questions. We also hope to find a single optimal technology solution that will meet all our business needs. This is as true in remote access as in any other technology area.

We have several network options—PSTN, ISDN, ADSL, cable, and wireless. Each of these options has unique advantages and disadvantages. They are also in different stages of technological maturity. On the other hand, our application needs are different for different users. Different user groups may find that more than one network may satisfy their functional needs at the least cost. Therefore, we need RACS/RAS models that support not one or two, but all the network options.

Large organizations who have diverse sets of users may implement hybrid solutions utilizing a mix of PSTN, ISDN, wireless, ADSL, and cable network options. The beauty of the Internet is that your organization need not worry about how your end users decide to connect to the Internet. They just need a strong enough firewall. Of course, if it is an Intranet solution, you do need to worry about support for new networks in your routers, and other intranetwork components.

12.7 Emergence of Integrated Network Providers

In Chap. 9, we mentioned the emergence of network providers who can take the headache of selecting the network option away from you. These companies have been formed as a result of mergers between large ISPs and Telcos. They provide an integrated set of network services, including network management. You could do an informed analysis of outsourcing your remote access needs to such a network service provider.

Summary

In this chapter, we compared various network options. If your needs are well defined, you may be able to select a network option that meets your functional needs at a reasonable cost. If, on the other hand, you have diverse needs, you may need a hybrid solution, consisting of multiple network services. As for exploiting an emerging network technology, such as ADSL or cable, measure the risk against your benefits. You could, of course, outsource your entire remote access requirements to a VPDN provider. No single solution fits all organizations—everybody has different needs.

PART 5

Remote Access Systems Integration

- Systems Design
- Application Design
- Security Design
- Bandwidth Management
- Network Management Issues

CHAPTER 13

Remote Access— Systems and Application Design Issues

Hardware and software infrastructure supplied by vendors will not perform well unless you apply good systems engineering and application design principles.

—*Chander Dhawan*

About This Chapter

In Part 3 of this book, we discussed various components of an overall remote access solution. Now we need to look at the systems design and application design of an end-to-end solution. This is where systems engineering and application design skills are called into play. The integration of components from different vendors imposes additional constraints that systems designers must also take into account. In this chapter, we study systems design and application design issues.

13.1 Systems Design Issues

We discuss the following topics under this heading:

- Remote access network design issues
- Capacity planning and response time calculations
- Improving effective capacity of a communications link through data compression techniques
- System availability design

13.1.1 Remote Access Network Design

13.1.1.1 THE DESIGN CHALLENGE   The design of a remote access network solution can be as simple as invoking remote access client software from Windows 95, and then dialing from your PC into the corporate RAS or your ISP. Alternatively, you may need to install an Internet browser, such as Netscape or Symantec's pcAnywhere software, in your remote PC. However, what we want to discuss in this chapter is the design of a remote access solution for a medium- to large-size organization, with hundreds to thousands of remote mobile users. This problem is somewhat different from the design of a conventional private network where you know your user population and their work patterns. Remote access network design presents a different set of challenges. Some of these challenges are:

- Lack of understanding about the number and type of remote users— in-house mobile professionals, telecommuters, field workers with handheld devices, small branch offices, business partners, customers, and suppliers

■ Difficulty in predicting traffic pattern of these nomadic users

■ Deciding which network to use, for example, PSTN, ISDN, xDSL, and wireless

■ If wireless networks form a part of the solution, there is a lack of standards for wireless networks

■ Use of public shared networks, such as PSTN, cable, and packet-switched wireless networks, including ARDIS or RAM Mobile Data, where you do not know the peaks of other users

■ Lack of knowledge about the backbone network, in terms of number of hops, capacity of the links

■ Use of virtual dial-up private networks (VDPNs) along with the Internet

■ Lack of integration between RACS/RAS and wireless computing MCSS

■ Managing response time expectations between a 10 Mbps Ethernet LAN and 33,600 bps PSTN

First of all, predicting the average number of remote users during the peak hour is not easy. However, we must start with some kind of estimate, and then refine it further as we gain a better understanding of their work profile. Even when it is possible to logically describe the tasks that mobile users perform, how does one translate application-level workloads into network traffic?

This need to start counting network traffic was not a problem before. Local area network professionals have had it easy: for the past ten years, they have had the luxury of working with ever faster LANs—first the 10 Mbps Ethernet; then the 16 Mbps Token Ring; and now the 100 Mbps Ethernet. Consequently, they have never had to worry about translating application traffic into network traffic.

In Chap. 12, we compared various remote access options. Your objective should be to match your application requirements with a network type that gives you the best performance at least cost. In this effort, you should keep the number of networks you have to support to a minimum.

With a public shared network, such as cable, you must ask the cable company how many users the network is designed for, and how many users they currently have signed up in a particular area. Remember that the cable network is like an Ethernet LAN, and if you have 200 users doing heavy-duty browsing, your effective bandwidth will be pretty low. Ask yourself and the cable provider—how many simultaneous users would they support on a 1.2 Mbps Ethernet LAN segment? Not only is it necessary to obtain information about one's own workload, but it is also necessary to worry about demand peaks created by other users.

13.1.1.2 CALCULATING THE NUMBER OF PORTS IN A RAS. There are formal mathematical tools based on the queuing theory that allow you to calculate the number of ports that you should configure in your remote access server for PSTN or ISDN. You can input the number of times remote users call in, and the average time of each connection in these models. Then they can predict that for no more than a specified percentage of times (typically 95 to 99 percent), you will not get a busy signal when you dial in. This will ensure that you will not get a busy port. You will also find suitable mathematical tables in telecommunications books such as *Systems Analysis for Data Transmission* by James Martin (Prentice Hall, 1972).

Instead of spending a lot of time in collecting information, which is based on guesses by planners, and then using very formal tools, you may simply use rules of thumb used by vendors in configuring remote access server ports. You should, of course, verify whether the metrics used by the vendor are close to your own situation. If they are not, adjust the numbers of ports up or down. Even if you use sophisticated mathematical models, you should validate the results by experience-based rules of thumb. As an example, the average length of a remote access connection by Fortune 500 users was 28 minutes, according to an IDC survey conducted during 1996. A ratio of 10 to 20 between the number of ports in the RAS and the number of remote users is another common rule of thumb in the industry. Please note that these ratios are for remote workers who call in for short periods of time at different times of day. On the other hand, telecommuters are an entirely different group of users who occupy the port almost continuously during the day.

13.1.1.3 WIRELESS NETWORK DESIGN ISSUES.   Implementing a remote access solution based on wireless networks is far more complex than wireline solutions. Network planners must deal with the following questions for wireless network—based remote access:

▪ Which wireless network out of ARDIS, RAM Mobile Data, circuit-switched cellular, CDPD, ESMR, narrowband PCS, and Metricom is most appropriate for the suite of applications that you want to implement?

- How does an application profile match a given network's capability?
- Should RNA technology be integrated with a wireless network?
- Does the network have good coverage and a minimum number of dead spots?
- Which mobile communications server switch should be selected?
- Should agent-based application development tools, like ARTour Web-Express or Oracle's Mobile Agents, be used?
- Should a private radio network be built (especially for public safety or utilities), or can a public shared wireless network be used? More and more, network planners find public shared networks give them more flexibility in the ever changing wireless world.

13.1.1.4 THE INTEGRATION OF RNA TECHNOLOGY WITH RADIO NETWORK INFRASTRUCTURE. As we stated in Chap. 10, the wireless industry and the RAS industry are moving in different directions. Ideally, from a network manager's perspective, a remote access server should support both wireless and wireline dial-in connections. Unfortunately, in practice, however, problems are still being experienced in integrating the two approaches. Wireless speeds are relatively low, and many LAN applications are not mobile-aware. Also, remote access servers from major vendors, such as Cisco, Shiva, and Ascend, do not support wireless WANs, such as ARDIS, RAM, or ESMR. CDPD is an exception by reason of its native TCP/IP support. The wireless MCSS, defined in Chap. 10, is functionally richer than anything the RAS industry offers.

13.1.1.5 COVERAGE AND AVAILABILITY ISSUES IN REMOTE ACCESS. Apart from PSTN, that has by far the maximum coverage in all parts of the developed world, emerging networks, including ISDN, but more particularly ADSL and cable have limited coverage right now. The situation is worse with wireless networks. You must do coverage analysis for all the networks you want to study.

Hybrid networks should also be considered for wireline and wireless networks. PSTN and ISDN should be supported in the same server. Similarly, in the case of wireless networks, QualComm has teamed with ARDIS to offer a hybrid solution for truckers who find that QualComm's satellite-based service provides excellent coverage on highways, but poor coverage in downtown cores.

13.1.2 Capacity Planning and Response Time Predictions on Remote Access Networks

A mobile computing application transaction traverses many hardware and software components before it reaches the destination server—and has to cover the same path again in reverse to complete the trip. There are many physical links (hops), wireless and wireline, between the end user's client application software and the information server. There are also several pieces of software involved, some of which feature queuing (i.e., they process these transactions in asynchronous fashion).

Thus, there are complex rules for scheduling priorities on a network. This makes it extremely difficult to build a mathematical model to estimate response times or transit time—and therefore, to plan reliable network capacities in advance. Network providers do give some estimates, using either complex queuing models or rule-of-thumb calculations, based on the experience of other customers.

The greatest surprise often comes in discovering the precise contribution of software to the total end-to-end response time. The average TCP/IP specialist may very well be unable to estimate how many packets will be shipped over a remote access network configuration, especially if it goes through Internet or if it employs a wireless network in its path with an application using a client/server design and middleware. While middleware may simplify application logic, there is no way of knowing just what is going on at any given moment. Many delays occur because the software configuration, the number of buffers, the number of physical ports, the number of parallel sessions, and software control blocks are not optimally tuned in the RAS and the application server.

It is important, therefore, to go back to the basics. We need to turn to fundamental engineering knowledge and mathematical modeling skills to come up with an initial estimate. This initial estimate can later be refined as understanding grows of the application's behavior on the wireless network with the communications software of choice.

We cannot overemphasize the fact that the greatest single cause of inaccuracies in response time calculations is the inability of systems designers to translate application traffic into network traffic. This inability too often stems from attempts to use crude mathematical models to represent very complex networks comprising multiple servers and queues. A second major contributor to the inaccuracy of estimates is the failure of many designers to allow for the holding delays in software queues at various points. Far greater attention needs to be paid to estimating this traffic, and to configuring software parameters.

One way of obtaining information about application-to-network traffic is to use a network probe inserted at an appropriate point while running the most typical transactions, and recording the network data flow. There are various software tools available to assist the communications designer in this task.

Once network data flow information is available, the service provider's network designer can use a mathematical formula similar to the one shown as follows to calculate transit times through the wireless portion of the network:

Transit time through remote access network (TT-RAN)
= queuing delay to access a port (PSTN) or access a channel (on wireless network) + transfer time between the end-user device and RAS, including errors and retries + transfer time between first ISP's router and Internet backbone + queuing plus transfer time between the Internet and second ISP + queuing plus transit time in the WAN link between the second ISP and destination RAS + queuing plus processing time within RAS + queuing plus processing time in LAN application server (or gateway to the mainframe) + queuing plus processing in getting information out of the mainframe information server

Also, note the following simple single-server queuing model—with a Poisson arrival time pattern and an exponential service time—as a guideline for rule-of-thumb type calculations:

Transit time in a server = waiting time + service time
= [(average service time in the server) ÷ (1 − server-utilization factor expressed in units of 1)]

For example, if server utilization factor $\mu = 0.5$ (i.e., 50% utilization), transit time including waiting time = $[S_t \div (1 - 0.5)] = 2S_t$ = twice the mean service time.

As can be appreciated, the number of hops in the path of the network could increase the transit time considerably. Remember that every new hop in the network path adds propagation time, plus link delay, however small it may be. (On any communications link, there are three response time components: queuing time, propagation delay, and transfer time.) (See Fig. 13.1.)

Note that end-to-end application transit times are much higher than network transfer times for packets of data. Application data might be segmented into multiple packets that have to be reassembled at the other end before the data is presented to the end user's PC monitor.

Figure 13.1
Components of total
response time in
a remote access
network.

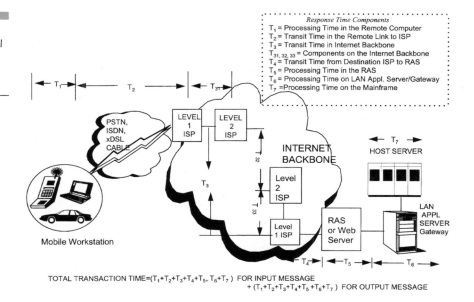

Response Time Components
T_1 = Processing Time in the Remote Computer
T_2 = Transit Time in the Remote Link to ISP
T_3 = Transit Time in Internet Backbone
$T_{31, 32, 33}$ = Components on the Internet Backbone
T_4 = Transit Time from Destination ISP to RAS
T_5 = Processing Time in the RAS
T_6 = Processing Time on LAN Appl. Server/Gateway
T_7 = Processing Time on the Mainframe

TOTAL TRANSACTION TIME=$(T_1+T_2+T_3+T_4+T_5+T_6+T_7)$ FOR INPUT MESSAGE
                      + $(T_1+T_2+T_3+T_4+T_5+T_6+T_7)$ FOR OUTPUT MESSAGE

Thus,

Application response time
= processing time in the mobile workstation + transit time through
the first remote link + queuing/processing time at the RAS/RACS
+ queuing/transfer time on the host link between the RAS and
the back-end DBMS server + queuing/processing time on the LAN
information server or mainframe + time for the response message to
traverse all the network components (described earlier) in the reverse
direction

For information on calculating the optimal number of ports in a RAS,
refer to James Martin's *Systems Analysis for Data Transmission,* published by
Prentice Hall.

13.1.3 Data Compression Considerations

Remote access network bandwidth is much less than user demands. Every
possible technique should be used to get the utmost service out of this
bandwidth. Compression of data is one such technique.

The most common place to compress data is in the modem. While a lot
of advanced compression work has been done on PSTN, ISDN, and cellu-

lar modems, the same efficiencies have not been achieved so far in proprietary packet radio modems, like Mobitex or ARDIS.

It is a good idea to go beyond modem hardware in reducing the amount of traffic on wireless networks. Intelligence can be built into client application programs so that short message codes can be used to indicate common data occurrences (e.g., an item description need not come from the database resident on the host computer, but instead can be generated on the mobile computer from an item number). Electronic forms and input screens for user interface should always be stored on the mobile computer. The network PC model where most of the application data comes from the network server may not work very well on slow-speed remote access networks, especially of the wireless variety.

13.1.4 Fault-Tolerant Design for Higher Availability

Remote access applications range from e-mail inquiries to mission-critical ambulance or fire truck dispatch for wireless networks. Whatever the application (but particularly in the case of dispatch applications for field force automation applications), the network is being relied upon to provide a service, and any lengthy downtime is unacceptable. Since a large number of users start relying on a remote network on a regular basis, a high degree of fault-tolerance is required.

Typically, most public shared networks have redundancies built into them, and network service providers guarantee an extremely high network uptime. Generally speaking, telephone switching centers, ISPs' router configurations, and other components have built-in hardware redundancy. Many large message switches typically employ fault-tolerant computer platforms, such as Tandem and Stratus. However, nothing should be taken for granted, and public shared network providers should be asked for details of their redundancies. Similar redundancies should also be built into your private remote access network configurations.

Another vital component that must have redundancy built into it is the RAS. Find out if the method of switchover from failing components to standby components is automatic or manual. If it is manual, you must ascertain whether the remote station is staffed, and how long a switchover takes.

A useful safeguard is to have two or more smaller RACS/RAS at different locations, rather than one large one. That way, if an RAS does fail, only a part of the network will go down. If the RAS has a reliability of 99 percent each, then the probability of two remote access servers failing simultaneously, and losing the entire network = $1 - (0.99 \times 0.99)$ percent, or 0.0199 percent. As well, excess capacity can be built into each of the RAS, so that if worst comes to worst, at least degraded service can be maintained to users—which is better than no service at all.

In this book, we advocate end-to-end integration and design. This means looking at all components—from the end-user device to the information server—and researching the likely effect of the failure of any of these components. Remember, the system is only as reliable as its weakest component, unless that component is duplexed.

13.2 Application Design Considerations— Mobile-Aware Design

The whole discipline of application development is subject to heated debate in our industry; a tremendous amount of literature exists on the subject. Many well-known experts in the field have spoken to countless packed seminars. What we would like to do here is to focus solely on those strategic issues that are relevant to remote access applications. We highlight the issues and discuss the pros and cons of different approaches. No specific solutions are offered. Readers must determine for themselves the strategies that best apply in their particular organizational contexts. It is the only way.

13.2.1 What Is Different About Mobile Applications and Users?

In order to appreciate the mobile user's requirements accurately, we should try to understand the differences between the ways in which mobile workers actually perform their work, and the ways in which they would like to perform their work, if technology were not a hindrance. Only with such an understanding is it possible to develop good solutions that make the most of existing technologies—even if technological restrictions do bar the way to a perfect, low-cost solution. The following requirements stand out:

- *Urgency.* Mobile workers are always in more of a hurry than office workers. They want to be able to work faster and more efficiently than inside the office.

- *Lower tolerance for error.* Mobile workers cannot afford to make mistakes in the mobile environment. A slip-up in a presentation, an erroneous piece of information, or an incorrect answer can all cost a sale.

- *Self-reliance.* Professionals on the road often have no access to administrative staff; nor in many cases can they take notes for later action.

- *Time is of the essence.* The value of a mobile worker's time outside the office (or the factory) is at a premium.

- *Low tolerance for technology faults.* Since mobile workers do not have access to external help, and technical support is far away, the technology configuration (software and hardware both) must work more reliably than inside the office.

- *Efficiency of use.* The application dialogue must be highly efficient, and lead to the completion of the task at hand in a minimal number of steps.

- *Essential only.* Outside the office, mobile workers do not have the time to tend to nonurgent tasks.

- *Criticality.* Many mobile computing applications (fire control, ambulance dispatch, law enforcement, etc.) are truly mission-critical, far more so than accessing a bank account for a withdrawal, or a travel agent for a ticket reservation.

- *Need for alternative inputting devices.* While the keyboard is generally accepted as a major way of interacting with computers, it is not as acceptable outside the comforts and convenience of the home or office. Pen and voice, or a combination of both, are often more intuitively useful.

- *Sporadic work bytes.* Due to the stop-and-go, multitasking nature of mobile workers' lifestyles, it is inefficient (and unnecessary) to make them wait for the completion of communications tasks or inquiries. It is better to be able to perform many of these tasks in the background while users are doing other things that do not require computer input or attention.

- *The personal touch.* Mobile workers are the ones who know their clients best. Thus, correct information about clients or subjects under review should be procured at the source, without intermediate processes or persons being involved, unless value is added or operational efficiency is thereby achieved.

■ *Optimal technological performance.* Hurried, harried mobile users deserve as good a technological performance as can be economically justified.

13.2.2 Making Applications Mobile-Aware[*]

We have stated that most remote access vendors simply enable remote users to access LAN applications. They expect that network providers will solve the bandwidth problem (10 Mbps LAN speed versus 33.6 to 128 Kbps for remote access, at least for now). The truth of the matter is that even when network providers do solve this problem, they will ask for a substantial premium. Therefore, it makes sense to also look for other ways of solving the problem. More important, as we described earlier, mobile workers have different requirements outside the office, on the road. To address these problems and unique requirements, we suggest a new approach—the *mobile-aware* design of infrastructure and applications. This is a term proposed by the author, in his previous book entitled *Mobile Computing—A Systems Integrator's Handbook,* as a descriptor for application designs and software products that meet the unique criteria described previously for mobile computing applications. Application designs and software products must meet the following criteria to be worthy of a mobile-aware label.

SYSTEMS AND APPLICATION DESIGN CONSIDERATIONS. *We are recommending that you should implement the following software, infrastructure, and application design features in order to make your applications mobile-aware.*

SYSTEMS SOFTWARE OR INFRASTRUCTURE DESIGN CONSIDERATIONS. First we shall address the criteria for systems design:

Criterion 1—Compression. Wireless networks will continue to be slower in speed and more costly to operate. Mobile-aware systems software should have as good a compression algorithm as possible. We are talking not only about modem-based compression of blanks and repeated characters, but also about phrases and common data elements being replaced by codes of shorter length.

---

[*]Note: Any similarity of the term *mobile-aware* to current products is unintentional. As far as we know, it does not violate any trademarks.

Criterion 2—Security.   Wireless networks pose greater security risks. The possibility of temporary connection failures should be minimized, and holes in error-recovery software that enable reconnection without full user validation in the event of a disconnection should be avoided. Mobile-aware system designs must provide encryption and authentication of users.

Criterion 3—Judicious Use of General-Purpose Software, Such as Middleware. Although middleware has a role in application development where network bandwidth is not an issue, mobile computing will continue to demand efficient design and special-purpose middleware, such as Oracle Agents, which optimizes the communications layer.

Criterion 4—Shortest and Fastest Logical Path to the Information Server.   In the past, many message switching applications were built with back-end applications acting as message switches. Since transaction response time is a critical performance requirement in many mobile applications (e.g., in public safety applications), there should be a minimum number of intermediary software components involved in communications switching. Ideally, messages should be directly switched to the destination information server by the MCSS (described in Chap. 11).

Criterion 5—Minimum Hops in the Network.   Many networks, especially the Internet backbone, are hierarchical in topology. Wireless networks have intermediate switches that are connected by wired line at 56 Kbps or T1. Nonetheless, they add delays. A mobile-aware infrastructure design should have minimum hops in the path.

Criterion 6—Agent-Based Client/Server.   An agent-based client/server design is more in line with the ways in which mobile workers work. In real life, they call to pick up their messages, check the status of customer orders, and ask administrative staff to do the errands until they return. Intelligent agent software (as in the Magic Cap Telescript paradigm) should know users' working preferences, and act on tasks in their absence so that they are ready with results when they call back. An advanced mobile-aware design should incorporate an agent-based client/server design.

APPLICATION DESIGN CONSIDERATIONS.   There are a number of application design criteria that must be considered to make an application mobile-aware.

Criterion 1—Fast-Track Dialogue for the User Interface.   An application's user interface should be similar to the one on the desktop, but it should also provide a fast track to the intended operation. Mimicry of CompuServe's GO concept or hot-button concepts in other applications could be incorporated.

Criterion 2—The User Interface Screen Should Be a Subset of the Desktop Menu Screen. Mobile users do not want to be burdened with large numbers of control buttons, multiple tool bars, too many menu items, too many entries in scroll bars, or a large numbers of options in menu items. Instead, mobile workers want a simple, customizable, uncluttered screen interface.

Criterion 3—Only Minimal Amounts of Information Presented, with Options for More Capabilities.   Mobile users should be presented with the minimum amount of information required to do the job. Only summary information should be presented initially. Any further details should follow, and only at the request of the user.

Criterion 4—The Data Input Interface Should Be Compatible with the Task.   Too many desktop keyboard-dependent applications are given to mobile users, even though these data input interfaces may be inconvenient for use outside the office. We should look at alternate input devices, such as pen and voice, that will become increasingly important requirements of mobile-aware applications.

Criterion 5—Abundant User Help.   A mobile user does not have access to help desks. There must be as much local application help available as is required to make users self-reliant.

Criterion 6—High Error Recovery at Network Levels.   The application interface to the communication layer must be extremely sturdy, and must recognize the roaming, coverage, and handoff issues of wireless networks. Mobile workers should not have to worry about being in or out of range, application errors, or network connection failures.

13.2.3 Data Replication and Synchronization on the Run

As end users increasingly take to the road, for a variety of reasons ranging from superior customer service, remote site auditing, repair calls, or meetings with branch offices, using a single source of the most current data becomes more difficult. Quite often, road warriors are working in a disconnected mode, entering data into the notebook that must be uploaded. In fact, the flow of data between corporations and their road field workers is now a two-way street.

One of the biggest problems mobile users face is keeping the notebook version of data files synchronized with the office version, irrespective of whether the primary file is on their desktop in the office or on a shared LAN server. This should be done on a regular and frequent basis. Examples of these files may be insurance rating tables, pharmaceutical price lists, product information, inventory information as of the previous day, or client information representing his/her most current investment portfolio.

What is important to realize is that an occasionally connected user has a different set of replication requirements than a permanently connected user. Mobile users want to minimize the time they spend connected to host systems transmitting and refreshing data, while ensuring that they have enough data and tools on their notebooks to function when they are disconnected. An ideal method of data replication should incorporate the following features:

- Automatic replication of all relevant files based on the profile of a user, in a single connection initiated by the user.

- The data replication and transfer of other information should be accomplished in the same session. As an example during one communications session, the user should receive his/her e-mail messages, faxes, sales reports (on demand), transaction queries that he had submitted, and application software update(s), if any. The paradigm that this communications session should emulate is one where the mobile professional calls into his/her office and his/her administrative assistant (represented by a software agent) gives to the user automatically all the important information, such as e-mail, and advises the user of other less-important information, which he/she may retrieve if the user so chooses.

- Bidirectional replication between the remote client and the central information server.

- Incremental replication, that is, only changes are sent. This is possible only with database technology, like Lotus NOTES, and not with word processing documents.

- Conflict resolution, in case of collision, through business rules.

- Checkpoint-restart capability for file transfer, in case of communications failure during the session.

- Ease of use à la synchronization feature in US Robotics' Pilot PDA, whereby simply by docking the pilot in its cradle, and pressing the *synch* button, the PC loads the latest version of relevant files into the Pilot.

Several database vendors, remote access vendors, or sales force automation vendors are trying to address this need from their individual perspectives. Sybase Inc., Emeryvale, California, offers Sybase SQL AnyWhere 5.0, which includes Sybase SQL Remote. SQL AnyWhere lets mobile users use their messaging systems, such as Microsoft Exchange, or Lotus NOTES, as a mechanism to update their databases. SQL AnyWhere and SQL Remote sit on top of the messaging systems.

SQL Remote, the replication component of Sybase's SQL AnyWhere DBMS, is designed to provide bidirectional replication between a central database server and remote databases. The product does not require direct connection between databases, but relies on messaging services, such as SMTP, MAPI, or VIM to replicate changes.

The consolidated database server arranges the data to be replicated into *publications*, to which remote users subscribe. Replication is handled by message agents located at the site of primary data store, and at each remote site. The messaging agent periodically packages any changes made to a publication, and sends them to subscribers using their e-mail systems. Meanwhile, any subscriber changes made to a publication are replicated back to the consolidated database. (See Fig. 13.2.)

Oracle has also released version 2.0 of its symmetric replication server. IBM offers a product called Lotus NOTES Pump and Data Propagator/Relational (DP/R), for the same function. During 1997, Novell released Novell Replication Services (NRS) version 1.0 which provides replications to NetWare users.

Figure 13.2
SQL Remote—
replication
component of SQL
Anywhere. (Source:
SYBASE.)

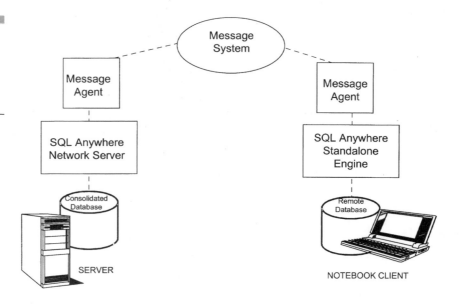

Remote Access Application Software Vendor Approach to Replication. The remote access application enhancement package vendor, Xcellenet, provides replication capabilities through its RemoteWare package by allowing you to develop scripts to perform this function. You can do similar replication with Lotus NOTES. Here are a few more examples of low-end solutions.

LapLink from Traveling Software. LapLink's latest version, called *Xchange Agent,* allows a salesperson to upload sales reports or call order sheets from a mobile PC at a set time, or at a set time performs the synchronization of file folders automatically. The job runs unattended: the laptop dials up the office, connects, transfers files, and then disconnects when the transfer has been completed. Traveling Software achieved this function by using Microsoft Window 95's System Agent scheduling technology. Symantec's *AutoXfer* provides a similar function.

Puma's IntelliSync. The IntelliSync for Windows CE software from Puma Technology enables you to synchronize your handheld PC directly with your desktop's PIM, contact management, and group scheduling applications, in one step. IntelliSync has the following features:

- Direct automatic synchronization
- Synchronization of multiple applications at one time
- Intelligent conflict resolution
- Complete customization

13.2.4 Legacy Database Access

During the last few years, we have seen a revival of the mainframe as a centralized information superserver. The data warehouse concept, the Internet revolution, and the total cost of ownership (TCO) of personal computers have prompted the IT community to fortify the mainframe's role as a keeper of corporate data. According to commonly accepted estimates, 50 percent of corporate business information continues to reside on the mainframe. Although LAN continues to be the connectivity vehicle of choice for desktops, and a keeper of departmental e-mail message stores, it does not have the glasshouse control for large masses of data.

As far as remote access is concerned, mobile workers need access to both sets of information—e-mail information on the LAN and operational information on the mainframe. We need an agent to consolidate this information on behalf of the mobile worker and ship it to him/her on connection. This is what the client-agent-server model provides. This concept is depicted in Fig. 13.3.

Figure 13.3
Client-agent-server
for legacy system
access.

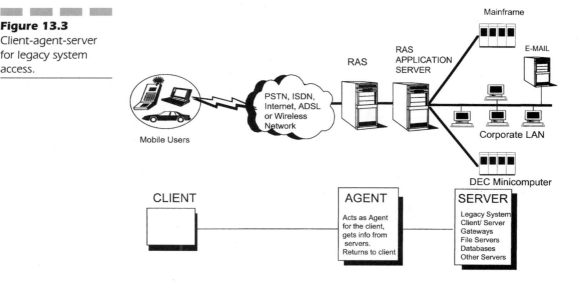

13.3 End-to-End Optimization

There are a number of factors that affect end-to-end optimization of a remote access connection. We discussed this issue in the context of PSTN in Chap. 6. Many of the factors mentioned there are relevant in other networks as well. Here, we mention those factors briefly once again, even at the cost of repetition, but include additional factors that are relevant from a global perspective.

- Processing speed of the remote computer
- Type of serial port on the remote computer
- Remote computer operating system
- Remote access client software
- Modem speed and features
- Telephone circuit characteristics
- Remote access server hardware
- RAS software
- LAN efficiency
- Type of network—PSTN, ISDN, ADSL, cable modem, or wireless
- Number of users sharing a public network, especially cable
- Protocol on the network

13.3.1 Tips on Optimization

In Chap. 6, we discussed the manner in which the aforementioned factors affect end-to-end performance. (Refer to this chapter for a more detailed discussion.) You may not have direct control of all of these factors. However, here is the list of things where you do have control, and what you can do to improve your remote access performance:

- Change your PC serial port to the newer 16550A serial controller chip, if it is not equipped with it; most notebooks and Pentium-class motherboards are, as long as they are not low-cost clones. It is worth the price.
- Switch to an ISP who supports 56-Kbps modems.
- Get a faster modem—move to a 56-Kbps modem from US Robotics or from Lucent-Rockwell, whichever your ISP supports.
- Get a faster PC—at least a Pentium class machine if you are a power Internet user.
- Install a faster graphics card that can display the graphics data received from the network in the shortest amount of time.

Summary

In this chapter, we discussed various systems design and application design considerations in order to make remote access solutions work from an end-to-end perspective, not just a communications perspective. We reviewed network design considerations, including capacity, data compression, and the overall availability of the system.

Then we introduced the concept of mobile-aware design and described the features that must be implemented in order to make our applications mobile-aware. We also talked about data replication issues, as well as access to legacy databases on the mainframe. Finally, we reviewed those factors that impact end-to-end design.

14

Remote Access—
Security*

Human beings have an intuitive sense of security against unde-
sirable events. The more we automate, the more we turn
human eyes away from what is going on around us. When we
make it easy for users to access secure information from remote
locations, we also expose ourselves to greater security risks.

—*Chander Dhawan*

---

*A major portion of this chapter is based on a white paper produced
by Shiva Corporation of Bedford, Massachusetts. The author thanks
Shiva corporation for their permission to use this material. The author
has made enhancements and changes to the source material.

About This Chapter

Computer security is an important topic in any information technology discussion. It is also a big and a complex subject. Many books have been written on computer security. One single chapter cannot adequately cover the subject. Our objective in including this chapter in the book is to provide you an overview of the subject from the remote access perspective. Topics covered include common threats and precautions, remote access security concepts, and enterprise security solutions. This chapter also includes information about the security features built into popular remote access server (RAS) products, as well as further levels of protection that can be achieved by integrating third-party security products into RAS hardware and software.

14.1 Remote Access Security— Introduction

In the days before remote access became widespread, organizations had predominantly closed, hard-wired networks, which provided physical security. Network access was limited to users physically located in the building. Requiring users to type in a name and password added another layer of security to the network.

Providing remote network access over telephone lines has added an entirely new dimension to the task of keeping business-critical information secure.

Conventional telephone systems are, by nature, public. Anyone can dial a number, and reach the *door* leading into a company's computer network. The primary concern of remote access security is to make sure that only known, authorized users can enter that door; the secondary concern is to ensure that data leaving or entering that door is encrypted to prevent unauthorized personnel from gaining access to this data. Before we discuss anything else, we must understand the threats and precautions at the door leading into a company's computing resources.

14.2 Security Threats and Precautions—On Everybody's Mind

We have seen that whereas technological advances have redefined the traditional workplace, these same advances have also created more opportunities for security breakdowns. These developments, such as the growth of the Internet and more powerful desktop systems, are placing a greater burden on those responsible for the security of an organization's computing resources. For example, the explosion of the Internet has compounded security issues by combining public and private network access, resulting in a new category of security requirements that must be addressed.

Security considerations have been further complicated by the multiplicity of functions and applications that are now accessible at the PC level. Because every application handles security in a different way, most users can't access all of their applications through a single password. The need to remember multiple passwords can lead to a natural temptation to repeat or simplify passwords. As a result, multiple levels of security can have the unintended effect of actually compromising security.

The good news is that solution providers are developing powerful and innovative methods for addressing these new threats, providing organizations with more methods of protecting themselves than ever before. In formulating a remote access strategy, companies need to determine what threshold of security works best for them. If highly sensitive information is generated only within certain areas of the company (e.g., by senior management or product development), precautions can be taken to partition networks accordingly.

14.3 Network Security and Ease of Use—Two Opposing Concepts

In general, improving a site's security will impact ease of use, and vice versa. While users want their data to be secure, they are not willing to navigate through a complex security system every time they need to access that data. However, there are many ways to protect your system without losing the advantages of connectivity. Here we discuss some of the possible means by which you can prevent intruders from gaining access to your system, while minimizing disruption to legitimate users.

Companies implementing remote access solutions need to attach a *value* to different kinds of information. One way to address this need during configuration is to have certain remote access components—for example, a specific remote access server—configured for a high security level, while other servers allow for unlimited access to insecure data, such as customer support knowledge bases. Other steps that organizations can take include requiring employees to change their passwords periodically, or requiring complex passwords that mix both letters and numbers, which are more difficult for an unauthorized user to guess. These internal procedures will have a definite impact on security and can greatly diminish the threat of security breaches.

14.4 Network Security— A Layered Approach

Network security can be seen as a series of layers of protection around the inner network. The levels are usually split between packet security, call security, and user security. The layers are shown in Fig. 14.1. While backbone LAN routers can implement the outer two layers, they cannot usually validate individual users; this has to be done by the host system. Terminal and communications servers, on the other hand, cannot usually verify at a packet level, but can, to some extent, implement the inner two layers. This chapter discusses all three security layers; however, the reader should bear in mind that a single system cannot usually implement all of them. It's also important to keep in mind that organizations have a responsibility for security once a user enters their system. This is the advantage of incorporating security features into individual applications, as well as networks. Companies also need to make the distinction between *authentication,* the process that determines how users access computing resources, and *authorization,* the process of determining different access privileges among the user population. The final consideration is privacy, ensuring that accessible data is viewed only by the intended audience.

Figure 14.1
Network security rings.

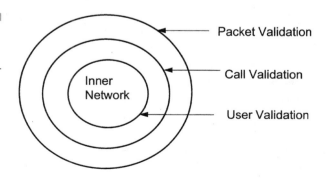

14.5 Security Concepts

There are several security concepts that are common across most systems. We describe those concepts here:

14.5.1 Authentication—You Are Who You Say You Are

The first step of authentication involves verification of the user's identity. Secure systems should verify, or validate, the user's name, password, and location. The use of third-party directory or naming services permits the use of a single database to store privileges for users on the local network, as well as for remote users. This helps minimize the number of user databases required to assign privileges such as remote dial-in access, use of hardware resources such as printers, application server access, and database access.

Authentication for remote access has distinct requirements and is more complex than general LAN security. The most critical authentication points are listed here and discussed in more detail later in this chapter.

14.5.2 Authentication Points—More Than One Door to Open Before You Get the Information

You should look at several points in end-to-end systems where authentication may take place. These points are:

- At the local remote PC level
- At the RAS
- At the network
- At the application level

At the remote device level, there are several physical and software security safeguards that hardware vendors provided. Many public safety agencies that have computers installed in police cars deactivate the device or application if it is not in use for more than a specified period of time.

At the remote access server, look for support of industry-standard name services, local security database support for smaller sites, and support for optional token-based security.

At the network, consider network-oriented solutions such as Novell Directory Services (NDS) or Microsoft NT Domains.

At the application level, password-protect your system's e-mail, personal information manager (PIM) applications, workgroup applications, and operational databases. Look for password protection features when evaluating productivity applications.

14.5.3 A Single Login—Simplifying Security Bureaucracy

One way to streamline security is by using a single login for accessing a number of systems, such as applications, printers, remote access, and network servers. This can eliminate the security risk of multiple passwords, which users often write down on a piece of paper and hide in obvious places. A single login can also reduce the administrative inconvenience that arises when adding and deleting users from multiple login configurations.

Although a single login is a desirable objective, it is still not feasible and practical for a number of reasons. First of all, until now there was no standard mechanism in the industry to define user-identification and passwords in the first place and a mechanism to pass this information to application programs in the second place. Second, there are thousands of legacy applications on different vendor platforms that have embedded security in the application subsystems. In the mainframe world of IBM, IMS/VS and CICS application subsystems have a security feature built into them. As a result, a more pragmatic objective is to have no more than two passwords—one at the network and the other at the application level. As for the network security, consider an integrated security solution from a third-party security hardware or software provider. Integrated third-party security solutions can provide the level of security required for the specific installation, while working seamlessly with the remote access server.

14.5.4 User Validation Authentication for Remote Access

There are several ways to ensure user validation for remote access systems. Password features such as expiration dates, grace logins, and allowing users to change personal passwords can provide flexibility to both users and security administrators. Integrating security features with your remote access system provides ease of management and simplicity, allows for configuration from a single location, and helps eliminate the possibility of security loopholes.

Organizations can also fortify validation through token-based authentication mechanisms. These protection schemes are discussed later in this chapter.

14.5.5 Password Protection— Avoiding Reusable Passwords

Given today's networked environments, the Computer Emergency Response Team (CERT), an industry association, recommends that organizations concerned about the security and integrity of their systems and networks consider moving away from standard, reusable passwords. CERT has documented many security breaches involving Trojan network programs (e.g., Telnet and login) and network packet sniffing programs. These programs capture information on host names, account names and passwords, which intruders can use to gain access to those hosts and accounts. This is possible because:

1. The password is used repeatedly (hence the term *reusable*), often over a period of months or even years.

2. The password frequently passes across the network in clear text, that is, without being encrypted in anyway.

Several authentication techniques have been developed that address this problem. Among these techniques are token-based challenge-response technologies that provide one-time passwords. The use of one-time passwords makes sniffing account information virtually useless.

14.5.6 Location Validation and Call-Level Security

Location validation employs various callback methods to confirm a user's location. One of these methods, CLI (caller line ID), validates an incoming call number against an internal database of approved numbers. CLI allows for the validation of ISDN calls by providing the called site with the ISDN telephone number from which a call is being made. This allows the called site to verify that the call is coming from an authorized location.

Callback can be used both as a security feature, and to take advantage of any asymmetries in call charges (where the service at each end is provided by different service providers, Telcos, or PTTs). It can also be used for centralized billing. PPP-based protocols may be used to negotiate callback. Alternatively, ISDN lines that support CLI may also allow callback to operate without involving PPP at the initial callback negotiation phase. The advantage of using CLI rather than PAP or CHAP (see the next section) to determine the callback address is the savings that result on the initial call.

Another callback method involves fixed dial-back numbers, in which the system dials a preassigned number. Fixed dial-back numbers help reduce costs, in that they enable the system to assume remote access line costs, thus allowing companies to centralize those expenditures. This method also allows companies to control dial-back access. Because the system dials a known number or location after the initial remote call request, an invalid user dialing in from another location who poses as the approved user will not be called back.

Whereas fixed dial-back numbers work well for users who require remote access from fixed locations, this method is not appropriate for mobile or field access. In these cases, organizations can use roaming dial-back, which allows a remote user to specify a number at which he or she should be called back. This is primarily a call- and toll-saving vehicle, not a security enhancement, since a user at any location can specify the callback number. Even though this method is not a security enhancement, it is no less secure than regular dial-in procedures.

14.6 PPP Security Authentication

A number of PPP-based authentication methods can be used to verify a caller's identity. The most common of these are the Password Authentication Protocol (PAP), and the Challenge-Handshake Authentication Protocol (CHAP). There are also a number of vendor-enhanced authentication protocols, such as the Shiva Password Authentication Protocol (SPAP).

PAP provides a simple method for the peer to establish its identity using a link establishment. Once this phase is complete, an ID/password pair is repeatedly sent by the peer to the authenticator until authentication is acknowledged or the connection is terminated.

PAP is not a strong authentication method. Passwords are sent over the circuit unencrypted, and there is no protection from playback or repeated trial-and-error attacks. In addition, the peer is in control of the frequency and timing of the attempts. Stronger authentication methods (such as CHAP) are recommended over simple PAP authentication.

CHAP is used to verify the identity of a peer using a three-way handshake. After the link establishment phase is complete, the authenticator sends a "challenge" message to the peer. This authentication method depends upon a "secret" known only to the authenticator and the peer. The secret is not sent over the link.

CHAP provides protection against playback attack through the use of an incrementally changing identifier and a variable challenge value. The use of repeated challenges is intended to limit the time of exposure to any single attack. The authenticator is in control of the frequency and timing of the challenges.

These methods can be used on any dial-up service, as well as on a point-to-point link or leased line. They can also be used on ISDN lines that do not support CLI. Even if the ISDN line does support CLI, the use of PPP authentication provides a greater degree of security than CLI alone.

Several vendors have implemented enhanced security methods. One such vendor is Shiva, which has extended CHAP standard with a third-party security server. Shiva's security software provides features such as virtual connection, dial-back, and support of third-party username directories.

14.7 Authorization Restrictions (Access Control)—Where and What Services

Authorization, or access control, refers to what parts of the network, resources, and services users can access. There are a number of remote access authorization models. For example, access can be controlled on a user-by-user basis, by administrator-defined user groups, by remote access server, by port, or protocol. These methods offer organizations the flexibility of defining control based on either equipment or people. Some common authorization privileges are described as follows:

- Dial-in, dial-out and LAN-to-LAN settings allow only specific users to gain access to features that they actually need, further closing security loopholes and minimizing exposure.

- Dial-back can be configured to authorize different users for either "fixed" or "roaming" dial-back (for example, office-based workers versus field workers). Roaming dial-back is less secure than fixed dial-back, but provides cost-containment.

- Administrator privileges are used to specify which administrators or groups can monitor usage, change user privileges, and administer remote access servers.

- Maximum connect times help ensure that idle users aren't staying connected for too long.

- Allowable connection times prevent off-hour access to network services. Organizations need to balance these restrictions with legitimate off-hour access requirements.

- Network and device filtering (ISDN, IP address, IPX SAP types), which are integral parts of firewall design, can prevent specific users from accessing particular networks, devices, or services on a network. (For example, the sales department might be able to access only sales servers, and engineers could access only development servers, whereas MIS would be able to access all servers.)

14.8 Security Accounting— Track What Is Happening

Security accounting involves tracking, auditing, and reporting security and usage activity. Accounting enables administrators to determine usage patterns and quickly identify unusual activity, either by authorized or unauthorized users. It can also detect attempts to access protected files and evaluate the integrity of a security system.

Accounting is most commonly used for billing, as well as for security auditing. Most companies, however, have many different accounting systems from multiple vendors, as well as multiple homegrown databases. Many legacy accounting systems that may have been implemented before the advent of remote access don't take remote applications into account, despite the fact that these capabilities need to be integrated with other accounting functions.

This problem can be addressed through a number of methods, including internal logs within the remote access server, SNMP trap hosts, RADIUS or TACACS+ accounting products, or applications designed specifically to log and monitor access attempts. Logging allows administrators to distinguish unusual occurrences from normal usage patterns such as remote login attempts by a user in the office, multiple concurrent dial-in sessions by the same user, or a significant number of failed login attempts.

14.9 RADIUS—An IETF Draft Standard

RADIUS, an IETF Draft, has recently emerged as a common standard for centralized security implementations. Almost all remote access server and security application vendors have announced RADIUS support due to its robust capabilities, commonality between vendors, and extensibility. The RADIUS Draft defines a protocol, but leaves client and server implementation up to individual vendors. The RADIUS "client" is the remote access server, which requests authentication against a RADIUS "server." RADIUS servers have been developed on a variety of platforms ranging from Macintosh, Windows 3.x, Windows 95, Windows NT, and Unix implementations.

14.9.1 RADIUS Features and Benefits

RADIUS can provide a single point of authentication and authorization. When a RADIUS server is used as a stand-alone server, remote dial-in users are authenticated against the server, with authorization privileges also controlled by the server. Accounting information regarding session time, protocols used, and utilization can be captured and output for billing and auditing purposes. (See Fig. 14.2.)

RADIUS server implementations from a number of vendors can act as "proxy clients" to other security servers, permitting seamless integration with other security implementations. RADIUS products currently provide proxy capabilities for NT Domains and Novell Directory Services. Future products will most likely be available to support evolving directory schemes, including X.500 and LDAP directories. Proxy support means that, in the case of NT Domain and NDS proxy, for instance, remote dial-in users can be authenticated directly against enterprise NT Domain and NDS directories. Separate user lists do not need to be maintained on the remote access server. When an administrator adds, deletes, or modifies user privileges directly in an NT Domain or NDS structure, these changes are reflected in the remote access server privileges, without administrator intervention.

Figure 14.2
RADIUS architecture.

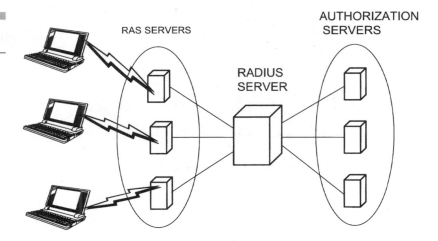

14.10 Interoperability of Security Across Remote Access Servers

RADIUS defines general parameters for remote access servers that are common across most products. Remote access server manufacturers have had to extend RADIUS to address additional device-specific capabilities not defined by RADIUS. Vendor-attribute dictionaries have allowed vendors to easily extend the RADIUS product to address these requirements. In a mixed product environment, multiple dictionaries have resulted in some confusion. Vendors are now working to extend common parameters within the RADIUS specification, guaranteeing interoperability, with only product-specific attributes maintained in vendor-defined attribute dictionaries.

Vendors have realized they need product interoperability, and have been working together to ensure that their products work seamlessly in a mixed-product environment. As remote access servers and RADIUS products mature, products will continue to be more interoperable, allowing full "plug-and-play" compatibility between RADIUS clients and servers.

14.10.1 Shiva AccessManager—An Example of RADIUS Implementation

Shiva AccessManager, a RADIUS implementation that can reside on Windows 3.x, Windows 95, and Windows NT servers, provides user authentication and authorization capabilities. In addition, extensive reporting capabilities allow the network or security management to review remote access server usage characteristics. With the included proxy support for NT Domains, remote dial-in users can be authenticated using Microsoft's NT Domains—providing a tightly integrated solution for NT shops. When announced support for third-party security token passthrough and NDS proxy is available, it would permit enterprise-wide remote access deployments to be secured and managed within a company's security infrastructure.

14.10.2 Blockade Systems Server as Centralized Third-Party TACACS+-Compliant Security Server

TACACS+ is an emerging protocol designed for third-party authentication, which complies with standards set by the IETF (Internet Engineering Task Force). Several vendors' RAS products, such as version 4.0 of the Shiva's Lan-Rover Operating System, provide support for TACACS+, enabling Shiva LanRover configuration to route user authentication requests to a centralized, third-party security server.

Blockade Systems has introduced such a server, which was created to address the needs of organizations with multiple RAS units and large numbers of users. Implementation of the TACACS+ server helps companies leverage the use of existing systems within the enterprise, and eliminates the need to define, audit, and maintain users in yet another security database.

The Blockade solution routes TACACS+ authentication requests from the LanRover to the Blockade Enterprise Security Server, which interfaces with IBM's RACF, CA-ACF2, or CA-TSS. Companies can then use their existing security systems and personnel to extend the existing corporate security policies to LanRover devices.

Optionally, the Blockade server also incorporates integrated support for dynamic password generators, such as those from Security Dynamics and AssureNet Pathways, to provide an additional level of security. When required, such devices can be used to enforce a level of user authentication which goes beyond the standard requirement for user ID and password. The Blockade system eliminates the need for the additional hardware usually required when using such devices.

A centralized TACACS+ server, such as the one offered by Blockade Systems, allows a company to retain a greater degree of control over its remote access servers, facilitates the enforcement of existing security standards, and eliminates the introduction of another database containing user ID and password information.

14.11   Enterprise Security

Larger organizations with global multipoint access requirements and those with greater security concerns will want to integrate their remote access capabilities with products specifically designed to enhance the security of information distributed over remote networks.

14.11.1 Centralized Security

One alternative approach is "centralized" security, which involves having the terminal or communications server authenticate a dial-in user's identity through a single central database, known as the *authentication server.* This server stores all the necessary information about users, including their passwords and access privileges. The use of a central location for authentication data allows a greater degree of security for sensitive information, greater ease of management, and a more scalable solution as the size of the network increases. Authentication servers can be configured in a variety of ways, depending upon the organization's preferred network security scheme. Common schemes for centralized security are based on RADIUS and TACACS+, discussed earlier in this chapter.

Authenticating a user includes the following steps:

1. A user dials into a network through a remote access server.
2. The remote access server forwards the user identification and password to the authentication server.
3. The authentication server validates and provides access privileges to the user.

14.11.2 Token-Based Security

Many of the most popular remote access security enhancements operate on the principle of *security by obscurity,* in which users must have a specific object in their possession to access the network. The "obscurity" of that object creates a much larger hurdle for attackers. One of the most popular methods for promoting security by obscurity is through a *token,* which is simply a credit card containing a small, built-in computer. Anyone who comes into possession of a token can't use it unless they have access to another piece of confidential information—the user password. Tokens also help solve the problem of having to remember multiple passwords (discussed earlier in this chapter).

The concept of a one-time password involves having the security server already know that a password is not going to be transmitted over insecure channels. When the user connects, he or she receives a challenge from the security server. The user takes the challenge information and uses it to calculate the response from the password. The security server then calculates the response and compares its answer to that received from the user. The password never goes over the network; nor is the same challenge used twice.

Racal's Guardata WatchWord II token is a similar product that offers convenient alternatives to passwords based on common names, birthdays, and so forth. When using WatchWord II, critical information is never entered in clear text. The operating principle is based on the challenge/response mechanism described in the ANSI X9.26 secure sign-on standard. The user enables the token by entering a PIN. The WatchWord Generate process takes a digital challenge from the host computer system entered into the token—which then generates a seven-digit response—a one-time password. The response is calculated from the challenge using the DES cryptographic process. There is a security controller or server at the host between the modem pool and the information server. SecurID is a similar product providing similar function.

Solutions such as these are driven by the fact that the electronic information that travels over telephone lines is subject to security threats and can be *stolen.* Integrating mathematical computations into the login process helps ensure that the information traveling across the telephone system is worthless after a single login, as opposed to the character string in a user password, which can be used repeatedly.

14.11.3 AssureNet Pathways Solution Based on Token-Based Authentication

AssureNet Pathways (formerly Digital Pathways) believes that user authentication forms the basis of a sound network security strategy. Compromised authentication may lead to other security functions being compromised, despite the fact that those other processes are well designed and intact.

Since reusable passwords are the weakest security link in most networks, many companies need to implement better authentication techniques. Token-based authentication and access control systems provide a reasonable balance among the goals of low cost, high security, and system simplicity. The basic premise of tokens is *two-factor* authentication: something you know and something you have. These systems eliminate many of the risks inherent in the use of static passwords by generating a changing (dynamic) password for authentication.

One of the most popular types of tokens is the challenge/response token, which is available in both hardware and software versions. Although both tokens perform the same challenge/response authentication, software tokens are much easier to use and represent the latest developments in token-based authentication.

When a login is attempted, the server challenges the client with a random number. This random challenge is then encoded using the token. Both the server and the client perform this encoding. The resulting value is used by the client as a one-time password. The server then compares the received response with its expected response, and if the two values match, the client is authenticated.

AssureNet Pathways products conform to existing communications APIs where appropriate, and work closely with partner companies where no standard APIs exist. This allows for easier compatibility when integrating AssureNet Pathways products with those of other vendors, such as Shiva.

14.11.4 Cylink's Remote Access Security Solution

Cylink is another vendor well known for providing remote access security products. Figure 14.3 shows a schematic of Cylink's SecureAccess System offering X.509 certificate-based security. It is designed for large Fortune 500 installations with thousands of mobile users.

14.12 Firewalls*

Firewalls are another method for reinforcing remote access security. As their name implies, firewalls can be used to restrict access between differ-

---

Source—Shiva's White Paper and an article by Kristina Sullivan in *PC Week,* dated October 7, 1996.

Figure 14.3
Cylink's secure access
system for mobile
workers.

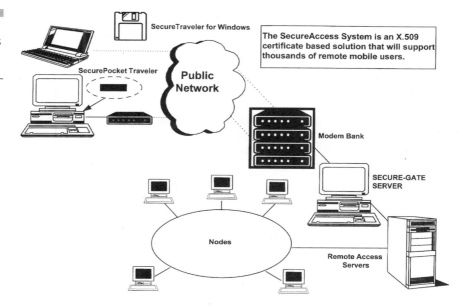

ent areas within a company's network. Firewalls are often used to help ensure separation between the public and private components of a company's computing resources, for example, the public information found on an external Web site versus the other information residing on an internal file server. Firewalls have a number of other applications, such as providing additional hurdles in the event that a break in authentication security occurs, or providing filters that let certain users and/or specific data pass through the firewall on a case-by-case basis.

There are a number of different approaches to firewall design. The various flavors of packet filtering and proxy agents employed by firewall vendors have their own advantages and trade-offs. Packet filtering and its variants, including state evaluation (also called stateful inspection) offer speed. On the other hand, proxy agents, application gateways, and circuit-level firewall models offer a higher level of security. (See Table 14.1.)

Packet Filtering. This feature is found in routers and is based on analysis of each IP packet at the network layer. The firewall determines whether to pass or block a packet based on a set of rules established by the network administrator. It can be used between internal and public networks or among internal departments. Packet filtering looks exclusively at the source, destination, and perhaps, the service type. The router determines if the source is allowed to talk to the destination, but it cannot determine

TABLE 14.1

Comparison of Firewall Technologies

	Advantages	Disadvantages
Packet filtering (routers)	■ Very fast ■ Application independent ■ Scalability	■ Complex access rules ■ Low security
State inspection	■ Can track connectionless protocols such as UDP, RPC ■ Highly secure	■ Must be supplemented with proxies to support application functions such as authentication
Application layer (proxy) gateways	■ Highly secure	■ New Internet protocols require new proxy ■ Low performance

SOURCE: *PC Week*, October 1996.

if it is legitimate. You need to do stateful inspection and look at the contents. Configuring routers for security is notoriously difficult.

State Evaluation.   This model was invented by CheckPoint Software Technologies Inc. This model expands packet filtering by adding state information as derived from past communications and other applications. It examines packets at the network layer and makes decisions in the context of previous transmissions. State evaluation boosts security by analyzing packets based on this state information, which is dynamically stored and updated.

Proxy Server/Application Gateway.   This model is highly secure and provides in-depth knowledge of IP protocols and allows application-level analyses. It examines each packet of information as it passes through the gateway.

Trends in Firewall Design.   Firewall vendors have already started implementing more than one model in their products. As an example CyberGuard and Firewall Plus utilize both proxy agents and state evaluation, allowing network administrators to decide which combination to use. CyberGuard also does packet filtering. To support application-specific functions, such as authentication, CheckPoint Firewall-1 adds proxy agents to its stateful inspection technology. There are situations where you want a proxy. When a user logs into an application, you need a proxy for that application.

Vendors such as CyberGuard expect other features such as additional security functions, such as antivirus software and encryption. CyberGuard has already entered into an agreement with Cheyenne Software Inc., of Roslyn Heights, New York. You should also see remote management and network management features.

14.12.1 Common Firewall/Security Features

You should look for the following features as a base set while selecting a firewall for your configuration:

- Support for various client devices operating under Windows 3.1, Windows 95, Windows NT, OS/2, and Macintosh
- Support various server applications running on NT, Unix, minicomputer and mainframe platforms, without any change
- Support inbound and outbound connection to and from the Internet, by incorporating integrated application servers such as e-mail, WWW, FTP, DNS
- Support for packet filtering, state evaluation, and transparent proxy models
- Network address translation all internal addresses into one single firewall IP address
- Encryption-based challenge and response authentication
- Audit trail and accounting
- Easy-to-use security administration facilities
- Support for domain name servers

14.12.2 Firewall Product Examples

There are a number of state-evaluation firewall vendors in the market. The following list names some of these vendors:

- AltaVista Firewall 97
- Ascend Communications's SecretAccess Firewall
- CheckPoint Software Technologies Inc.'s Firewall-1
- Global Internet's Centre-Firewall
- IBM's Internet Connection Secured Network Gateway
- NetGuard's Ltd.'s Guardian 1.3
- Secure Computing Corp.'s BorderWare Firewall
- SideWinder Security Server
- SunSoft's Sunscreen EFS Firewall

Since firewall concept is relatively new, there are more vendors in the marketplace than the market can support. Market rationalization has not occurred. Therefore, some of these vendors will either leave the market, get acquired by larger players in the remote access arena, or merge. While selecting a firewall, you should take into account the staying power of the vendor and strong customer acceptance of the product, as well as its functional features.

14.13 NT Domains, NetWare Bindery, Novell Directory Services

Security databases can also exist as part of a network operating system (NOS). The primary advantage of these mechanisms is that they provide a single point of authentication and authorization. Users can enter a single password and be automatically authenticated into the remote access server; a general-purpose server running NT Domains and NDS; or even multiple servers, automatically. Network administrators can define, control, and administer users across an entire network through a single point, greatly simplifying this task. Strength of security across the network is based on the robustness of the operating system security. In the case of a fully implemented version of NT Domain security, this can provide security levels as high as U.S. Government C2 protection level. Access to these services by the remote access server can be achieved either natively in the router's operating system, or through another standards-based protocol, such as RADIUS. Both implementations can provide the same level of functionality and security.

14.13.1 NDS and NT Domains versus RADIUS, TACACS, TACACS+

RADIUS, TACACS, and TACACS+ can provide a single point of authentication and authorization, similar to NDS and NT Domains. The difference, however, is that whereas NDS and NT Domains are proprietary, RADIUS and TACACS+ are open IETF standards that have been adopted by many organizations other than Microsoft and Novell, the respective owners of NT Domains and NDS. The advantage of these open standards is that they can be used between multiple vendors and shared between many products.

Both RADIUS and TACACS+ provide the ability to pass security data to a variety of databases. For example, a remote access server that implements the RADIUS client component can use RADIUS to authenticate a user via an NT server, a Novell server, or another host, as shown in the following sections.

14.14 Encryption for Security and Data Integrity

Encryption involves scrambling digital information bits with mathematical algorithms and is the most potent protection available against security intrusions into wireless and wireline communications. Different encryption schemes have been proposed and implemented. The Data Encryption Standard (DES) is one algorithm that has held sway since the 1970s. RSA, based on public key cryptography and named for the three MIT professors—Rivest, Shamir, and Adleman—who developed it, is another. Pretty Good Privacy (PGP) is a public domain implementation of RSA available for noncommercial use on the Internet in North America. All these schemes use some variation of private and/or public keys as shown in Fig. 14.4.

Encryption focuses on two areas: password protection and data integrity. Password encryption prevents unauthorized users from capturing valid users' passwords. Because an encrypted password cannot be entered into a password prompt field, superior password integrity and system security are maintained.

Figure 14.4
Remote access security via public and private key combination. (Source: *Open Systems Today,* October 1994.)

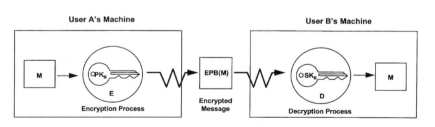

Sending A Secure Message

1. User A wants to send a secure message to User B. User A encrypts (E) the message (M) using User B's public key, PK_B.

2. The encrytped message (EPB(M)) is sent across the wire.

3. User B gets the encrypted message and decrypts (D) it using his secret key (SK_B), resulting in a decrypted message (M). Encryption and decryption are inverse processes.

Legend
PK_B -- Public Key Of User B
SK_B – Private (secret) key of User B
EPB(M) - Encrypted message for B

Data encryption is of great importance to financial institutions, public safety organizations such as police, and any other organizations with highly sensitive data. With data encryption, users need an encryption key to access data. Without this key, the data is meaningless. This "listen-proof" data offers good protection for network access, but it does cause significant CPU overhead as the server encrypts the data.

There are three types of keys used in encrypting data:

1. A private key known only by the sender and the recipient

2. A private/public key combination

3. A one-time key

In private-key systems, the two parties have a secret key which they use to encrypt and decrypt data. The private/public key combination is more secure, however. In this scheme, the recipient's public key—available to all who need it to send encrypted data—is used to encode information for transmission. The recipient uses a private key associated with the set to decode the information. (See Fig. 14.5.)

The one-time key method is based on the generation of a new key every time data is transmitted. A single-use key is transmitted in a secure (encoded) mode and once used, becomes invalid. In some implementations, the central system will not issue a key for a new connection until the user supplies the previously used key.

14.14.1 Vendor Key Administration Solutions—Nortel's Entrust[*]

Nortel has introduced a public-key management infrastructure that allows privacy of data and authentication functions performed on corporate-wide distributed networks. Entrust has received very good reviews by the technical trade press and has been endorsed by a number of leading information technology vendors, including IBM. (See Table 14.2.)

Although there are a number of very good technologies in the market for creating encrypted data, Entrust addresses several inherent requirements of key administration that any large organization faces; for example, how do you validate a public key given to you, how do you get your data back if you lose your private key, and finally how do you get to the

---

[*]*Source*—Paper published in Nortel's *Telesis* magazine during 1996 and *Network Computing,* November 1, 1996.

Figure 14.5
Encryption of Internet e-mail messages.
(Source: *Data Communications Magazine*, May 1996.)

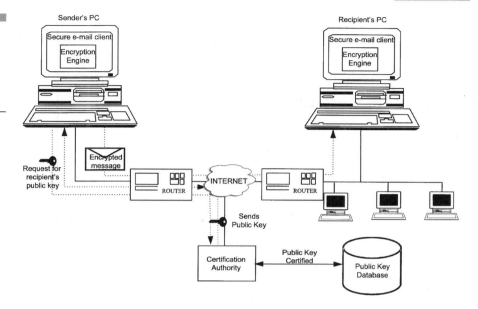

secret data that employees guard after they leave the organization? Nortel's Entrust is a file-signing and encryption package with an intuitive client interface, so that users will have little excuse for not using it.

An important aspect of Entrust is the way it divides ownership of keys among security officers, administrators, and users. Personal signing keys are held exclusively by the individual; they are not kept in a centralized

Figure 14.6
Nortel's Entrust supports balance of power among security officers, administrators, and users.
(Source: *Network Computing Magazine*, November 1996.)

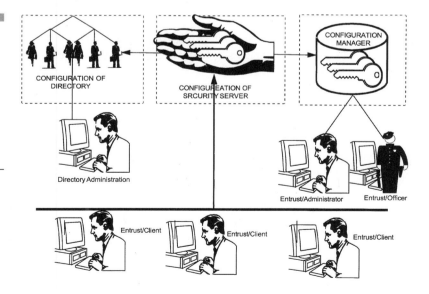

TABLE 14.2

Nortel Entrust
Features

Feature	Description
Client platforms	Windows 3.1, Windows 95/NT, Macintosh, Unix
Server platforms	Unix (HP-UX, Solaris, AIX)
E-mail client interface	MAPI, MS Mail, cc:Mail (VIM)
Encryption algorithm	Nortel's CAST, U.S. DES (FIPS PUB46-2/ANSI 3.92)
Digital signature/hashing algorithm	RSA Digital signature (PKCS#!) RSA MD5, MD2, US NIST DSA, US NIST SHA
Certificate formats	ITU-T x.509
Directory protocols	X.500 directory access (DAP) X.500 Directory System (DSP) IETF RFC 1777 Lightweight Directory Access Protocol (LDAP)
Application programming interface	IETF RFC 1508/1509 Generic Security Services (GSS-API) Nortel's Entrust/Toolkit API
Price	Client—approximately $159 (qty one), cheaper with larger quantities Servers: Entrust Manager—$4000, Entrust Server—$2000 and X.500 server—$5000

SOURCE: Nortel and article by David Willis in *Network Computing*, November 1, 1996.

database. Encryption keys, however, are kept within Entrust/Manager, a centralized key repository, so that encrypted data files may be decrypted after the administrator releases a new pair of keys. Thus, an organization retains ownership of its data, but it doesn't own an individual identity. (See Fig. 14.6.)

Nortel's Entrust recognizes that users, administrators, and security officers have different views of security. Users tend to shy away from security procedures that make their jobs more difficult. Security officers are often too restrictive and sacrifice user access to diminish risk. Finally, security administrators often get stuck in the middle, trying to make it work and yet unable to please everybody. Entrust provides a system of checks and balances to assist these three groups.

Using the Entrust/Security Officer application, security officers can define policies for their certification authority (CA) security domain. There may be multiple CA domains within an organization. They can set certificate lifetimes (a certificate contains signing and encrypting key pairs). They can set up cross-certificates across other trusted CAs if there is network connectivity between servers. Security officers can assign security administrators for day-to-day activity of using Entrust. The administrator's main responsibility is to enable users, recover keys when they forget them, and revoke access when necessary. User lists are maintained in an X.500 directory, whether it is Nortel's or some other vendor's. Security officers can audit administrators' activity using an Audit Record Viewer.

Other features of Entrust are:

Standards-Based Security. Entrust utilizes many industry standards, protocols, and algorithms, including those from RSA, National Institute of Standards and Technology (NIST). It is based on TCP/IP, Lightweight Directory Access Protocol (LDAP), and X.500 directory. One piece of proprietary algorithm it uses is Nortel's CAST encryption. Of course, you can substitute CAST with DES.

High Performance. Entrust sends minimally sized requests to Entrust Server, which then contacts the directory and Entrust/Manager applications on the client's behalf. Creating a client profile moves approximately 8 KB of data between client and server. Actual encryption/decryption times are quite low when compared with similar products.

Entrust/Toolkit. Besides IETF's RFC 1508/1509 generic Security Services (GSS-API) Entrust provides a toolkit of its own for developing secure applications.

14.14.2 Vendor Encryption Solutions— Security Dynamics Inc.

Security Dynamics and several other network security vendors provide security solutions that positively identify users and protect the integrity of network resident data, while providing some form of encryption. For example, all communication between a protected client and server can be encrypted, along with the database of user records residing on the server. This ensures that private access information is not readable as it travels across a network. In 1996, Security Dynamics acquired RSA Data Security, Inc., one of the recognized leaders in the encryption arena.

14.15 Electronic Signatures

Electronic signatures can also be used to ensure that users are who they claim to be. With the appropriate hardware and software—PenOp from Peripheral Vision in the United Kingdom—a system can literally demand a valid signature. While the primary use of such software is in contract-related applications (mortgages, loans, etc.) there is no reason why it cannot also be used as a substitute for a password.

PenOp is based on a biometrics signature-verification technique. It supports a variety of signature capture methods, ranging from low-cost digitizers attached to desktop PCs, through to handheld PDAs or pen computers.

14.16 An End-to-End Encryption Scheme with a Constantly Changing Public/Private Key Set

Although each of the encryption schemes described earlier provides a certain amount of security in and of itself, we believe the best scheme is one based on end-to-end encryption using private and public keys, where not even the network provider's control center knows what information is being transferred. To achieve this, the remote client device and the information server must each perform encryption/decryption as appropriate, depending on the direction of the transmission. Several PCMCIA cards provide encryption capabilities and—even though hardware cards are certainly the fastest way of achieving DES and RSA encryption—software-based encryption is also available. In fact, with Pentium-class client devices, software-based processing of encryption and decryption with 40-bit keys does not add significant overhead to the total transaction time. Moreover, it is a much cheaper solution. With 256-bit keys, you should evaluate which data elements should be encrypted, because processing overhead may be significant.

14.17 Network Security Policy, Procedures, and Tips—More Important Than Tools

We have spent a lot of time on security concepts and tools in this chapter. These tools can never give you complete peace of mind unless you adopt and implement good security policy within your organization. Without spending an undue amount of time on this issue, we briefly mention nontechnology-related issues for enforcing security.

Obtain Senior-Management Awareness and Commitment.   Without this support, you will not get the resources you need to implement a good security solution.

Continuous Risk Assessment.   Many organizations do not know accurately the amount of losses the organization may be suffering. Also, risks to corporate information will change over time. You should continuously assess risk to information. Only then can you develop security plan to offset these risks.

Enterprise-Wide Security Policy.   To be effective, security policy must be a dynamic document that identifies the potential threats to computers, networks, and information within the organization, along with the procedures and measures to be taken by employees to address these risks.

Employee Awareness.   There should be an ongoing campaign to make employees aware of security requirements and their responsibility in enforcing it.

Physical Security.   It is as important to enforce physical security policy for protecting access to notebooks, desktops, and servers as it is to employ good security software tools.

Security Policy Monitoring.   The organization should institute regular audits of the security policy and procedures. These should be spontaneous and random for a subset of the users done on a more frequent basis. An organization-wide audit should also be done, even if it is on a less-frequent basis.

Summary

In this chapter, we discussed various security threats, concepts, and techniques to allow only authorized access to corporate information. Our primary focus was from the point of view of remote access, though some concepts go beyond the network. We also looked at encryption techniques, including private and public keys, for protecting data as it moves over the network. We described several product implementations to illustrate many of the security requirements. Finally, we emphasized nontechnology security policy issues. We hope that we achieved our objective of highlighting the need for analyzing security requirements in any remote access solution.

15

Remote Access— User Issues

When we hand over technology to our users without understanding their training, ergonomic, and social needs, we ignore basic principles of how people change their habits— business or social.

—Chander Dhawan

About This Chapter

Many network professionals spend a lot of time analyzing technical aspects of a remote access solution, but not enough analysis is done regarding user concerns and issues. A perfectly good technical solution may not result in user acceptance, if many of these concerns are not addressed. In this chapter, we deal with some of the more common user issues that remote workers complain about. We discuss the importance of remote device characteristics, ergonomics, backup, software upgrade, and technical support issues. Our objective is that network planners and implementers will become more conscious of these requirements.

15.1 User Issues

There are a number of end-user-related issues that determine the success of remote access projects. Although these issues may not directly impact the communications solution that is the focus of this book, they should be analyzed, nonetheless, by the project team handling remote access. It is in this context that we include this topic in our book. We consider the following issues:

1. End-user-device-related issues
   - How many remote devices to support
   - Which is your primary PC—notebook or desktop?
   - Processing power and capacity
   - Size and weight
   - Notebook screen type and size
   - Mobile computer screen resolution
   - Keyboard size
   - Peripheral input devices
   - Touch screen, as a special input device
   - Packaging—the need for ruggedization
   - Logistics of installation in vehicle for in-car applications

2. End-user device configuration issues

3. Client software choices and interface issues

4. Workplace ergonomics

5. Safety and health considerations

6. Human factors and organizational issues
7. Data backup issues
8. Application software upgrade and version control issues
9. User training issues
10. Technical support issues

Let us review these issues one by one.

15.1.1 How Many and Which Remote Devices Should You Support?

Nothing is more visible and more important to the end user than the device he/she holds. It is through this device that the end user perceives the benefits of the mobile computing and remote access. In general, end-user devices should be highly portable. Application requirements might well affect shape, size, and weight, in which case, usability pilot testing should be carried out with different models and devices. Intended users themselves should be involved in the selection of the device. A trade-off may be necessary that should be brought to the users' attention (e.g., functionality and ruggedness over elegance and superior ergonomic design).

With so many remote devices in the market, you should determine which devices you are going to support and for what function. This is especially true for e-mail application. While an Intel-compatible notebook may represent the workhorse of the road warriors, MAC Power-books, PDAs, WIN CE-compatible organizers, two-way pagers, and personal communicating devices based on digital PCS telephones are all vying for the attention of the remote worker. Each device has its own unique operating procedures, software compatibility, and technical support requirements; you should try to keep this variety to a manageable level.

15.1.2 Which Is Your Primary PC— Notebook or Desktop?

Downsizing—figuratively and literally—has become a major phenomenon of the '90s. It is a trend that is reflected in the types of personal computers sold. According to PC manufacturers, the sale of notebook PCs will overtake those of desktop computers in 1997. In fact, according to market researcher BIS Strategic Decisions, not long thereafter, notebook PCs will become the primary PC of choice. The graph in Fig. 15.1 shows the relative percentage of notebooks that will be used as primary PCs. According to this graph, by the year 2000, 80 percent of the 21.6 million PC notebooks sold will be used as primary PCs.

If notebook is the primary PC used by the mobile workers, master files will have to be resident on the remote device. It will also affect data replication strategy and business rules to be used in conflict resolution.

15.1.3 Processing Power and Capacity

The power of a remote device must meet the requirements of the remote access applications. The processor type and its megahertz rating, the amount of RAM installed, the speed and the capacity of the hard disk, the type of graphics card, the CD-ROM access time, the operating system software, and the application software design all contribute to the overall performance from an end user's perspective. As far as performance of remote access on the chosen network is concerned, the type of serial port has a significant effect on the transfer time. (We discussed this point in Chap. 6, while analyzing end-to-end optimization of PSTN.)

Other factors being equal, the configuration of the mobile PC should also match the user's desktop configuration, particularly if the mobile worker spends time both in the office and in the field.

Figure 15.1
Total notebooks installed and percentage of notebooks used as primary PC.

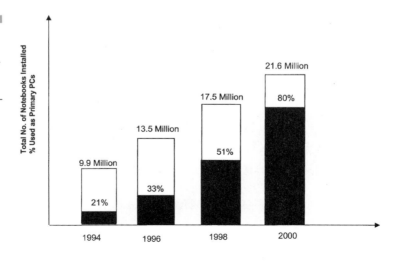

15.1.4 Size and Weight

Two of the most important considerations in selecting a mobile device are the size and weight of remote devices, especially when they are to be carried from place to place during the day. Fully loaded, ultraportable notebooks small enough to fit inside a briefcase, and weighing less than 4 pounds (2 kilograms), battery pack included, are available today. An extra battery is desirable for road warriors, especially on long plane trips, though a few airlines have started providing charging facilities on the plane.

15.1.5 Notebook Screen Type and Size

At present, there are three types of screens commonly available for notebooks—monochrome, dual-scan color, and active-matrix color. Monochrome screens are like black-and-white TV—almost obsolete. Active-matrix color screens are more difficult to manufacture and are more expensive. From the end user's perspective, especially where sales applications are concerned, active-matrix color screens are more desirable than dual-scan screens. Unfortunately, however, color screens do not function well in direct or reflected sunlight, though recent developments in monitor technology will hopefully address this shortcoming.

The size of the screen and its resolution are important factors in terms of ease of viewing and crispness of graphics. Standard screen sizes for notebooks currently range from 11 to 13 in (29 to 34 cm) which is slightly smaller than the average 15-in (38-cm) desktop monitor.

15.1.6 Mobile Computer Screen Resolution

Most notebooks on the market today have an SVGA resolution of 600 × 800 pixels. Some notebook manufacturers are now producing XGA (768 × 1024 pixels) screens (standard on many desktops). Since screen sizes, by nature, are fixed, higher resolutions result in reduced font sizes. To compensate for this, SVGA notebook screens are being made in the 13 to 14 in (30 to 33 cm) range. Although the fundamental requirement that a notebook fit in a briefcase does limit the maximum screen size possible, the combination of these higher resolutions and accompanying larger screen sizes does reduce the strain on the eye.

15.1.7 Keyboard Size

The keyboard will continue to function as the primary mode of input for another few years, at least until such time as speech recognition becomes mainstream. Space availability, especially in airplanes, cars, or customers' offices, dictates that keyboards be at once small and compatible with the desktop variety. The latter requirement is important for end users who spend time in the office, as well as on the road. Compatibility can be compromised, sometimes, in favor of smaller application-specific keyboards with specialized mobile devices if the function keys and numeric pad normally available are not needed.

15.1.8 Input—Keyboard, Mouse, Pen, Touch, or Speech

Whether a keyboard should be used or not depends to a large extent on the amount of data to be entered. Although the keyboard is not a natural method of communication, it has been the predominant method of information input and dialogue manipulation for typing and computer-based applications. For GUI applications, the mouse supplements keyboard input. However, both can be awkward to use in the field when no desk or other physical platform is available. To overcome this problem, supplemental input peripherals, including trackball, pen, touch, and voice, are becoming common. Even though a trackball may be acceptable to sales and professional mobile users, it is not considered favorably by field personnel, or by public safety agencies. In such cases, a pen-based device is more acceptable for manipulating GUIs and modern mobile applications.

A pen-based device can be either active (electronic) or inactive. An inactive pen works with a touch screen, where the pen is used instead of a finger. Both the low cost of inactive pens and the frequency with which pens tend to be mislaid should be considered when selecting a method of data input.

Unavoidable keying requirements, on the one hand, and the degree to which pen-based application manipulation is possible, on the other, should determine which device is assigned the greater weight. In some applications, the pen can be the only input device. In such implementations, minimal keyboard input can be provided through a soft keyboard on the screen.

Voice as a method of input has significant potential for future mobile applications. It is expected that as this technology matures, it will find its way into mobile computing end-user devices. Voice recognition is enabled by way of a microphone connected to a sound card, along with a hardware adapter which converts analog voice information into a digital format. In some cases, this conversion can be accomplished by software. Several speech recognition software packages (such as Kurzwell, Holvox, Dragon, and Verbex) are available under Windows and OS/2.

15.1.9 Touch Screen as a Special Input Device

Touch is intuitive and is commonly used in many tasks in the daily work. Dialing a phone, writing, using a calculator, and thumbing through an address book are all dependent on touch. Capitalizing on this fact, many field and industrial mobile applications have been designed with touch screens that complement pen-based computing. Often this allows for the total elimination of the keyboard, which results in units that are more compact and easier to use in the field.

For simple applications, such as selecting menu items in a restaurant, touch has three major advantages when used on a pen-based computer:

1. It is more natural and intuitive.

2. It is more convenient, even if the pen is tethered to the computer, because the pen does not have to be located or manipulated.

3. It is quicker.

The intuitive nature of simply touching or pointing with a finger is well known and demonstrated by the universal acceptance of touch-screen technology in public kiosks where there is no opportunity to train users to interact with the computer.

In the pen-based computing world, the stylus or the pen serves as a natural method for entering text. However, touch input is more natural and intuitive—not to mention convenient—for many other actions, including pushing keys, selecting menu boxes, or manipulating radio buttons and sliders. This is especially true in the case of many vertical applications—such as a route person making deliveries or a stock clerk checking for stock outs—where input is often intermittent.

15.1.9.1 TECHNOLOGY BEHIND TOUCH/PEN-BASED DIGITIZ-ERS. There are three touch/pen technologies. The first uses resistive membrane. This touch screen is, in fact, a transparent plastic switch that is activated by a finger or a pen. It offers good resolution, low power consumption, and a high digitizing rate. It is also low-cost, proven, and easy to integrate into products. On the down side, these sensors suffer from a scratchable surface, diminished light transmission due to multiple layers of plastic, and poor hand rejection (the ability to reject signals that result from the hand resting on the screen while writing).

The second technology, employed by IBM, uses a glass sheet with discrete conductive lines. The pen is electrostatically sensitive, while the finger is capacitively sensed. Because both finger and pen signals are separately digitized, this system offers good hand rejection. Other advantages are good resolution, high point speed, and a more durable surface. However, the image quality is reduced due to the visibility of the grid lines.

The third technology is used in MicroTouch products. This technology employs a single continuous conductive coating on a glass sheet, and senses both the finger and the pen capacitively. This technique offers excellent optics, low power consumption, high resolution, and high data speed. It also offers automatic hand rejection through a capacitive sensor in the lower barrel of the pen that activates the switch, puts it in pen mode, and invokes hand rejection.

15.1.10 Packaging—Do We Need Ruggedized Notebooks?

There is no doubt that mobile equipment is subject to far more rough handling and movement than its desktop counterpart, and as a result, mobile users, telecommuters, and road warriors have demanded ruggedization for some time now. Consequently, most components are of higher quality, and have to meet tighter specifications for shock absorption, temperature and humidity resistance, waterproofing, and overall packaging integrity. Many notebooks designed for industrial use are encased in a magnesium shell to withstand shock and movement.

In law enforcement and heavy industrial applications, such as environmental inspection applications, more stringent requirements for ruggedization, such as the following may be required:

■ Withstand both cold temperatures (to −35°C) and high temperatures (to 45°C)

■ Withstand humidity levels in the range of 80 to 90 percent

■ Shock resistance of up to 400G

■ Withstand drop test from 3 feet (1 meter) onto a concrete floor

15.1.10.1 HOW MUCH RUGGEDNESS?—ECONOMICS IN AN ENVIRONMENT OF CONSTANT INNOVATION.

Until recently, some users have insisted upon highly ruggedized devices because of rough handling, as well as hostile environmental conditions, under which they work. However, these units cost twice as much as other units to purchase and service, because of the low volume of production and perceived added value. They are also difficult to upgrade. We believe that notebooks will have to be upgraded more and more frequently as technology innovation continues its exponential growth. As such, business-case justification for long-lived ruggedness will increasingly be questioned. Already, senior management has started questioning the need for the high level of ruggedness, and the price tag that comes with it. Instead, they are evaluating the feasibility of buying off-the-shelf semirugged devices. Panasonic has recently introduced a unit (Panasonic CF-25) which meets most ruggedness requirements at an economical price tag.

15.1.10.2 WIRELESS POWER—LIFE OF THE BATTERY PACK.

A remote computer permanently installed in a vehicle can be powered by the vehicle's battery. However, the device's own rechargeable battery is often the only source of power. In such cases, the duration of the charge is an important consideration in selecting the right device. While we should see continued improvements in this area, current limitations constitute an inconvenience factor that has to be endured. The discipline of keeping a fully charged spare battery close at hand is a good one to cultivate.

Power management software is a common feature of most notebooks. A recent innovation in power management was introduced by SystemSoft Corporation. It is the MobilPro suite of BIOS and system control software. When used in notebooks powered by smart batteries, MobilePro enables users to request the battery life needed for a specific job. The product automatically slows down components, as needed, in order to meet the requested time. MobilPro supports hot-docking, and is available in notebooks from Compaq, Hewlett-Packard, and others.

15.1.11 Portable or Fixed Device in the Car— Logistics Problems

In pondering an overall mobile computing solution for field dispatch, courier, law enforcement, and similar applications, a decision has to be made as to whether or not to have a portable end-user device, get it permanently mounted, or one of the available variations on the theme. Several companies specialize in building in-vehicle mobile-device mounts. There are also portable slave units available that work in conjunction with permanently mounted devices. Another alternative is to install a portable-unit docking station in a vehicle.

With the arrival of dual airbags as standard equipment in post-1995 cars in North America, there is now minimal space inside vehicles for the installation of a computer. Experimental new fixtures and cabinetry have been prototyped by IBM and specialty contractors like Johnson & Gimbel. However, the reaction of users to the prototypes has not always been positive because of their tendency to restrict movement on the one hand, while putting the computer out of easy reach, on the other. Contractors are looking at the possibility of completely redesigning dashboards to enable more flexibility with respect to in-vehicle computer installation.

15.1.11.1 LIGHTING SOURCE FOR THE SCREEN FOR NIGHT USE IN VEHICLES.   A screen may either be backlit or use reflective lighting. For public safety applications where ruggedized notebooks are used at night, backlit screens are preferred. If this option is not available, special lighting attachments for external lighting may have to be installed.

15.2 Incompatible Software Configuration Issues

One of the major problems from a technical support point of view is the amount and type of software that remote workers install on their notebooks. A lot of it is often done without the knowledge of the head office staff. While the users may question the need for control by the central support organization, this uncontrolled software installation does make the life of the help desk staff more difficult. We suggest that you should use the following guideline in controlling software configuration. If the organization is paying for the hardware and software, they should put some controls in place for the software configuration. If, on the other hand, the mobile worker is relatively independent, is computer-literate, and pays for his/her own equipment, the help desk should provide support on a best-effort basis; that is, if the problem is possibly due to incompatible software, they should advise the user accordingly, but need not go out of the way to diagnose the problem.

15.3 Ergonomics, Health, and Safety Considerations

There are a number of ergonomic design considerations you should keep in mind. No longer is it acceptable to spend money and time on software development and technology infrastructure only. You should institute ergonomic studies as well. Inconvenient standing position in airports, uncomfortable sitting posture, having to stretch your arms to use the keyboard, lack of arm rest in a car or in a plane are only some of the many problems that may sound as irritants that mobile professionals may accept initially. However, sustained use of devices in an inconvenient fashion manner is not desirable, and the user community should ask the industry to provide solutions. Other ergonomic considerations when selecting a notebook are:

- Does the unit have an adjustable screen angle to accommodate individual preferences?

- Is there an armrest? This can be an important consideration where extended use is anticipated.

- In the case of a vehicle installation, is the notebook mount adjustable (swivel sideways and vertically) with respect to the user's seat?

- Is there a light available for the keyboard, if you are using the computer at night in the car?

As the need is felt and users demand solutions, more and more ergonomic accessories in the form of portable laptop support, nonskid pads, and wristpads have started appearing in notebook stores. Industry researchers feel that more help is on the way, but users had better become vocal and demand notebooks with both portability as well as ergonomics. We also recommend industry magazines, such as *Mobile Computing* and a feature series by George Hobica who writes on the subject frequently.

As mobile computing becomes a norm, and labor unions get involved, they will start talking about possible health issues as a result of extended hours of using notebooks in less-than-perfect mobile environments. While there may not be solid evidence, some recent studies have associated cellular and radio frequencies with health risks, such as cancer. Work in mobile environments is not ideal and can lead to fatigue and stress as well.

15.4 Do You Know Where Your Mobile Workers Are?— The Organizational Issues

Mobile computing has changed the work environment, with the result that telecommuters, mobile workers, and road warriors are no longer under constant supervision. Interesting organizational, ethical, and supervisory issues emerge. Now, with increased independence of the work force, and consequently, less on-the-spot supervision, we need a thinner layer of middle management. On the worker side, we need to raise the standards of professional ethics, whereby the work reports you receive from the staff show genuine progress of the work, and the amount of time spent on various tasks. More and more, supervisors have to develop work appraisal techniques that allow them to document tasks, and measure the results by quantifiable criteria—the real results that count. No longer is being in the office for eight hours and not producing measurable results acceptable performance.

15.5 Data Backup Issues

Remote workers are in hurry. They do not always back up their configuration data and applications on a regular basis. This can be a source of serious problems, in the case of hard-disk failure or computer viruses. If it is a highly centralized organization, with important operational data on remote notebooks, or if the central IS department provides this as a service to the field force, you should back up daily and weekly on an incremental basis. Alternatively, the users themselves should buy external backup drives like Zip drive from Iomega, and back up their data on a regular basis.

15.6 Application Software Upgrade and Version Control Issues

Legacy mainframe application development did not have application software upgrade problems as long as they relied on dumb or semi-intelligent terminal interface. That was before the client-server paradigm came along. Applications and data got partitioned among the two tiers, where application logic on the PC was very much dependent on server logic to complete a unit of work or a transaction. The client software version has to be synchronized with the server portion. Now, we have the Internet-centric model with network computers, Java applets loaded from the network at the time of execution. While this model has tremendous potential for resolving application software upgrade and version control issues, the reality is that the majority of applications being used by remote workers are of the first two varieties where application software must be upgraded frequently. It is not easy to do this with mobile workers who log in from the road, home, or customers' offices, where they do not want to wait for half an hour of software download. To complicate this problem, incremental upgrade for application software is not a common practice with software developers because software distribution is not on their mind when they are concentrating on business functionality. Also, it is more complex.

We recommend following practical approaches to reduce the intensity of this problem:

1. Schedule and mail full upgrades on a diskette or CD on regular basis—monthly or quarterly.

2. Provide a Web site from which the mobile users can pick up new software at their convenience.

3. Send emergency software updates (for fixes) when users log in.

4. Develop server applications, such that they work with multiple versions of client software à la Internet model where Web applications work with multiple versions of the Netscape browser. With this approach, remote users do not get additional functionality unless they are at the latest version level, but do get old functions even if they have not updated their software.

15.7 User Training Issues

Irrespective of how friendly the user interface is and how computer-literate your users are, you should always plan to provide training. Of course, the length of the training program will depend on a number of factors, including how much business process reengineering change you are introducing. The remote users do not visit your offices on a regular basis. They also do not have ready access to technical support personnel. Therefore, you should provide training with an objective of making them self-sufficient. Training should be provided by somebody who has worked in the same department as your users. Only those who have faced those problems that you are trying to solve through automation can understand and explain the new technical solutions to their colleagues. Do not forget to include a budget for training at the time of obtaining project funding.

15.8 Technical Support Issues

Technical support is both necessary and expensive. Problems will occur as you introduce a new application. Situations that you never recognized during formal testing will arise and result in application failure during rollout. The hardware is going to need maintenance services. Remote users should be trained to handle these situations and to get help from the technical support group. Please remember that mobile users become dependent on their notebooks, and would feel more exasperated when computer applications do not work out to their satisfaction. They may lose a sale or not be able to complete a service request. When they start trying to fix the problem on their own, they may create more problems. The loss in productivity, opportunity cost of time spent in fixing problems, software upgrade costs, and technical support costs are only some of the components that constitute the total cost of ownership of PCs.

In case of hardware failure, special arrangements should be made to send a loaner unit, fully loaded with application software, by overnight express service, such as FedEx or UPS.

Of course, the help desk should have suitable tools to debug problems remotely, download fixes, and provide timely support. Only then will user issues no longer be issues.

Summary

In this chapter, we have described some of the user issues with remote access. We have done this to highlight some of the concerns that remote access project teams must attend to. The foremost among these is the suitability of end-user devices and user interface to business applications. In this evaluation, the ergonomic design of the device should also be considered. Data backup, software upgrade capabilities, and user training requirements should not be among low priority tasks that get done if there is enough time and money. While evaluating remote access hardware and software, the team should ask the vendors how their solution addresses these issues. Network management capabilities of the proposed solution will certainly assist the help desk in solving technical problems expeditiously.

16

Remote Access— Tariff and Bandwidth Management*

Give the mobile users remote access without control; they will use it to maximize their personal convenience first and organizations' costs second. Productivity will improve but telecommunications costs will skyrocket. You better learn how to manage bandwidth efficiently and apply conservation principles. Inexpensive bandwidth will always be in short supply.

—*Chander Dhawan*

---

*This chapter is based on a white paper produced by Shiva Corporation of Bedford, Massachusetts. The author has made appropriate changes and thanks Shiva corporation for their permission to use this material.

About This Chapter

In this chapter, we discuss the need for managing bandwidth and network costs so that we utilize the network for useful data traffic only. We
must understand the need for keeping system overhead traffic to a minimum, especially when we use LAN protocols over a WAN. We discuss analyzing various network services and selecting an optimal set to keep our
telecommunications costs under control. We explain compression and
spoofing techniques in order to achieve higher network efficiency.

16.1 Need for Tariff and Bandwidth Management

Network managers have been surprised by telecommunications costs for
remote access. There are a number of reasons for this unexpected rise in
these costs. First of all, there are a number of carrier services that have different tariffs. As a result of deregulation in the telecommunications
industry, you can negotiate special deals with Telcos and network service
providers. Mobile and remote users' requirements can vary from time to
time. In fact, these requirements are quite difficult to estimate at different
times of the day, even though some ISPs can now provide reliable statistics
on usage patterns. The time zones when these services are used have an
impact on these costs. While your telecommuter programmer may prefer
to use network services during off-peak hours, your sales professionals do
not want to wait for the cheaper tariff during the evening. You may want
to use additional bandwidth for videoconferencing during certain periods. In an environment like this, you need sophisticated techniques and a
supporting business discipline to manage both the tariff and the bandwidth usage.

16.2 Bandwidth All the Time versus Bandwidth on Demand

Until recently, most corporations based their wide area networks (WANs)
on leased lines. Traditional OLTP applications have a relatively determinis-
tic usage pattern. Therefore, you can determine network bandwidth

requirements more easily and satisfy this demand through a private network. Leased-line networks provide constant bandwidth all the time, meaning the organization pays for the telecommunications lines regardless of the time they are in actual use. While this technology is sufficient for enterprises that need consistent connectivity with fairly constant usage, it does not meet the performance and cost control needs of organizations that have a large number of users and offices accessing client/server applications remotely. Mobile users, telecommuters, and many small but remote offices have changed the network picture significantly during the past few years. The Internet and Web applications based on graphics and video make high-bandwidth requirements at different times during the day.

Branch offices or remote users that need access for only a few minutes to a few hours a day are better served by WAN-switched services such as ISDN and frame-relay. In the ISDN environment, fast call-setup times and other attributes of switched services enable bandwidth on demand, meaning bandwidth is available when it is needed and charges are incurred only when data is actually being transmitted over the line. With switched services such as ISDN, it is cost-effective to connect even the smallest remote or home office.

Despite the many advantages of switched services, however, they must be managed properly in order to keep spiraling telecommunications costs under control and realize the maximum benefits for the organization. This requires that remote access vendors and network service providers provide solutions where bandwidth can be increased or decreased based on changing demands during the day. Ideally, a network provider should be like an electric utility from which industrial consumers can draw as much electric current as they need. They pay for the amount of power used, not for having a voltage with a capacity to provide this current. Bandwidth management in an information network is a unique set of technologies designed to help companies minimize the cost of using switched services.

16.3 Tariff Management— Main Areas to Control

Tariff management is based on three technology areas:

1. Bandwidth control
2. Connection control
3. Data control

Bandwidth control reaps maximum network efficiency at minimum cost by deploying flexible and dynamic bandwidth-on-demand techniques.

Connection control provides the most efficient way of connecting remote locations. This is based on taking advantage of different tariffs and on prioritizing connections. Connection control also provides fast and efficient recovery from failure.

Data control makes the most efficient use of available bandwidth by using spoofing and triggered routing update techniques. It ensures that usage-sensitive LAN-to-WAN services such as ISDN are not left on, when there is no data to send. Data control also utilizes data compression to squeeze as much data as possible into the available bandwidth.

When these three innovative concepts are integrated under a tariff and bandwidth management umbrella, network managers are able to gain the greatest possible monetary and competitive value from remote network access.

16.4 Bandwidth Control

LAN-to-LAN traffic is inherently sporadic. Bandwidth control ensures that WAN services are used only when required and closed down when there is no user data transmission. This is critically important when services are being paid for, regardless of the amount of traffic being transmitted across the network. It also ensures that optimal services are used for particular applications and/or particular remote sites, and that extra bandwidth can be made available when there are unexpected bursts of traffic.

Only by combining these bandwidth control features can network managers be confident that WAN costs are minimized and the most flexible service is available. There are four key areas of bandwidth control:

1. Bandwidth on demand
2. Minimum call duration timer
3. Bandwidth aggregation and augmentation
4. Switchover

16.4.1 Bandwidth on Demand

In the ISDN environment, fast call-setup times and other attributes of switched services enable *bandwidth on demand,* meaning bandwidth is available when it is needed and charges are incurred only when data is actually being transmitted over the line.

With bandwidth on demand, a call is opened only when there is data to send and then closed as soon as the data is sent. This is totally transparent to users on the network. For example, when users are running a Web browser to access a remote Web server via ISDN, they cause an ISDN connection to be opened at the point of first access to the Web. While they are reading the data they have received, the connection times out because no data is being sent or received. As soon as they access the next page of information, the connection is reopened. Since the time to make the ISDN call is so rapid, the users appear to have been connected all the time.

The time-out parameters are usually configurable on the ISDN access devices and the most suitable values will depend on carrier tariff policy and the applications being used.

16.4.2 Minimum Call Duration Timer

The minimum call duration timer is an extension of the bandwidth-on-demand time-out. Many carriers have a minimum call time that is different in length (and possibly in tariff rate) from subsequent call times. For example, the minimum call time may be three minutes and thereafter, the tariff is per minute. Having a separate configurable timer to handle this creates the flexibility needed to successfully manage costs.

16.4.3 Bandwidth Aggregation and Augmentation

With the availability of combined bandwidth, extra data channels—an ISDN B channel, a leased line, an X.25 virtual circuit or a dial-up circuit— are used only when the existing channel capacity is saturated. Channels are shut down when the extra bandwidth is not required. Bandwidth can be increased by combining channels of the same network type, such as two ISDN B channels, or by combining channels of different types, such as an ISDN B channel and a leased line.

Combining the bandwidth of two or more channels of the same type, on the same interface of across interfaces, is termed *aggregation*. In this scenario, when a router receives the first packet for transmission, a channel is opened to the remote router. A further channel is then dynamically opened when the number of packets or bytes queued exceeds a certain value, which is normally user-defined. After each new channel is opened, there is a short delay before a subsequent channel is opened, allowing the existing queue to be emptied.

When the measured data throughput indicates that fewer channels are needed, data is no longer transmitted on the channel that was opened last. If both ends stop sending data, the channel is closed after a user-specified time-out. This latency is used to accommodate bursty traffic patterns. (See Fig. 16.1.)

Channels from different interfaces can also be combined. For instance, one channel on an interface is specified as primary while another is specified as secondary. Channels on the primary interface are used before channels from the secondary interface. This technique is used to combine bandwidth from interfaces of similar speed.

Adding bandwidth from a different type of interface is known as *augmentation.* Using an ISDN B channel as on-demand bandwidth for a leased line is a common application of combined bandwidth. This allows a 64-Kbps leased line to be used for average load, while an ISDN B channel is added when the leased line is saturated.

16.4.4 Switchover from Less Expensive to More Expensive Link, When Needed

Switchover enables traffic to be moved from one circuit to another, depending upon the traffic rate. A slow-speed leased line running at 19.2 Kbps can be linked to a 64-Kbps ISDN B channel. When the traffic rate on the leased line reaches saturation, the ISDN link is opened and traffic

Figure 16.1
Bandwidth requirements during the day. (Source: Shiva White Paper.)

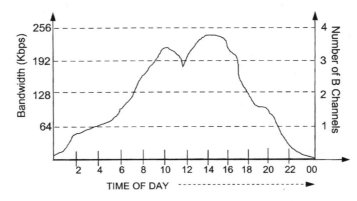

moved to it. Once the traffic rate drops below that of the leased line, the ISDN link is closed down and traffic diverted back to the leased line. The threshold at which traffic switches can be defined by the user. Switchover ensures that the most cost-effective circuit is always used, and provides a very cost-effective solution for networks with changing bandwidth needs throughout the day.

16.4.5 Bandwidth Control in RAS Products

Many RAS products support bandwidth control features either manually or automatically. As an example, Shiva's *Integrator* line of products supports all the bandwidth control features described previously. Dynamic bandwidth aggregation and augmentation are achieved through a dynamic multilink mechanism.

16.4.6 Bonding—Method of Aggregating Data Channels

When data channels are aggregated together they provide a wider pipe through which you can send the data. This is analogous to adding additional lanes to a single-lane highway. More traffic bandwidth is made available. There are various techniques which can be used to manage the data down these combined channels. These range from the simplest round-robin way of sending one data packet down each channel in turn, to splitting the packets of data into fragments and sending each down a different channel. Both PPP Multi-link and bonding use the more complex fragment approach.

The PPP Multi-link Protocol (RFC 1717) is an extension of the PPP (Point-to-Point Protocol-RFC 1661) standard. It describes a standard method of combining channels to ensure packet ordering and compatibility between manufacturers of internetworking equipment. It employs a method known as *packet chopping,* wherein individual packets are chopped into smaller fragments of a uniform size. These fragments are then distributed among all the channels in use. Because it is a software solution, PPP Multi-link is limited in the number of channels that can be combined at any given time. However, it does allow internetworking products to combine channels of any type, not just ISDN.

Unfortunately, there are currently no standards for dynamically aggregating additional data channels using PPP, for ISDN or any other service.

Bonding allows channels to be combined at the physical framing level. It gets its name from the Bandwidth On Demand Interoperability Group. It is independent of the framing protocol used. Bonding is a very efficient method of combining channels because it is normally performed in hardware without any software packet handling overhead. Bonding is particularly effective when used with primary rate ISDN (PRI). With PRI, up to 63 B channels can theoretically be combined to provide a very high speed link. This bandwidth can be used to provide the primary method of connection between sites, or as a convenient backup to high-speed leased lines.

Bonding is based on an open standard so it provides interoperability between vendors. It is independent of any higher-layer protocols. Bonding can only be used with ISDN. Also, there are various modes of operation for bonding and the most common mode does not support the dynamic addition and removal of circuits on demand.

16.5 Connection Control

Connection control provides the most efficient way of linking remote locations. This is based on tariff parameters or prioritization. Connection control also provides fast and efficient failure recovery.

16.5.1 Time-of-Day Tariffs

WAN services are subject to different tariffs at different times of day. Usually, these tariffs are lower during the evening than at peak usage times in the day. Since networks operate continuously and often replicate data at night, it is important to be able to take advantage of these lower tariffs.

Network managers can take advantage in several ways. For example, they could use X.25 or frame-relay during the day when interactive traffic is high, and employ ISDN at night for data replication and backup. Or, they could make sure that ISDN was not being used at peak times at all, by preventing any remote access during certain periods.

In some circumstances, it is useful to switch off the ISDN link for a period of time, to make sure that applications are not stuck in a loop, using valuable bandwidth.

16.5.2 Callback

Callback is another tariff-based technique. Tariffs between two remote locations frequently vary, depending on which site initiates the call. This is particularly applicable to domestic and international long-distance calls. With *callback,* when a remote site calls a second site, the second site closes the call and dials the first site back. When this is done via ISDN, CLI (calling line identification) can be used for the callback number. This is very efficient as the initial call will be refused, incurring no charge at all in this direction. In the case of ISDN and dial-up, the PPP-based PAP or CHAP can be used by network managers to identify which remote site should be called back. Callback can also be used for security purposes or for centralized or decentralized billing, by an Internet provider, for example.

16.5.3 Prioritization

Circuits can be prioritized to ensure that low- or high-priority calls are opened or closed down depending on their importance. For example, consider a router with one basic rate ISDN interface. The two B channels on this router can be aggregated to provide 128 Kbps of bandwidth to another router. If router one needs to call a third router, but another call is in progress, circuit prioritization could give priority to the router one call. In this event, router one would close down the second call to router two and dial up router three. The call to router two would continue with reduced bandwidth. (See Fig. 16.2.)

16.5.4 Failure Recovery: Transparent Backup and Multiple Static Routes

Transparent backup and multiple static routes are two techniques that offer cost-effective resilience against communication failure. Communication failure can be caused by either an internetworking device failure or a network failure. Both of these potential problems—and the right internetworking product to solve them—must be considered when designing a resilient network.

Transparent backup is applied when two WAN interfaces on the router provide different paths to the same destination. One interface may be

Figure 16.2
Prioritization and
bandwidth manage-
ment. (Source: Shiva
White Paper.)

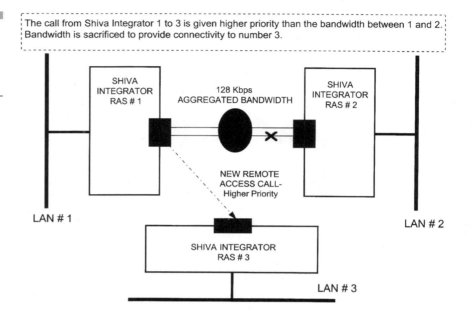

The call from Shiva Integrator 1 to 3 is given higher priority than the bandwidth between 1 and 2. Bandwidth is sacrificed to provide connectivity to number 3.

used as the primary circuit, while the second acts as the secondary or backup circuit. If the primary circuit fails, the secondary circuit is automatically activated. When the primary circuit comes up again, traffic is transferred from the secondary circuit back to the primary circuit. If, for example, the primary circuit is a leased-line service with the backup circuit running across ISDN, the user would pay for just the ISDN backup when it was needed. (See Fig. 16.3.)

It is vital that routers be capable of switching from one circuit to another without requiring routing protocols such as the Routing Information Protocol (RIP). Routing protocols often take up to three minutes before implementing the desired topology change. A delay of this length can cause serious problems in a network running critical applications. Data can be lost if there is no known route. With transparent backup, switching to a backup link is done transparently when a link failure is detected. As a result, backup is provided immediately and with no data loss.

Multiple static routes are another cost-saving feature that avoids running routing protocols such as RIP over WAN links while still providing a backup service. When a failed static route is detected, a router employs the next best route. If the original best route returns to service, it replaces the alternative route.

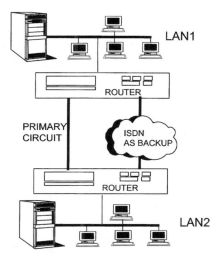

Figure 16.3
Transparent backup.
((Source: Shiva White
Paper.)

As the router determines which route to use, there is no need to run RIP over the WAN links. This leads to significant WAN savings as running RIP can be expensive.

It is important to understand the difference between transparent backup and multiple static routes. Transparent backup is used to provide two alternate circuits to the same remote router. Only one route will ever be advertised via the routing protocols, regardless of which circuit is in use. This mechanism is transparent to all devices—except the two communicating routers—whether a primary or backup circuit is providing the connectivity. Multiple static routes can be used when connecting multiple routers. The routers will advertise the new route locally, an action that will be detected. Multiple static routes are therefore not transparent.

16.5.5 Support for Connection Control

Several RAS products such as Shiva *Integrator* support all these Connection Control features. Services can be dynamically changed depending on the time of day. Callback can save considerable money as well as provide flexibility. Circuit prioritization means that key remote locations are always guaranteed bandwidth in any circumstances. Multiple static routes and transparent backup give added levels of backup and security that are vital for an efficient network.

16.6 Data Control

The third element of bandwidth management is data control. It involves controlling the data sent over usage-sensitive services such as ISDN. There are three primary data control components:

1. Data compression
2. Triggered routing protocol updates
3. Spoofing responses to service or housekeeping messages

16.6.1 Data Compression

The issue of compression in a routed network is a complex one, and there is no definite solution. However, it is clear that data compression in many networking environments will increase user response time and increase the volume of data transmitted in a given time. As a result, network costs will be reduced.

The right choice of tools and techniques for any particular network will depend on:

- Traffic bandwidth requirements
- Traffic protocols
- Nature of traffic
- Network topology
- Application latency requirements

16.6.1.1 WHY USE COMPRESSION? Here are some reasons why using the right form of compression can improve your network. It provides:

- Continued use of a legacy low-speed line despite increasing bandwidth requirements. Getting an average 2:1 compression ratio across a 9.6-Kbps leased line gives it the bandwidth of a 19.2-Kbps line.
- Improved latency across a low-speed line. If the routers at the ends of a slow-speed leased-line connection can compress the data sufficiently, any latency improvement may be observed.
- Reduced network costs on a time-based tariffed service such as ISDN.

16.6.1.2 TYPES OF COMPRESSION. There are three forms of compression: header compression, body compression, and link compression. Each has its advantages and disadvantages, and should be employed in different circumstances. It is worth noting that in internetworking, all compression algorithms must be *lossless* (the packets must look the same following compression/decompression as they did initially). Therefore, in data communications, different algorithms are used than those employed in the fields of voice or video compression, where *lossy* algorithms are used with the expectation of a drop in signal quality. This would be unacceptable in data communications.

HEADER COMPRESSION.   In any protocol, a packet consists of header information which defines where the packet is to go, and what type of information it contains. This is followed by the information itself. In a link dedicated to one type of traffic between the same two hosts, this header information does not change, yet is duplicated in each message. Where the body content is small, the header information forms the larger percentage of the bandwidth generated, even though it serves no purpose.

Header compression removes these duplicated headers before the packet is sent over the link, and regenerates them at the remote end. This technique can be employed to great advantage for interactive protocols such as Telnet and X-Windows, where typically the packet content may be only one byte.

The most common form of header compression in the Internet world is Van Jacobsen Otilde's header compression, which is defined by IETF's RFC 1144.

BODY COMPRESSION.   Where applications are communicating using protocols with large body contents, compressing the body will achieve a greater effect than compressing the headers. Although many forms of compression algorithms may be employed, the choice should be made based on the memory or speed requirements. Both sides of the link must, of course, agree on the algorithm to be used.

A good example of successful body compression is the transportation of Microsoft LAN Manager packets across a WAN. These packets typically contain a large percentage of repeated characters and empty space, and are therefore ideal candidates for compression.

LINK COMPRESSION.   On point-to-point links, the entire data stream may be compressed and regenerated at the remote end. This is a protocol-independent mechanism and may be implemented in devices separate from the internetworking equipment used for the transmission.

Claims of compression ratios vary considerably from different manufacturers. Many advertised claims are for best-case scenarios for preconfigured data. This is often a poor reflection of real data compression that you may achieve in a real-life operation. The best method of testing a compression algorithm is to try it out on a specific data stream that your own application suite may generate.

16.6.2 Compression Techniques and Standards

Apart from Van Jacobsen's header compression, there are currently no other RFCs in this area. There are, however, a multitude of Internet Drafts. One Internet Draft defines the compression control protocol used to negotiate, at link initiation, which compression protocol, if any, should be used. The others define the use of each of the compression protocols within the PPP framework.

16.6.2.1 THE PPP COMPRESSION CONTROL PROTOCOL.   This protocol is a mechanism which allows two PPP peers to negotiate which of the compression protocols, if any, they will use. Technically, this draft has been ready to progress down the Internet standards track for some time.

16.6.2.2 THE PPP BSD COMPRESSION PROTOCOL.   This uses the widely implemented Unix Compress compression algorithm, the source for which is widely and freely available. The PPP BSD algorithm has the following features:

- Dynamic table clearing when compression becomes less effective
- Automatic compression shutdown when the overall result is not smaller than the input
- Dynamic choice of code width within predetermined limits
- Effective code size that requires less than 64K bytes of memory on both send and receive

16.6.2.3 V.42bis.   This has been proposed by the CCITT (now the ITU-T) as a compression standard to work in association with the V.42 error-correction protocol for modems. It uses a variant of the Lempel-Ziv-Welch

(LZW) compression algorithm and can be implemented in hardware or software. It has the benefit of automatically turning compression on/off depending on the compressibility of the data stream.

16.6.2.4 THE PPP PREDICTOR COMPRESSION PROTOCOL. PPP Predictor compression protocol is the algorithm intended to be the vendor standard. According to the Internet Draft which defines it, "*Predictor is a high-speed compression algorithm, available without license fees. The compression ratio obtained using Predictor is not as good as other compression algorithms, but it remains one of the fastest algorithms available.*"

16.6.3 Triggered Routing Protocol Updates

In an IPX environment, routers use RIP (route information protocol) to broadcast their current routing tables every 60 seconds. Also every 60 seconds, servers use SAP (service access point) to broadcast their currently available services. Routers also use RIP to broadcast their routing tables every 60 seconds. When these broadcasts are allowed to traverse usage-sensitive WANs, costs soar.

However, there is a problem with stopping these broadcasts: If they are eliminated, the routers and servers cannot communicate with each other and do not have a complete view of the network. There are at least several ways—some of them flawed—to minimize these broadcasts.

Timed updates are a flawed method of transmitting network changes at predefined intervals. It is a simple method that unfortunately sends messages and opens ISDN calls when there are no changes, thereby increasing traffic with no accompanying benefit. If the timer is set to a small interval, the network load can be increased by about 15 percent, but if the timer is set to a high interval, updates are often transmitted too late to be useful.

Piggybacking is another flawed technique. It allows some traffic types to be designated as transmittable across the WAN if the link is already open, but it does not allow the link to be opened solely for the purpose of transmitting this traffic. When the link is opened for some other user traffic, the routing update information is allowed across. The exception to this behavior comes when a routing link fails, and the routing updates are forced across. With piggybacking, updates are not acknowledged.

Triggered RIP (IP and IPX) and SAP (IPX only) routing updates are more efficient. They only broadcast across the WAN when the available services or network topology changes—an infrequent occurrence in a well-behaved, stable network. In addition, these updates must be acknowledged—meaning that the remote routers have successfully received the updates. By running Triggered RIP and SAP, network managers can realize the benefit of the protocols without the network overhead costs.

16.6.4 Spoofing

The term *spoofing* is used differently by WAN internetworking vendors, and is defined as a set of techniques to keep service packets or network housekeeping information off the WAN link while fooling the network into thinking the frames have been sent.

Spoofing is most applicable to Novell NetWare networks. The Novell protocols were written assuming that devices were connected to LANs where bandwidth was not an issue. Many types of service or housekeeping packets are sent between devices. These include:

- IPX keep-alive (*watchdog*) packets
- SPX keep-alive (*probe*) packets
- NetBIOS over IPX keep-alive packets

16.6.4.1 IPX KEEP-ALIVES.   NetWare is a client-server operating system. When using IPX, each client (user) logs on to a server on the network, and for the duration of the login session, the server PC sends keep-alive packets to reassure itself that the other side of the session is still live. The period of the keep-alive packets is configurable but is typically measured in minutes. Keep-alive packets are also known as watchdog packets.

Even though a user may not be doing any live work, the session will be considered as live until the user logs out from the server. A typical NetWare user may choose to be permanently logged into one server, rather than logging in and out every day.

16.6.4.2 SPX KEEP-ALIVES.   SPX is similar to IPX, but is a connection-oriented protocol which sits above IPX and is used by applications to provide guaranteed delivery of packets in the correct sequence. The SPX header includes the IPX header of 30 bytes, and then adds another 12 bytes for sequencing, flow control, connection, and acknowledgment information. The SPX protocol also uses keep-alive—also known as probe—packets, but in this situation, they are sent both from the client and the server. Lotus NOTES is one application which uses SPX/IPX.

16.6.4.3 SPOOFING FOR NETBIOS OVER IPX. Corporate internetworks are increasingly built using NetBIOS over IPX as the common transport protocol; this is the standard configuration in Microsoft Windows 95. Microsoft used to standardize on the LAN NETBEUI protocol, which relies on frequent broadcasts to communicate between hosts in the same way as IPX does.

NetBIOS over IPX carries on this practice of communication via broadcasts, so in order to spoof a NetBIOS/IPX network successfully, the WAN router must be able to handle spoofing at both levels simultaneously.

SPOOFING EXAMPLE—IPX/SPX. Spoofing of IPX and SPX keep-alives and NetBIOS over IPX is essentially the same. Consider the following SPX example.

In a typical client-server SPX interaction, one station (the client) requests services from another station (the server). Before submitting any requests to the server, the client must establish a connection. When the client has completed its interaction with the server, the connection is closed down. Once a connection is established between client and server, there may be periods of inactivity when no data is sent by either workstation. During this idle period the server will send keep-alive or probe packets to the client to ensure that the workstation is still up and running. When both the client and the server are on the same LAN this presents no bandwidth or cost issues.

If the client and server are at either end of a switched WAN service such as ISDN, the ideal solution would be to keep the connection closed, ensuring the periods of inactivity. However, the keep-alive or probe packets keep the connection open. This is where spoofing comes in.

Spoofing reduces network costs significantly because keep-alive packets are responded to locally by the router and not sent across the WAN link. This means the WAN link remains closed during the periods of inactivity. This technique is called spoofing because, by responding locally to keep-alive packets, the router *spoofs* the client and server, making them act as though the WAN connection is still active. (See Fig. 16.4.)

SPOOFING TCP/IP. The TCP/IP protocol stack was designed 20 years ago with WANs in mind. It employs few broadcasts. One example is directed broadcasts, wherein the broadcast is automatically restricted to the LAN on which it originates. The only broadcast common to TCP/IP is the RIP broadcast, used by routers to advertise the routes known to them. Since

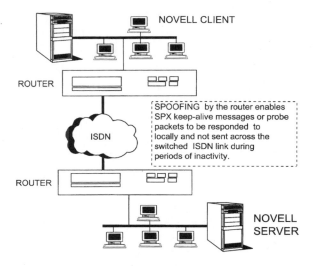

SPOOFING by the router enables SPX keep-alive messages or probe packets to be responded to locally and not sent across the switched ISDN link during periods of inactivity.

the release by Microsoft of a TCP/IP stack for Windows, internetworks built on NetBIOS/IP are frequently used. Spoofing at the IP level is unnecessary because IP is more WAN-friendly than IPX. However, spoofing is still required for the NetBIOS protocol.

16.6.5 Example of Cost Savings with Spoofing and Triggered Updates

Both spoofing and triggered updates prevent calls being made to send data that is not necessary. These features can save enormous amounts of money. Consider the following simple example which shows the high costs of sending unnecessary data.

You have a long-distance, ISDN-based Novell Network between two locations. You probably would not have considered using ISDN if these locations required access for more than two hours per day, so assume that this is the period over which real data is actually being sent. Assume a call charge of $0.03 for the first minute and $0.01 for each additional minute. Assume the average call is in progress for five minutes, so there are 24 calls per day for real data.

16.6.5.1 COST SAVINGS DUE TO SPOOFING OVER ONE DAY

■ Assume 24 calls of five minutes each. The first minute costs $0.03 and four subsequent minutes cost $0.01.

- 24 * ($0.03 + (4 * $0.01)) = $1.68 per day. However, if you do not have spoofing and/or you are using normal, not triggered RIP, there are many additional calls to send service frames or housekeeping data. This could easily result in one call per minute to send this unnecessary data. This amounts to one call every minute of every additional hour that no real data is being sent—22 hours in this example.

- 22 * 60 calls of a few seconds each, so it costs $0.03 for every call.

- 22 * 60 * $0.03 = $39.6 per day.

In this example, since the initial call charge is more expensive than the per minute charge, it would actually be cheaper to keep the call open all day.

With spoofing and triggered RIP/SAP, these calls will not be made. This is a simple example, but it does illustrate the enormous savings that these features can produce.

Summary

Before network managers can truly optimize their networks, they must have a firm command of the many tools and techniques associated with bandwidth control, connection control, and data control.

The concept of bandwidth control is initially simple to grasp, but complex to implement. It is based on the notion that WAN services should be used only when they are needed, and paid for only when they are used. Through the use of techniques such as aggregation, augmentation, switchover, and minimum call duration timing, bandwidth control is well within reach.

Connection control is part art and part science. The art involves having the imagination to devise strategies that optimize the same WAN elements that are available to all users. The science involves taking the time to thoroughly understand WAN elements so they can be employed to maximum user advantage.

Data control requires an in-depth knowledge of network protocols and how spoofing can be used to optimize them. Primarily applicable to the widely installed base of Novell NetWare networks, spoofing can eliminate wasteful, expensive network misuse. At a time when every expense is scrutinized, this kind of data control can make a significant corporate contribution.

There are many costs to consider when implementing a network solution. The cost of transferring data across the WAN is quantifiable and is the largest cost associated with managing a network.

Source

This chapter is based on a white paper produced by Shiva Corporation of Burlington, Massachusetts, one of the pioneering organizations in the remote access industry. The white paper was written by Paul Gowns, a product development manager for the Shiva *Integrator* range of products and Val Wilson, a product marketing manager in Shiva Corporation. The author thanks Shiva Corporation and the original authors of the white paper for their permission to include this material in this book. The author has edited the original white paper to suit the needs of the book.

17

Remote Access— Network Management

The world is full of paradoxes. We create a solution to satisfy a need. In the process, we create sub-problems which we must solve and manage for the good of the bigger problem that we started to solve in the first place. Network management falls into the same category when it is reactive—an afterthought.

—*Chander Dhawan*

About This Chapter

In this chapter, we discuss network management requirements and tools for remote access. Since many of the tools in question are still under development, we concentrate our discussion on operational management requirements and the work of standards-setting bodies such as the Mobile Management Task Force (MMTF).

17.1 Justification for Remote Access Network Management

There has been a lot of attention in the trade press on the total cost of ownership of PCs, especially the mobile devices. Irrespective of the controversy surrounding the actual cost, there is no doubt that managing mobile devices and providing technical support to remote users costs more than fixed desktop devices inside the office. Both the number of users wanting to access corporate information servers and the traffic from remote devices are increasing at a much faster rate than the capability of the central infrastructure (such as remote access servers or back-end network) to handle them. Network downtime is prohibitively expensive. According to some studies done by Infonetics, the average revenue loss for one hour outage in certain financial services companies can be as high as $78,000 per hour. You can very quickly estimate a similar dollar number for your organization. You may note that your mobile workers are, quite often, crucial human resources in the organization—field service people, sales staff, or senior business executives.

It is relatively easy to build a business case for network management of remote networks.

17.2 Remote Access Network Management Requirements

Operational management requirements of a remote access application are in many respects no different from any other IT application. We still want to diagnose problems when they occur, measure performance when users complain, and sometimes we do this for planning and avoiding future problems. However, when it comes to network management techniques, mobile users and remote access networks do have certain unique requirements.

Apart from anything else, the use of different network technologies (some of them still in a nascent stage) and connections that are temporary and often tenuous makes remote access networks intrinsically more complex, on the one hand, and less robust, on the other. Being on the move most of the time, mobile users are often not easy to locate or contact, their only telephone connection busy transmitting data. Additionally, security is a far bigger problem with mobile user devices than with fixed desktop PCs.

Unfortunately, to make matters worse, mobile workers are also among the most demanding of users and expect everything to work properly 100 percent of the time—especially in front of customers, or at night in hotel rooms when there are only minimal (if any) support staff on duty. To top it all, this expectation is often exacerbated by a tendency on the part of these same workers to tinker with hardware and software configurations themselves.

There are several tasks that technical support staff have to perform to keep a system operating smoothly, including:

- Network problem management
- Network asset management
- Network change management
- Network software upgrades
- Network performance monitoring
- Application status monitoring

17.2.1 Network Problem Management

In large organizations, network managers are preoccupied with multivendor internetworking (especially LAN-WAN integration) problems, ATM, Ethernet switching, call center implementation, and with the integration of different network management schemes (increasingly based on the SNMP standard) into a single enterprise-wide solution. Now, in the midst of all this preoccupation, users are starting to demand better remote access technical support. These demands can quickly—and only too often do, judging by the experiences of early implementors—become major stumbling blocks for network managers as they struggle to cope with emerging technologies and vendors who concentrate their own efforts on providing functionality first and standards-based network management second. (Or third, or fourth! In the absence of generally accepted industry standards, vendors tend to provide only a limited amount of proprietary implementations of network management capabilities.)

Problem management deals not only with recording of a problem, but also with diagnosis and assigning the problem to a specific hardware or software component failure. The following list describes a typical set of problem-resolution requests that a hot line receives from the mobile workers:

Component	Type of Problem or Change Request
Mobile-device-related	Setting up new user IDs and passwords
	Installing new hardware, software drivers, and software applications
	Recovering and downloading mistakenly erased data files
	Hardware failures
	Application hangs
	Out-of-memory messages
	Virus alerts
Operating-software-related	Missing drivers
Application- or database-related	Application hangs
	Older versions of application software encountered
	Data files corrupted
	Problems running multiple applications
Communications-related	Communication sessions interrupted
	Failures during file transfer
	Slow network (poor response time)
	Cannot connect—weak coverage (poor signal strength)
	Network modem failures
Security	Can not log in
	Want to change password
Server-related	Server hangs
	Backup/recovery problems
	Databases down

In order to diagnose the cause of the problem, the network help desk staff needs tools to monitor the status of various components. The first challenge they face is that they do not always have monitoring capability or status information about the health of their temporary network connections—be it switched dial-in link, ISDN, Internet, ADSL or wireless network. In most cases, it is a public shared network. The second challenge to the network staff comes from the mobile workers' tendency to experiment with their machines more often on the road than in the office. Many enterprising users try to be self-reliant in computer networking, despite not having had the necessary training to make these changes on their own. Such well-meant, but ill-advised, experimentation often leads to more problems, and ultimately to more calls for help.

17.2.2 Network Asset Management

How to manage network assets when user locations are not known? How to ascertain user configurations when the users themselves have not been seen for days or weeks? How to upgrade user software when there is no way of ascertaining available disk space? Which management software to use at the central site when there are thousands of mobile users? Is the software scalable? Is there a system administrator robot that constantly polls remote mobile users to retrieve stored status information, even if the MMTF's Mobile Management Information Base (MIB) is implemented across the enterprise? These are some of the open-ended questions that have to be answered if accurate information is to be kept about mobile users' computer configurations.

Whenever possible, the same network management software should be used as is used by the organization for the wired network, especially if it is SNMP-compatible. Once a mobile MIB is available for end-user client devices, it can be included in the network asset-management software.

17.2.3 Network Change Management

Remote access networks are very dynamic, with new users being added every day and existing users constantly adding and changing hardware.

When new users become a part of the network, they must be assigned network addresses and given user IDs and initial passwords. As well, it is often necessary to create custom boot diskettes for new users containing information that cannot be sent over the air link because of security policies.

When existing mobile users upgrade hardware or send hardware in for repair, additional demand is created for new user IDs and passwords.

All this in turn results in a need to constantly update the configuration database.

17.2.4 Network Performance Monitoring

One of the most common complaints from mobile users is that remote access networks are slow compared to wired LANs, and that performance quality varies with geographic area, especially in public shared networks. Understandably, network managers prefer to be thoroughly conversant with network performance in a proactive manner prior to any onslaught of user complaints.

To be able to discuss problems with a network provider, a network performance-monitoring capability is needed so that network transit times can be identified in various components (see Fig. 13.1), starting with application processing times in the client, network, and information-server components.

As far we can tell, not many tools are currently available for measuring network transit times in remote access networks, but sample transactions could be time-stamped and measurement data collected. This data should be used to prepare service-level reports on a regular basis. Unlike a private WAN where you can predict load characteristics more predictably, a shared network is more volatile and you may never know how many other users the network service provider has started servicing.

17.2.5 Software Upgrade Requirements

Users are constantly changing their operating systems software. Today they might be using Windows 3.11; tomorrow, Windows 95. As users upgrade operating system software, they need help in upgrading software drivers and OS-dependent configuration files. While self-loading diskettes or CD-ROMs can and should be issued for larger files, the fastest method of upgrading software is via dial-up (or via wireless networks in some cases, where you have very small changes) transmissions from a central site, provided the upgrades are not extensive.

At the present juncture, the use of wireless communication bandwidth for upgrades should be restricted to transfers of application data. On-line systems software upgrades should be made only via faster and more reliable dial-up links using V.34 modems.

17.3 The Mobile Help Desk

The center of action for all these activities is the customer service call center, otherwise known as the *help desk*. This is where all requests for help with problems, upgrades, and changes come. Help-desk staff have access to operational staff, database administrators, and network specialists—the second-line technical support staff.

It is the responsibility of the help desk to record all problems and monitor them until they are resolved to the satisfaction of the users.

17.3.1 Do We Have Remote Access Network Management Solutions?

Leading network management vendors such as HP, IBM, Tivoli, 3COM, Cisco, and Bay Networks are presently focusing their attention on solving the wired network problems. They are not paying too much attention to remote access related network management. Perhaps justifiably so. Accordingly, there is a patchwork of solutions that address specific problem areas, but no single toolset capable of effectively addressing all of remote access's many unique requirements. A few vendors are creating a standard so that the industry can build standards-based solutions, chief among them being the MMTF's SNMP-based MIB, which we mention along with a few other tools in the next section.

17.4 Mobile Network Management Tools

We describe the following initiatives, products, and approaches to the development of standards and tools for mobile network management.

- The MMTF initiative
- Remote control software
- Symbol's Spectrum24 NetVISION
- Plug-and-Play with Windows 95
- Individual hardware vendor approaches
- Individual software vendor approaches
- A remote access server or wireless MCSS as a network management hub

17.4.1 The MMTF Initiative*

Recognizing remote access's unique network management problems, several vendors, headed by Epilogue Technology and Xircom, have joined together to create the Mobile Management Task Force. Epilogue Technology is a pioneer in providing SNMP-based network management development software to hardware, software, and chip vendors. Xircom is a leading vendor of PC Card—based solutions for wireless LANs and remote access. Other active members are Compaq, Fujitsu, IBM, LanAire, Motorola, National Semiconductor, and Zenith Data Systems.

17.4.1.1 THE BASIC TERMS AND CONCEPTS OF SNMP.   While we do not intend to enter into a detailed discussion of SNMP in this book, we nevertheless explain a few terms and basic concepts. (For more detailed insight into the topic, the reader should refer to *Network Management Standards* by Uyless Black.)

SNMP is the most popular network management protocol for multivendor environments. Just as TCP/IP has been adopted as a de facto transmission protocol standard, SNMP has assumed a similar role in network management. SNMP architecture was designed with simplicity in mind to support the remote management of network resources to the fullest, independently of host computers or gateways.

SNMP has three components—the network element, the network management (NM) station, and an SNMP agent. The network element is the resource that is being managed (mobile computer, PDA, information server, etc.). The NM station is where management control information is requested and received. The SNMP agent is the software that collects control information on behalf of the NM station from the network entity.

The agent may in fact be a proxy agent in situations where the network element cannot be reached by conventional management protocols (low-function devices such as modems and bridges, for example, cannot interact with the NM station). Proxy agents can serve many useful features: first, the managed entity does not have to concern itself with network management; second, it can perform protocol conversions; third, it can provide security.

Management control information is described and stored in a MIB, which resides in the managed entity. The SNMP protocol defines the way

---

*Information available on Xircom and Epilogue World Wide Web homepages.

an NM station accesses information in MIBs using one of the three methods shown in Fig. 17.1.

An SNMP manager can be programmed to send periodic polling messages to the managed devices, the intervals established through an SNMP MIB. This facility is important because of the need with wireless networks to keep management traffic to an absolute minimum. Managed objects can send interrupts under certain conditions as well. Interrupts are not preferred by designers, however, because of the unpredictability of traffic.

Finally, there are SNMP traps, based on restricted parameters. A trap is an interrupt handled by a filter that conserves precious network bandwidth by reporting back only if a certain threshold value has been reached, or a combination of different parameters reaches a specified value (when message queues and the number of transmission errors combined exceed a threshold value, for example).

In the past three years, equipment vendors have made significant progress in implementing the first version (v1) of the SNMP standard, as well as its enhancements (v2), in a few cases. Conformance to the SNMP standard is now a common requirement in networking RFPs.

17.4.1.2 THE EPILOGUE/XIRCOM MOBILE MANAGEMENT SOLUTION. Xircom has developed the initial structure for a mobile MIB

Figure 17.1
SNMP operations.
(Source: *NM Standards* by Uyless Black.)

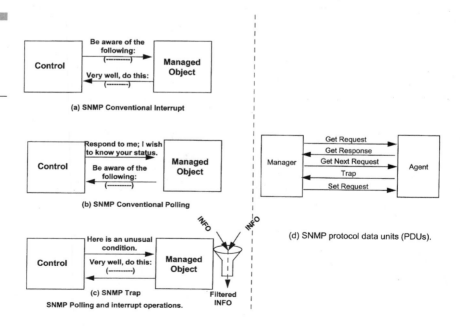

(a) SNMP Conventional Interrupt

(b) SNMP Conventional Polling

(c) SNMP Trap

SNMP Polling and interrupt operations.

(d) SNMP protocol data units (PDUs).

extension to the current SNMP v2 specification. Comments and contributions from industry vendors, network administrators, and end users are being solicited. These will be refined and submitted as part of a draft MIB document. Epilogue Technology will publish an informational request for comment (RFC), and will present the draft MIB to the Internet Engineering Task Force (IETF) for consideration.

17.4.1.3 **PROPOSED MOBILE MIB EXTENSION.** The following is the initial structure for the mobile MIB extension being proposed by Xircom and Epilogue:

mobilePlatformInformation.   An object that identifies the mobile platform being used, including the manufacturer and model, BIOS date, operating system (including vendor and version number), and PCMCIA client driver identification data (including driver ID, vendor, and version number).

mobileLocale.   Although it is not clear where the information will come from, this object is designed to give the network administrator local access information, specifically the country, city, local time, time zone, and in the case of dial-up users, the local telephone number of a dial-up connection.

mobilePower.   An object that identifies the version of the POWER.EXE driver resident on the user's computer, the version of the Advanced Power Management (APM) ROM BIOS, and the current power-saving configuration (none, low, medium, high, or the time of the last critical-suspend event). Traps related to power management could include "suspend" and "resume."

mobileResourceAllocation.   This object includes information about memory, IO, IRQ, and related parameters used by mobile devices, and could be expanded to track the availability of various resources, such as adapters, slots, bus extensions, ports, and so forth.

mobileCardServices.   An object that identifies the version of Card Services.

mobileSocketServices.   An object that identifies the version of Socket Services.

mobileBattery.   An integer that would determine whether a device is running on AC or battery, and, if a battery, determine the number of minutes of battery life remaining. It is set to 0 if unknown.

mobileConnectionMedia.   An integer that identifies the connection media being used: 0 = unknown; 1 = serial port; 2 = Ethernet; 3 = Token Ring; 4 = Wireless; 5 = 100 MBit.

mobileConnectionSpeed. An object that identifies the connection speed of the media being used (9600, 38,400, 10 MB, etc.).

mobileConnectionMechanism. An object that identifies the connection mechanism being used. (Examples include pcAnywhere, CarbonCopy and NetWareConnect.)

Other traps being considered for inclusion in the Mobile MIB extension include insert/extract PCMCIA devices and docking/undocking.

17.4.1.4 TYPES OF MIBS FOR MOBILE COMPUTING AND REMOTE ACCESS.

MMTF group has defined four MIBs that define the objects for mobile computing. The System MIB provides information about the Mobile software drivers, card and socket services, power status and related system conditions. The adapter MIB gives information about the network adapter. The Link MIB provides information about the Mobile network link to help network managers troubleshoot and optimize the connection. The extended System MIB contains information about system extensions. This is shown graphically in Fig. 17.2.

17.4.1.5 SCENARIOS WHERE MOBILE MIB COULD BE USED

Problem Scenario	Mobile MIB Solution
Company wants to standardize on one brand of PCMCIA cards because older nonstandard card and socket services cause intermittent problems.	With the support of a specialized Mobile MIB extension residing on mobile users' laptops, the network manager can identify and upgrade the CS/SS to a newer version known to fix the problem.
Different users connect to the network using many different methods, making it difficult to standardize on one method.	By monitoring connection methods, the network manager can identify which is the most popular.
Company's auditors want an inventory of laptops.	SNMP provides a means of tracking and controlling capital equipment.

17.4.1.6 PROGRESS WITH IMPLEMENTATION OF MOBILE MIB.

Xircom is shipping SNMP Mobile MIB with the network management software included in its PCMCIA network adapters. Epilogue has incorporated the MIB as an extension of its Emissary MIB compiler software. Shiva Corporation has also announced plans to support Mobile MIB.

Figure 17.2
MMTF's mobile
MIBs. (Source:
*Communications
Week,* January 1996
and MMTF page on
Internet.)

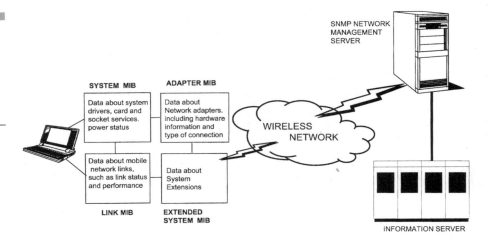

17.4.2 Remote-Control Software

The following remote-control software is often used for problem diagnosis:

- CarbonCopy from Microcom
- NetModem from Shiva Corporation
- pcAnywhere from Symantec
- ReachOut (from Ocean Isle Software, Vero Beach, Florida)
- IBM's Netfinity 3.0

There are other utilities, such as Novell's Netware utilities, Frye's utilities for NetWare, and Intel's management utilities that enable the diagnosis of notebook-related problems.

Of course, problem management and software upgrades take on a whole new dimension if the software has to be upgraded in hundreds and thousands of notebooks, and there is no way of scheduling the transfer of data. The most pragmatic solution is to have users pick up upgrade diskettes from the nearest convenient location.

17.4.3 Symbol Technology's NetVISION Tools

Symbol Technology's NetVISION Wireless LAN Management Tools simplify the configuration and management of a wireless LAN. This easy-to-install program immediately initializes and configures the Spectrum24 (Symbol's wireless LAN) radio networking environment with minimal user intervention. Once installed on the network server or host system, NetVISION uses autoconfigured defaults to boot Spectrum24 remote nodes, including access points, radio terminals, and devices with wireless LAN adapters.

NetVISION's GUI provides network managers with the tools to administer the Spectrum24 radio network nodes. Included are tools for:

- *Booting.* NetVISION boots Spectrum24 mobile units automatically with no intervention required.
- *Configuring.* NetVISION is a configuration manager. It configures Spectrum24's range of network parameters, remote mobile units, access points, and access to security features.
- *Monitoring.* NetVISION sets limits for acceptable operations and monitors the network for noncompliance, issuing exception reports for departures.
- *Diagnostics.* NetVISION tests the status and operation of network nodes. Run-time system errors are monitored via NetVISION diagnostics and logged into an error file located on the host machine.

17.4.4 Plug-and-Play with Windows 95

Notebook manufacturers have started using the Plug-and-Play technology in Windows 95—along with the widespread integration of the PCI bus standard in docking stations—to provide IT managers with help in keeping track of laptops used on LANs. Unfortunately, this still does not address the wider problem of mobile users who do not visit their offices for days and weeks. However, it is expected that the PCI bus standard will soon be implemented in notebooks themselves. Central network management systems will be then able to retrieve information directly from mobile computers.

17.4.5 Individual Hardware Vendor Approaches

MMTF's MIB implementation by the industry at large will take several years. Meanwhile, a few vendors are providing their own utilities to provide partial answers to the bigger problem of overall network management for mobile computing and remote access. Intel is providing management utilities to provide system information remotely. Similarly, Compaq includes intelligent manageability utilities in some of its notebooks, notably the Contura and the LTE Elite.

17.4.6 Individual Software Vendor Approaches

In our limited research, we found that Xcellenet's RemoteWare product offers good practical solutions to some of the network management issues that we have discussed in this chapter, including:

- Reasonably effective but inexpensive security control where full-fledged encryption is not warranted
- Automated data/file distribution and synchronization in case of failures
- Store-and-forward transmission of large files, using agent-client-server design
- Comprehensive history and control of changes
- Remote control, with system activity alerts (e.g., software updates)
- Application-configuration management

If the mobile solution is based on RNA using PSTN or ISDN only, Xcellenet's RemoteWare software bundled with Shiva's hardware platform could be a very effective combination from functionality and network-management perspectives. Shiva has already announced its intent to implement MMTF's Mobile MIB in future products.

17.4.7 A RAS or MCSS as a Network Management Hub

We have focused on remote devices so far. We also need network-management information about the network, the physical and logical network connections, the TCP/IP sockets, and the logical flow of data in the network. One possible source of this information in a remote access network is the remote access server itself. In case of wireless networks, it could be MCSS described in previous chapters. As an example, Shiva's NetRover products provide SNMP-compatible information that can be viewed by anybody in the network control center or the help desk.

Most of the popular RASs make this type of network management information available—either locally to the network staff or remotely to the enterprise network control center staff or the hardware supplier's support staff, who can diagnose problems and apply software patches or upgrade complete software modules. You should ask your vendor to provide an SNMP v2 MIB that can be integrated with enterprise network-management software, such as IBM's NETVIEW or HP's OPENVIEW.

17.5 The Integrated Network Management of a Mobile Infrastructure

Mobile computing solutions consist of many components, and individual component manufacturers provide network management information about their products. There is not a single vendor, however, that we know of who provides an integrated, systemwide view of the various components in a mobile computing solution. The closest anyone ever comes is the occasional systems integrator who builds a limited capability for a specific customer.

We propose that network managers should ask for this integration of information at the central site where the rest of help desk resources reside. We believe that the centralized network management of an enterprise is the most pragmatic approach for two reasons. First, very high skills are required for remote access network management—skills that are also very expensive to acquire. Second, for logical network management (application status), the necessary information is likely to be available only at the central site.

With present-day call-center telecommunications networks increasingly being centralized, or even outsourced, we can learn from the experiences of the airline industry. In the case of airline reservations systems, it does not matter where a call is physically serviced—the call-center representatives know the status of the network elements wherever they are.

Of course, simple asset management or server-status information can be provided by regional network administrators, if such individuals are available at these sites.

We make the following suggestions for the enterprise-wide network management of remote access systems:

- We should use both a reactive approach to solve problems as they occur as well as a proactive approach of monitoring the network regularly and continuously in order to forestall potential problems in future. This is especially true for response-time problems.

- You should try to solve as many problems as possible remotely by using software probes. However, in some cases, there is no alternative to dispatching a service representative to the site.

- Ensure that most of the network elements, including mobile devices, have SNMP v2-compatible MIBs.

- For mobile devices, MMTF's Mobile MIB should be requested, if it is available. If it is not available, the vendor should be asked to indicate its intentions regarding a future implementation.

- Remote access server and wireless network service providers should be asked to provide SNMP-compatible network management information to the central network management console through a logical port into enterprise network management system.

- We should use software-based approaches of dispatching a network robe.

- There should be standardization on remote diagnostic software such as CarbonCopy, pcAnywhere, or any other similar package selected by the organization for the RNA solution (see Chap. 10).

- For the sake of overall integration, we should use whatever network management solution the enterprise has chosen as a standard. In most cases, a new standard for remote access is not warranted.

17.6 Status of Remote Access Network Management Capabilities

Network management for remote access servers lags behind network management capabilities of mainstream network components such as LANs, routers, switches, and hubs. On the other hand, in the case of the wireless industry, the situation is even worse. Elemental, nonintegrated, and proprietary, network management in the wireless network industry appears to be evolving very much as an afterthought, lagging behind the RNA industry, which has made substantially more progress with network management software. There are two main reasons for this. First, network management is more easily addressed in the RNA industry. Second, there are not many vendors who are interested in a standards-based network management solution for the enterprise, or in promoting and licensing a de facto standard.

We have started seeing some coordination and a movement toward standards through the efforts of the National Portable Computing Professionals Association (NPCPA). However, only when there is concerted demand from all users for standards-based and integrated network management will real progress be forthcoming.

17.7 Policy and Procedures to Support Technology-Based Network Management

There is no magic wand for network management. No amount of network management automation will solve problems created by a lack of standards and well-observed procedures. We strongly recommend, then, that network management tools be augmented by appropriate policies, procedures, and standards. The following suggestions are but a few of the many guidelines that can be institutionalized in an organization:

- Standardize, if appropriate, on a brand/model of notebook and other mobile devices to be employed for remote access. This is especially recommended when a central organization funds such an acquisition.

- Standardize on a base of permissible software configurations.

- Private software packages should be restricted and allowed only if they do not violate software-configuration standards imposed by a central organization.

- For a tightly controlled organization, consideration should be given to purging renegade software after checking hard disk directories.

- Insist on regular virus scans, backups, and software refresh procedures.

17.7.1 Who Should Be Responsible for Mobile Computer Management?

Network management can be performed by either a centralized group, or it can be distributed at a departmental level. This decision involves both technical and organizational issues. While centralized network management is more cost-effective, a hybrid solution may be more responsive from a users' perspective. Under such a scheme, responsibility for simple network management tasks can be given to department-level systems administrators, and more complex problems can be assigned to a skilled group in the central network organization. The division of responsibility can be handled in such a way that the user sees only a single problem-management interface with call/trouble ticket switching between the departmental and the central organization automatic and transparent. The current support-organization structure should be considered as a model. In most cases, no new support organization is warranted for mobile computing.

Summary

In this chapter, we have examined network-management tools and related issues for remote access. We stated that network management in remote access is in the early stages of development, with individual vendors creating proprietary and narrowly focused approaches. We discussed the MMTF's work on creating SNMP standards-based MIB. We should see a more unified, integrated, and standards-based approach in the years to come.

Chances are, however, that a single network management tool is still a long way off. Rather, it may be necessary to use a combination of tools from various vendors. For now, the more important network diagnosis, asset management, and standardization issues unique to remote access should be the focus of attention. For other issues, current network approaches should be used.

Source and Reference

Network Management Standards by Uyless Black, published by McGraw-Hill: an excellent reference book for understanding SNMP, MIBs, and network management in general.

APPENDIX **A**

Remote Access Products and Services

It is not possible to provide a comprehensive list of remote access products in a book of this size. Therefore, we shall cover only a subset to give you a general idea of what is available in the market. In some cases, we have described specific products. In other cases, we have mentioned the names of the vendors only with a brief description of their product line in a given category. Moreover, the limited space does not allow us to describe the features and functions of the listed products in any great detail. However, we have described a few of the popular products in the body of the book in order to illustrate the underlying technical concepts.

This industry is evolving at a fast pace. Accordingly, the products are changing regularly also. The industry will go through a rationalization process during the next few years. Some of the vendors may not survive this process as a result of either voluntary exit or forced acquisition. Therefore, the reader should contact the vendors or refer to trade magazines to get more current information.

A.1 Remote Access Product Categories

For the sake of simplicity, we have divided the products into the following categories:

- Remote access server (RAS) products
- Remote access concentrators and switches
- SOHO or personal RAS products
- RAS software products
- Remote access client software products
- Application-specific remote access products
- Firewall products
- Value-added network services

A.2 Remote Access Server (RAS) Products

Remote access servers represent the main infrastructure component of this industry. The basic functionality of these products has been described in Chap. 4. (Refer to the description of the architecture and components of these products in that chapter.) Most of the products described here follow this architecture.

A.2.1 Vendor Scene

First, we shall list the major vendors that provide RAS products. Although the following list is incomplete, it does represent most of the larger players in the market:

- Ascend Communications
- Bay Networks/Xylogics
- Cisco
- IBM
- Microcom
- Multitech
- Novell
- Perle
- Shiva
- 3COM/US Robotics
- Xyplex

A.2.2 Product Features, Functions, and Performance Criteria for Evaluation

The following features, functions and performance considerations are often used for comparing products:

- Server features, including variety of protocols supported, type, and speed of network options supported—PSTN, ISDN, ADSL, wireless, and so forth
- Client software features
- Network management features
- Security functional features
- Scalability as measured by the number of analog and ISDN ports supported on client and server side
- Type of modems, adapter boards, and their speeds supported—both on client and server side

- Effective, not rated, RAS performance, measured in terms of effective bytes/sec on popular LAN protocols, such as IPX and TCP/IP
- Ease of installation
- Ease of use by remote users for the supported platforms
- Compatibility and integration with rest of the network infrastructure
- Open architecture—number of client and server platforms supported
- Technical support
- Price/performance, typically represented by average price per port

A.2.3 Brief Description of Common Products

We now briefly describe some of the vendor products. You should obtain current and more complete information from a specific vendor or from the World Wide Web.

A.2.4 Ascend Remote Access Product Line

Ascend is one of the leaders in the remote access market. It has a full suite of products. These products can be divided into two distinct families of products—Pipeline family for low-end to medium-size organizations and MAX family for high-end organizations, carriers, and network service providers. We describe Ascend's MAX family under Sec. A.3, since it is based on concentrator/switch architecture.

A.2.4.1 ASCEND'S PIPELINE FAMILY OF REMOTE ACCESS PROD-UCTS. Ascend's family of Pipeline products represents an extensive range of remote access devices for Internet access, remote offices, home offices, and telecommuters. Key applications such as corporate LAN access, Internet/intranet access, management of large numbers of tele-commuters, and mission-critical remote networking benefit from the Pipeline family.

Pipeline 25-Px, 50, and 75 models support high-speed ISDN BRI communications for stand-alone PCs and PC/LANs. Analog devices are also supported on the Pipeline 25-Px, 25-Vex and 75 models. High-performance Pipeline 130 models support growth from a 56-Kbps frame-relay dial-up or leased-line connection at up to T1/E1 speeds. The Pipeline 130 provides a dedicated WAN connection and an integrated switched connection for mission-critical backup and overflow applications.

A.2.4.2 ASCEND'S PIPELINE FAMILY FEATURES AT A GLANCE

- Multiprotocol bridging and routing
- Dynamic Bandwidth Allocation (DBA), inverse multiplexing
- Multilink Protocol Plus (MP+)
- Data compression on Pipeline 50, 75, and 130
- Remote management
- Extensive standards-based security
- Integrated Secure Access Firewall optional on Pipeline 50, 75, and 130

Pipeline products are convenient solutions with modem-sized footprints and no external wiring. The members of this family, as of mid-1997, are shown in Table A.1.

A.2.5 Attachmate's RLAN v4.1

Attachmate's RLAN product has been in the remote access lineup for a long time, much before the current influx of remote access products. Although Attachmate has impressive domain-oriented authentication and configurable client software installation, *Network Computing* magazine reported in its June 1996 issue the poor performance of the server, even with approved serial I/O hardware. Notwithstanding these test results, Attachmate does have superior integration with IBM back-end infrastructure. The node server can be purchased either as a turnkey hardware solution or with vendor-approved hardware and software.

A.2.6 Cisco Product Line

Cisco is a major vendor in remote access market. It has a full suite of products from the low-end to the high-end enterprise solutions. (Refer to Chap. 4 for a more complete description of AS5200 universal server.)

A.2.7 IBM 8235 Remote Access Solutions

IBM offers the 8235 family of remote access servers or switches that can support from 2 connections to more than 71 connections simultaneously. The smaller remote offices can be served by IBM 8235 Models 051 and 052 which have two ports for modems or ISDN attachments, respectively. The larger IBM 8235 Model 140 switch supports both individual modems and

TABLE A.1

Ascend Product
Line

Product	Pipeline 25-Vex	Pipeline 25-Px	Pipeline 50	Pipeline 75	Pipeline 130
Bridging	Yes	No	Yes	Yes	Yes
IP routing	No	Yes	Yes	Yes	Yes
IPX routing	No	No	Yes	Yes	Yes
PAP and CHAP security	Yes	Yes	Yes	Yes	Yes
POTS	2 RJ-11	2 RJ-11	No	2 RJ-11	No
ISDN S/T Interface	Yes	Yes	Yes	Yes	Yes
ISDN U Interface	Yes	Yes	Yes	Yes	Yes
Number of users	Up to 4	1	Limited by user profile	Limited by user profile	Limited by user profile
PPP	Yes	Yes	Yes	Yes	Yes
Multilink PPP (RFC 1717)	Yes	Yes	Yes	Yes	Yes
MP+	Yes	Yes	Yes	Yes	Yes
SNMP	No	No	Yes	Yes	Yes
Data compression	No	No	Yes	Yes	Yes
Availability	Now	Now	Now	Now	Now

aggregated connections such as channelized T1/E1 and primary rate ISDN to accommodate a large number of users.

All 8235 models have the following features:

- Support analog and digital connections, including ISDN
- Support IP, IPX, AppleTalk, NetBIOS, and LLC protocols
- Come with advanced function client software for OS/2, DOS, and Windows 3.X at no charge
- Can also be accessed from Windows 95, Windows NT 3.5, SLIP, PPP, and terminal dialers
- Have money-saving communications and common management functions designed to lower the two largest remote access costs (i.e., line charges and user support)
- Offer a variety of security options

IBM ISDN support uses virtual connections that minimize connect time by terminating the connection when it is not being used and restarting it when it is needed again. This is done in such a way that the applications using the connection are not aware that the connection is broken. Without virtual connection support, ISDN can be too expensive for many applications. Automated installation aids lower support costs by automatically detecting the client configuration, and by allowing network managers to build a sample configuration and rapidly replicate that configuration to large numbers of similar clients.

A.2.8 Microcom LANexpress 4000, rel 4.0

Microcom's RAS is easy to set up, has analog modem and ISDN BRI support, gives good network performance, and provides good management functions—called expressWATCH. It has an integrated remote node and remote control (CarbonCopy) Win95 client that allows for customized installation and use. The client software includes scripting capability to automate installation, dialing, network login, and application execution. This release does not support Multilink PPP (MPP) or suspend and resume feature. Third-party security support like TACACS or RADIUS is also lacking. Watch for these features in a future version. Please check with the vendor.

A.2.9 Microsoft Windows NT Server v3.51

It is a natural candidate for no-hassle introduction of remote access in Microsoft-exclusive shops. Close integration into the operating system and ease of installation are worth considering, if you are committed to MS Windows NT for other applications. It has a 256-port capacity from software point of view, though hardware implementation might present some configuration challenges, especially if it was based on low-speed analog ports.

Monitoring statistics are available through NT's system monitor and event log. However, network performance of NT-RAS as software server is average—middle-of-the-road in its class of RAS. Decision to adopt MS RAS is generally made on software integration and affinity issues, rather than performance issues.

Citrix WinFrame and US Robotics' Network Hub are only two of several different RASs that utilize Microsoft NT 3.51 RAS as the software under the hood.

A.2.10 Novell's NetWare Connect 2

This product is a preferred platform for those shops that have some level of NetWare installed, because of its close integration with the network operating system. It has good usability, true call accounting, good speed, and an attractive GUI manager called *Connectview.* (See Chap. 4 for further information on this product.)

A.2.11 Shiva's LanRover/E Plus

Shiva's popular LanRover/E remote access solution is meant for small and medium-size organizations. LanRover has won several awards in performance, ease of use, and network management. It has superior configuration setup and monitoring.

LanRover remote access servers provide economical turnkey remote access solutions for users who only want to purchase the functionality that they want. All LanRover Power base configurations feature IP, IPX, and AppleTalk dial-in capabilities. In addition to supporting Microsoft dial-up networking for Windows 95 and Windows NT clients, Shiva includes Shiva Remote for Win95 and Shiva Remote 3.x. Supported authentication methods include LanRover user lists, LanRover user list server, and NetWare bindery. STAC compression, virtual connection, MLP, and IP filtering are also supported. Modular LanRover/Plus configurations include integrated mix-and-match V.34 modems and ISDN BRI cards to minimize external cabling and maximize server reliability. (See Fig. A.1.)

Figure A.1
Shiva LanRover
access switch.

A.2.11.1 LANROVER CONNECTIVITY KIT. The LanRover Connectivity Kit enhances information access with dial-out capabilities to on-line services, bulletin boards, the Internet, and fax machines.

A.2.11.2 LANROVER SECURITY KIT. LanRover security kit restricts unauthorized access to information by centralizing network access control through the support of the TACACS, TACACS+, and Radius protocols and enabling communications with third-party challenge-response authentication servers such as Security Dynamics and AXENT. The Security Kit includes Web-based comprehensive network monitoring capabilities with the Shiva Access viewer with Shiva Access Alert. Shiva Access Viewer for Windows NT 3.51 or 4.0 (workstation or server) gives you all the flexibility and power to monitor, in real time, just what is happening on your network via a Web browser. Check on device status and usage patterns, and be alerted to network break-in attempts.

A.2.11.3 LANROVER PERFORMANCE KIT. Speed into the latest advance in remote access technology. Slow performance is a common problem for remote users, given bandwidth limitations, latency constraints, and the design of *chatty* file-system-based LAN applications. The LanRover Performance Kit featuring "PowerBurst" caching software is unique in accelerating remote node performance. Fully integrated client and server agents make deployment of this technology easy and result in lower communications costs.

PowerBurst accelerates application response time performance by caching frequently accessed data into the local hard drive. The use of intelligent algorithms and an integrated refresh function that all the requested data is current and that only the changes are transferred over the bandwidth-intensive telecommunications link.

A.2.11.4 SHIVA LANROVER'S TECHNICAL SPECIFICATIONS. Shiva LanRover's technical specifications are briefly summarized in the following paragraphs in a list format:

Dial-in Connectivity

- Single number IPX, TCP/IP, AppleTalk Communications protocol
- ShivaRemote for Windows 95 and Windows 3.1
- MS dial-up networking client
- IPX dial-in with NetWare VLMs and NETX

- Telnet from asynchronous terminals for dial-in
- DNS domain name server support

Management

- Centralized management of multiple LanRovers
- SNMP, including MIB II and other extensions over TCP/IP, IPX, and AppleTalk
- HP OpenView link to Shiva Net Manager
- TCP/IP addresses can be assigned via DHCP, per user, an internal address pool, or selected remotely
- Maintains a detailed activity log for accounting, billing, and troubleshooting
- Bootable from nonvolatile memory or BOOT/TFTP server
- Interactive on-line documentation via any standardized browser

Security

- User name, password, and dial-back security (fixed and roaming)
- CHAP, PAP, and SPAP support
- Centralized authentication via Shiva user lists, Shiva user list server, NetWare bindery
- Dial-in password expiration and user change password options
- IP device filtering and network filtering
- AppleTalk zone and device filtering
- Win95 security pack

Performance

- STAC data compression for ShivaRemote 5.0, Windows 95, and LAN-to-LAN connections—as much as 400 percent faster in ISDN and X.25 environments
- Header compression and broadcast filtering
- Broadcast filtering in IPX, TCP/IP, NETBEUI, 802.2/LLC, and AppleTalk environments
- Minimize update traffic with Shiva "Delta" technology (Delta SAP, Delta RIP, and Delta RTMP)

A.2.12 US Robotics Total Control Enterprise Network Hub

The US Robotics Inc. Total Control Enterprise Network Hub provides remote access by implementing a hardware configuration consisting of a large chassis, a single board processor running Microsoft Windows NT 4.0 Remote Access Server (RAS) software, ISDN PRI card, and a quad-modem module. Many vendors have implemented proprietary software for RAS; instead, USR's Total Control hub utilizes existing software platform that many network administrators are already familiar with. This hardware-software combination allows NT's RAS software to take advantage of the hardware features in the chassis. (See Fig. A.2.)

The large chassis of USR's Total Control hub can support up to 60 modems and 10 ISDN PRIs, two T1 connections, or a combination of both. The Total Control hub also supports frame-relay. The cards are hot-swappable, and you can remove and insert them without affecting other connections. A redundant power supply can also be added to prevent downtime of mission-critical applications. The chassis is rack-mountable, and all the wires connect to the rear of the unit. The USR hub supports V dot modems with new X2 capabilities (discussed in Chap. 6).

The Total Control Management software runs on a Windows 95 workstation. It offers an easy-to-use GUI for configuring the server. You can even configure the modems from within the management software—a feature not found in server management software of too many RASs. This environment has a low learning curve.

The product does not currently support dial-out—the modems are for inbound calls only. This capability may be available in future releases, because it is a common requirement in remote access applications.

Over and above support for NT-based user security, the Total Control hub supports PAP and CHAP as well as RADIUS. A detailed event log and call accounting keep track of dial-in statistics. This information is viewable in both textual and graphical form on the management console.

The Total Control Enterprise Network hub provides flexible management, scalability, density, security and other features, described as follows:

Scalability and Ease of Upgrades.   You can easily, cost-effectively add ISDN feature when needed. US Robotics' modular design and software-based architecture allow the Total Control Enterprise Network hub to migrate to other technologies easily.

Figure A.2
(a) 3COM/US Robotics
total control network
hub; (b) network
administration for
total control hub.

(a)

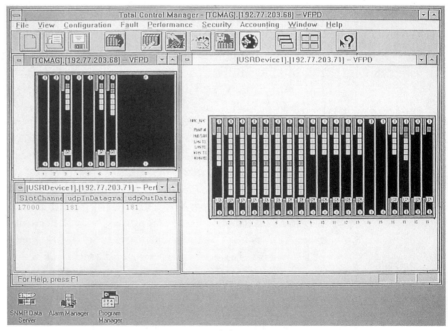

(b)

Density.   A single chassis accommodates up to 10 PRI spans for 230 simultaneous digital connections, density that is among the highest of any product on the market. High density is important to organizations that lease space by the square foot for data centers or points of presence.

Universal Connect.   The product automatically handles both analog and digital calls over incoming PRI trunks.

Integrated Network Access Servers.   Total Control provides access to Ethernet, frame-relay, Token Ring, or X.25 networks through integrated remote access servers.

Management and Security.   Options include Total Control Security Server and Accounting Server; SNMP management systems; and the Auto Response intelligent, automatic management system.

Design for 24-Hour Uptime.   Fault tolerance, hot-swap, a front-loading design and available redundancy provide 24-hour uptime for mission-critical applications.

Worldwide Homologation.   The Enterprise Network Hub is approved for use in a number of countries, including Australia, Austria, Belgium, Denmark, France, Germany, Israel, Italy, Japan, Singapore, Spain, the United Kingdom, and many others. US Robotics is currently pursuing worldwide approvals for its ISDN PRI Access System.

Price.   The price of a typical configuration is in the $20,000 to $80,000 range, depending on the type and number of cards installed. Price for a PRI-ISDN card is around $5995.

A.2.13 3COM AccessBuilder 4000

Refer to Chap. 4, section 4.4.2.5 for information on AccessBuilder family.

A.2.14 Xyplex RAS Product Line

Xyplex Networks offers a complete family of products in both chassis-based and stand-alone configurations. The Xyplex Networks' RAS solutions include Network 3000, Network 6600, stand-alone RAS solutions, and the high-end Network 9000 Remote Access switch. We describe the latter product in Sec. A.3.

A.2.14.1 XYPLEX NETWORK 3000. The Xyplex Network 3000 has the same software base as the Network 9000, but is designed for branch offices. Its scalable design provides full routing, bridging, and hub functionality. The modular Network 3000 family of integrated branch office bridge/routers supports IP, IPX, AppleTalk, ISDN, and Frame Relay protocols, in addition to concurrent 802.1 bridging. DECnet, X.25 and OSI are offered as options.

Every Network 3000 product is supported by the unique WANScape WAN-independent software architecture. This architecture, in combination with support for the industry-standard Multilink Point-to-Point Protocol (MLPPP), provides support for extensive bandwidth management features, such as inverse multiplexing, intelligent bandwidth-on-demand, leased-line backup, link aggregation, time-of-day routing, and data compression.

Providing simplicity, interoperability, flexibility, and scalable management, the Network 3000 family provides solutions for today's changing business environments.

A.2.14.2 XYPLEX STAND-ALONE RAS SOLUTION. Xyplex's high-end Network 9000 switch and Network 3000 solutions are complemented by a line of stand-alone solutions for supporting smaller workgroups. Xyplex Networks' stand-alone solutions support the same hardware, software, and management implementations as their chassis-based Network 9000 and Network 3000 counterparts, making them easy to manage and integrate into an existing network.

Xyplex Networks' stand-alone concentrators include support for 10Base-T and FDDI networks and range from low-port-density, fixed-configuration, nonmanageable units to SNMP-managed units that support over 100 ports per stack.

The Xyplex 6600 Series of LAN switches provides a wide range of solutions from small workgroup-sized switches and scaling up to high-performance collapsed backbone solutions.

Xyplex Networks remote access server solutions provide a variety of LAN access methods from terminal-to-host connectivity using LAT, TCP/IP, TN3270, or XRemote to remote node operation via TCP/IP, IPX or AppleTalk. Xyplex Networks' award-winning 1600, 1620, and 1640 Access Servers provide a complete solution to address the transition from host-based to client/server architectures.

Network Management. Xyplex Networks' products can be managed by a rich array of local and remote management tools for configuration, diagnostics, operating statistics, network load status, and on-line help information. Xyplex Networks' equipment can be managed locally via an out-of-band management console port or from any locally attached terminal, PC, or workstation on the network that supports Telnet or Digital's Remote Console Protocol (RCP). Xyplex Networks' products also support full SNMP agent functions, including support for standard MIBs such as MIB I and MIB II. In addition, Xyplex Networks has been recognized as having the most extensive private MIB extensions, significantly enhancing SNMP central management and system configurability. Any SNMP management system, including Xyplex Networks' *ControlPoint* network management application, can be used to manage Xyplex Networks' devices.

ControlPoint is designed to manage Xyplex Networks' networking equipment in a multivendor network of geographically dispersed users. ControlPoint provides an extraordinarily robust graphical toolset that allows network administrators to meet the challenges posed by today's open, distributed, client/server networks.

ControlPoint offers consistent look and feel and functionality across the leading network management platforms, including SunNet Manager, HP OpenView for Windows, and HP OpenView for Unix.

With ControlPoint, Xyplex Networks' hubs and routers support the RMON Alarms, Events, and Statistics Groups for distributed fault management. For establishing a network baseline and for trend analysis, Xyplex Networks' routers additionally support the History Group. These four RMON groups provide the information most network administrators need to operate and manage their networks.

A Microsoft Windows—based, point-and-click graphical user interface (GUI) configuration tool is available to step users through the configuration process for the Network 3000 family and the *RouteRunner* bridge/router products. This easy-to-use tool includes an exceptionally intuitive ISDN setup window with default settings for many of the ISDN access arrangements available worldwide, dramatically reducing total setup time.

A.2.15 Sample Feature Comparison for RAS Products

We show in Table A.2 various features of RAS products that you may evaluate on functional basis. You should also compare other factors listed at the beginning of this section.

TABLE A.2

Sample Comparison of RAS Products

	Novell NetWare Connect 2	Shiva Corp. LanRover/E Plus	3Com Corp. AccessBuilder 4000	Xylogics/Bay Networks Remote Annex 4000
Server features				
Processor type	Intel 386 DX	68020	i960	80486 SX
LAN topologies supported	LAN-independent	Token-Ring, Ethernet	Ethernet, Token-Ring	Ethernet
WAN topologies supported	Async, ISDN	Async, ISDN	Async, ISDN BRI	Async
Max async/ISDN BRI per server	128/64	8/4	16/8	72/0
Serial protocols supported	PPP, TTY, Proprietary	PPP, SLIP/CSLIP, Multilink PPP	PPP, Multilink PPP, SLIP, ARA	PPP, SLIP/CSLIP
PPP control protocols supported	IPXCP, IPCP, IPXWAN	IPCP, IPXCP, NBFCP, CCP	IPCP, IPXCP, BCP	IPCP, IPXCP, ATCP
LAN-to-LAN routing	No	Dial-up	Dial-up, Dedicated	Dial-up, Dedicated
Autodetection at port of inbound transport	PPP, SLIP, ARA, Telnet	PPP, SLIP, ATA, Telnet	PPP, ARA	PPP, ARA, IPX Proprietary, Terminal
Auto Port Pooling for dial-in, dial-out, LAN-to-LAN routing	Dial-in, Dial-out	All	All	Dial-out
Busy-out port with modem failure	Yes	Yes	No	No

Client features

	DOS, Win3.x	Win3.x	DOS, Win3.x, OS/2	DOS, Win3.1
Provided client's OS	DOS, Win3.x	Win3.x	DOS, Win3.x, OS/2	DOS, Win3.1
Client license	Unlimited	Unlimited	Unlimited	Unlimited
Remote control client included	No	No Remote Control	ReachOut	Funk Proxy
PPP client included	Novell	ShivaPPP	AccessBuilder Client	Funk FastLink II
IPX stacks included	VLM	NETX, VLM	No	Netx, VLM
IP stacks included	Novell LANWorkPlace	No	FTP	No
Dial-back/roving dial-back	Yes/Yes	Yes/Yes	Yes/No	Yes/Yes
Scripting of logon/install	Yes/No	Yes/Yes	Yes/Yes	Yes/Yes
Dial-on-demand	Yes	Yes	Yes	No
ICONization of connection	Yes	Yes	Yes	Yes
WAN execution warning	Yes	No	Yes	Yes
Auto reconnect from client	Yes	Yes	No	Yes

Management

Operating system platform	NetWare 3.12, 4.1	DOS, Win3.x, WinNT, Mac, OS/2	Win3.x	Unix
Connection time limits	Yes	Yes	Yes	No

TABLE A.2

CONTINUED
Sample Comparison of RAS Products

	Novell NetWare Connect 2	Shiva Corp. LanRover/E Plus	3Com Corp. AccessBuilder 4000	Xylogics/Bay Networks Remote Annex 4000
Management				
Inactivity timeouts for users	Yes	Yes	Yes	Yes
View users/port status	Yes	Yes	Yes	Yes
Management interface GUI/Char.	Yes/Yes	Yes/Yes	Yes/Yes	Yes/Yes
Reporting outputs format	TXT	TXT	TXT	TXT
IP address assignment	Static per Port, Client Defined, Pool on Server, BOOTP, DHCP	Static Per Port, Per User, Pool, Client assigned, BOOTP, DHCP	Static Per Port, Pool on Server, Client Defined, BOOTP, DHCP	Static Per Port, Client Defined
Security				
Authentication supported	PAP, CHAP NWCAP	PAP, CHAP, SPAP	PAP, CHAP	PAP, SPAP, CHAP
Authorization services supported	NDS, Security Dynamics, Enigma Logic, Digital Pathways	Security Dynamics Digital Pathways, Defender, RADIUS	Security Dynamics, TACACS, WinNT, NDS, NetWare Bindery	Kerberos, Enigma Logic
Password aging	Yes	Yes	Yes	No

SOURCE: *Network Computing Magazine, June 1996.*

A.3 Remote Access Concentrators and Switches

This category represents high-end remote access concentrators, also called remote access switches that are utilized by large corporations, telecommunications carriers, and network service providers to extend their backbone networks to support remote office access, telecommuting, and Internet access. (See Fig. A.3.)

Table A.3 shows several carrier-class switches from five different vendors.

Now, we briefly describe the following products in this category, only to make you familiar with the product line. Our intent is not to provide a comprehensive product lineup.

A.3.1 Ascend's MAX Family of Remote Access Switches

Ascend's MAX products are powerful remote access concentrators that help corporations build remote networks of any size, with virtually any combination of analog and digital carrier services. All MAX products combine the functionality of a router, a terminal server, an ISDN switch, and a frame-relay concentrator in a single box. MAX products represent

Figure A.3
Remote access
switching.

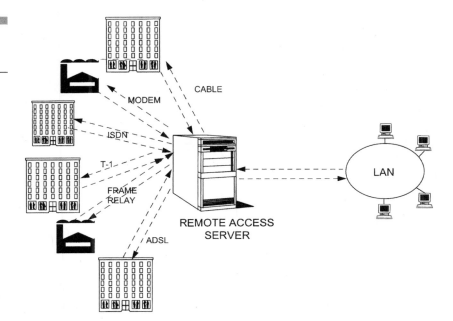

TABLE A.3

Remote Access Switches

Vendor	Product	Capacity/Max Serial Rate	Features
Ascend Communications	MAX TNT	672 users	Multiprocessor architecture, hot-swappable modules
Advanced Computer Communications (ACC)	Tigris	672 modem users; 1080 ISDN or leased-line users	Hot-insertable interface cards, audio and visual alarms
Bay Networks	Remote Annex 6100	One T1/115 Kbps	AppleTalk, TCP/IP, IPX/SPX protocols
Cisco Systems	AccessPath TS	720 analog or ISDN users	Redundant paths, parallel processing
RAScom	2900	720 users	Redundant hot-swappable components, real-time fault-identification
Shiva	LanRover Access Switch	Over 100 users, 72 modems	Round Robin call distribution; Tariff management, multiprocessing, PowerBurst caching agent
US Robotics	Total Control Enterprise Network Hub	2 T1 or 10 ISDN PRI/6 Mbps/64 maximum modems	IP, IPX, SPX support Random call distribution
3COM	AccessBuilder 5000 LAN/WAN Switch	10 T1 or 30 ISDN PRI/2048 Mbps 240 integrated modems	Source routing, transparent spanning tree, IP, IPX, AppleTalk

SOURCE: *LANTimes*, July 1997 issue, upgraded by the author.

the industry's most adaptable, secure and interoperable solutions for remote networking. No wonder there are more than 2 million digital ports and 1.2 million digital modems installed worldwide.

With industry-leading manageability and plenty of room for expansion, there is a MAX product to precisely fit your needs—whether you are a company with several offices across town, a corporation with thousands of users around the world, or a service provider looking to consolidate multiple carrier services on a single, high-density platform.

Ascend's MAX family features include:

- Integrated support for digital, analog, and cellular services
- Dynamic Bandwidth Allocation™ (DBA), inverse multiplexing
- Multilink Protocol Plus™ (MP+)
- Remote management
- Scalable and stackable units
- Digital modems
- Extensive standards-based security
- Optional integrated Secure Access Firewall

Table A.4 presents an overview chart of the MAX family of products.

A.3.1.1 ASCEND'S MAX 4000 REMOTE ACCESS SWITCH. The MAX 4000 is the high-end multiprotocol WAN access switch in Ascend's MAX family. By consolidating analog and digital access lines over high-speed digital trunks, the MAX 4000 enhances call performance by maximizing line utilization and minimizing equipment problems. (See Fig. A.4.)

The scalable architecture provides an easy migration path from analog-based solutions to the next generation of networking using Hybrid Access for ISDN, frame-relay, or leased-line solutions. The MAX 4000 allows remote users to move from analog systems to digital connectivity, using existing carrier E1/T1/PRI circuits and supporting up to 120 digital users. Ascend Max TNT, the top-of-the-line switch, can support up to 672 digital sessions.

As users migrate from PSTN to ISDN, high-density digital modem cards can be added to eliminate the need for costly analog modem banks and the downtime that can result from inefficient connectivity.

A.3.2 Shiva's LanRover Access Switch

This is Shiva's entry in the high-end (carrier-class RAS switch) market. With its $25,000 price tag, it is a cost-effective solution for midrange

TABLE A.4

Ascend's MAX Product Line

Product	MAX 200Plus	MAX 1800	MAX 2000	MAX 4002	MAX 4004	MAX TNT
E1	No	No	1	No	No	28
T1 with integrated CSUs	0	0	1	2	4	28
ISDN BRI(S/T)	Up to 4	8	0	Up to 32	Up to 32	0
ISDN BRI(U)	Up to 4	8	0	0	0	0
T3 (channelized)	No	No	No	No	No	1
Frameline	No	No	No	No	No	150
T3 (unchannelized)	No	No	No	No	No	1
Ethernet	1	1	1	1	1	16
High-speed serial	No	1	1	1	1	30
Hybrid analog and digital access	No	Included	Included	Optional	Optional	Optional
Frame-relay	0	Optional	Optional	Optional	Optional	Optional
Software digital modem	8*	16	24	48†	96	672
Capacity concurrent digital sessions	0	16	24	48‡	96	672
Expansion slots	8 or PCMCIA	2	2	6	6	48
IMUX ports	0	2	2	6	6	0

SOURCE: Ascend documentation available on the Internet, July 1997.
* Supports analog PCMCIA modems
† Up to 72 with upgrade options
‡ Up to 96 with upgrade options

438

Figure A.4
Ascend MAX 4000
remote access server.

remote access requirements. This type of remote access switch may offer a better network design to some organizations who do not want to have a single point of failure by centralizing all remote access in one switch.

LanRover Access switch offers as many as 72 analog ports and 4 PRI ports in a rack-mountable chassis. All the cables connect at the chassis, leaving the front clear. Of course, this design requires unrestricted access to the rear of the rack.

Configuring the LanRover is easy—in fact, easier than most servers in its class, according to independent testing by test labs and trade press evaluation reports. Configuring the hardware and software, including the management software, may take less than an hour for a small number of ports. After that, it is just a matter of repeating the process for remaining ports. The management software installed on Windows or Macintosh searches the network to locate Shiva products. If it discovers one or more, you use the management utility to add users, configure ports, and route protocols.

Please review other features of LanRover/E Plus described earlier. It is perhaps the only product with a client- and server-side caching agent, called PowerBurst, that boosts network performance. Caching is persistent, meaning that cache is continuously updated. However, PowerBurst works only with IPX.

A.3.3 Xyplex 9000 Remote Access Switch

According to the vendor, Xyplex Network 9000 is the industry's first software-based hub, around a chassis that accommodates an array of present and future technologies. The Network 9000 integrates media concentration, internetworking, remote access services, and switching in one scalable architecture. With the Network 9000 you can support media from 10 Mbps Ethernet to ATM. The Network 9000 supports a wide range of WAN services in a single chassis, and Xyplex Networks' software tightly integrates them together. And because the Network 9000 is available in 3-, 6-, and 15-slot chassis, you buy only what you need.

Xyplex 9000 has a flexible chassis design. With its midplane design, processor and I/O functions connect at the midplane of the chassis. I/O modules can be mixed and matched with processor modules. For example, a new ATM module can easily be added to the 9000 without changing the routing processor. And the 9000's software easily integrates the new module. Other hubs require major changes or additional boxes to accommodate new technologies.

The Network 9000 was the first hub to support advanced switching and ATM capability. With the existing midplane, the Network 9000 can support interconnection of high-speed switch modules in adjacent slots. Equipped with the *SwitchPlane*, Network 9000 supports interconnection of up to seven high-speed switch modules over an 8.4-Gbps switching fabric.

A.4 SOHO RAS Products

Since the SOHO market has become a sizable market in its own right, most of the RAS vendors have introduced inexpensive products to meet this need. Most of the products have an ISDN feature built in. These products are also called *personal ISDN routers* in the trade press. In terms of size, these products are about the size of a modem and cost in the range of $600 to $1200. The following vendors offer these solutions:

- Advanced Computer Communications (ACC) Congo Voice Router
- Ascend Communications Inc.'s Pipeline 25-VEX
- Bay Networks Inc.'s Clam
- Cisco's Cisco 752
- Develcon's Orbitor 2000
- Digi International Inc.'s Retoura Version 2.0
- Flowpoint Corp.'s Flowpoint 100
- Gandalf's XpressConnect 5242I
- ISDNet's NetRouter 1080
- Motorola's Vanguard 311

- Network Express's NE Nomad 750
- Proteon's Globetrotter 70
- RNS's NetHopper
- SBE's NetXpand SOHO
- Shiva's ShivaIntegrator 150
- 3COM's AccessBuilder RemoteOffice 500, 520, Remote User 400, and so forth
- Xyplex's RouterRunner ISDN

A.5 Application-Server-Based RAS Products

There are three application-server-based products worth mentioning. These are:

- Citrix's WinFrame/Enterprise Application Processor
- Citrix's WinView for Networks
- Xcellenet's RemoteWare

We describe each of these in the follow subsections.

A.5.1 Citrix's WinFrame/Enterprise Application Processor (Using Microsoft NT 3.51 RAS)*

Citrix has a unique way of extending Windows applications beyond the LAN to remote users. In order to do that, Citrix offers two products— WinFrame/Enterprise and WinView for Networks.

Description. WinFrame/Enterprise from Citrix Systems, Inc. is a unique way to extend Windows applications beyond the LAN. It is a multiuser Windows application server software for enterprise application deployment. Based on Windows NT Server under license from Microsoft, Win-Frame/Enterprise extends the reach of business applications beyond the LAN to remote users over the extended corporate network and public

---

*Information based on the material obtained from Citrix's homepages on the Internet.

Internet. The application server's main function is multiprocessing (time-slicing of multiple Windows remote control sessions) and administrative management of multiple server-executed applications.

A.5.1.1 APPLICATION CHALLENGES FACED BY ORGANIZATIONS. Organizations seeking to broadly deploy line-of-business applications across the extended enterprise face a diverse set of challenges. Some of these challenges are:

LAN-Locked Applications. Most business applications, such as two-tier client/server, are designed for the LAN and are not optimized to run over high-latency phone or WAN connections, which run 100 to 1000 times slower than a local LAN segment.

Servicing New Group of Users. The older corporate computing infrastructure was built for internal users only, not always for company's customers, prospects, business partners, and suppliers.

Heterogeneous Clients. Not everyone uses or needs a PC on their desktop. Some use non-Windows systems, like OS/2, Unix, or Macintosh. Some need low-cost, fixed-function devices, like terminals. Others need new computing devices, such as wireless tablets and personal digital assistants (PDAs).

Escalating Cost of Remote Network Management. Managing remote access (security), version control (maintenance), system configuration (moves, adds, changes, deletes), and support (help desk) are very costly, particularly for distant users.

A.5.1.2 BENEFITS OF WINFRAME APPROACH. Citrix believes that WinFrame/Enterprise satisfies the performance and universal application access needs of remote users, while equipping I/S organizations with the economic benefits of single-point management and control.

Extending the Reach of Business Applications. I/S organizations can securely and cost effectively deploy line-of-business applications to employees over the extended corporate network as well as to customers, prospects, and suppliers over the public Internet.

Performance. I/S organizations can provide users with consistent application performance over any type of network connection.

Management. I/S organizations can economically distribute, support, and manage line-of-business applications from the server.

Security.   I/S organizations can deliver broad access to corporate information with minimal security risk in data transmission and information storage.

A.5.1.3 TECHNOLOGY SUMMARY.   WinFrame/Enterprise is a multi-user Windows application server software based on a network-centric computing model, which includes Citrix's ICA-based universal, thin-client software. Citrix has source and distribution licenses to Windows NT Server from Microsoft. The company has added two core technologies to Windows NT Server to produce WinFrame/Enterprise.

ICA WINDOWS PRESENTATION SERVICES PROTOCOL.   ICA is a general-purpose presentation services protocol for Microsoft Windows. Conceptually similar to the Unix X-Windows protocol, ICA allows an application's user interface to execute with minimal resource consumption on a client PC, while application logic executes on the WinFrame multiuser application server. Because ICA is implemented at the system level (GDI), it is highly efficient and compact. This architecture allows applications that usually consume the largest bandwidth to run at near-LAN speed over low-bandwidth phone lines.

MULTIWIN MULTIUSER ARCHITECTURE.   MultiWin is a Citrix-developed technology that extends the Windows NT operating system into a true multi-user system. This means multiple users can execute applications on the same machine at the same time; each user runs as a virtual session on the WinFrame/Access server. This architecture provides the following end user benefits:

Fast Application Access.   The WinFrame/Enterprise universal, thin-client gives remote users fast access to any type of application including DOS, 16- and 32-bit Windows, or client/server programs. The performance of remote control session is quite good because of local caching and efficient though proprietary PPP software.

Acceptable Network Performance.   According to a test conducted by *Network Computing* magazine, text file transferred at 4.7 Kcps, the compressed file at 2 Kcps, and bit map file at 5.3 Kcps.

Local/Remote Transparency.   WinFrame/Enterprise provides the familiarity of a local LAN desktop. Remote users have complete access to all local system resources such as notebook drives, remote printers, and clipboards. Users can also cut and paste between local and remote applications and drag-and-drop to copy files in the background while they continue to work.

Integrated Desktop.   From a single desktop, remote users can run applications locally from the notebook PC or remotely from the WinFrame/Enterprise server for best performance.

Ease of Setup.   In spite of the complexity of the way Citrix WinFrame is implemented under the hood, it is relatively easy to install, configure, and use. With its Windows 95—like installation and setup wizard, Win-Frame/Enterprise clients are quite easy to implement for various environments—Windows 3.1, Windows for Workgroups, Windows 95, and Windows NT. The wizard guides users through all the necessary installation steps and automatically detects the PC's available modem. The vendor provides an evaluation copy of the software to test-drive on its own system by dialing a 1-800 number.

32-Bit Windows Application Availability.   Remote users gain immediate access to Windows 95 and Windows NT applications, regardless of their client hardware. WinFrame/Enterprise enables even DOS-based 286 systems to run Windows 95 applications at near-LAN speeds over low-bandwidth connections.

A.5.1.4 CORPORATE I/S MANAGEMENT BENEFITS.   Assuming that the unique architecture of WinFrame fits your organizational needs, you may be able to achieve the following benefits:

Economical for Lightly Loaded Clients.   WinFrame supports multiple concurrent users on a single processor and offers free, unlimited client software licensing, making it a cost-effective solution for enterprise-wide application delivery. Of course, the application server is licensed for specified number of *concurrent* users.

Enterprise Scalability.   Symmetrical multiprocessing (SMP) hardware compatibility enables WinFrame/Enterprise to support hundreds of concurrent users.

Extensive Connectivity.   WinFrame/Enterprise connects users to the network through standard telephone lines, WAN links (T1, T3, 56-Kbps, X.25), broadband connections (ISDN, frame-relay, ATM), or the Internet. The remote user's access to the remote access sessions on the server can be over asynchronous, PPP, or LAN connections.

Single-Point Application Management.   With WinFrame/Enterprise, all application upgrades and additions are made only once at the server and are instantly available to all remote users.

End-to-End Management.   Using WinFrame/Enterprise's extensive Windows-based utilities, network administrators can set up applications, view active sessions, monitor system performance and events, troubleshoot problems, and create reports from the server. WinFrame/Enterprise also allows administrators to use popular network management tools, like Microsoft Systems Management Server. It also provides compression ratio for file transfer.

Remote Administration.   System administrators can dial up to the Windows NT—based WinFrame/Enterprise server for remote administration and management.

Remote Support and Training.   Administrators can connect to a remote user's session to visually see what is on the screen and interact with the user, making WinFrame/Enterprise a valuable remote support and training tool.

Seamless Network Integration.   WinFrame/Enterprise integrates into NetWare, Windows NT, Novell, and other PC networks, allowing administrators to quickly set up users from existing domain or bindery information.

C2-Compliant Security.   WinFrame/Enterprise incorporates Windows NT's C2-level security, including multilevel passwords and privileges, roving callbacks, encrypted login and data, and file-level security to protect data privacy and network resource integrity.

A.5.2 Citrix's WinView for Networks

With WinView for Networks, your NetWare users can enjoy fast Windows and DOS performance, regardless of their remote location. With WinView, most of the Windows-specific tasking is done at the application server. Only the user interface is executed at the client workstation, allowing remote users to run Windows and DOS applications with local performance. Therefore, it reduces the amount of data that travels over the wire. As a result, bottlenecks are eliminated and performance is improved. Some of the features and benefits of WinView are listed here:

32-Bit Processor for Performance.   WinView for Networks is built around a high-performance 32-bit operating system. As a result, WinView provides the reliability, flexibility, security, and manageability you need in order to extend your LAN to a large number of mobile users and remote branch offices.

Reliability.   WinView runs in protected mode, so misbehaving applications cannot affect system integrity or other users' work. This supplies you with a reliable platform from which you can deploy mission-critical applications.

Supports Hundreds of Remote Users.   Unlike remote control software, which requires one dedicated PC for every dial-in line supported, WinView is a scalable, affordable remote computing solution for users requiring access to the network—anytime, anywhere. A single WinView application server supports up to 10 concurrent Windows users—actual performance is application-dependent. So it's easy and cost-effective to distribute remote Windows throughout even the largest organizations.

WinView's Security and Ease of Use.   WinView is a secure, reliable, and manageable remote computing solution. Its advanced multilevel system security, including dial-back, protects your network while maintaining tight integration with NetWare security. Featuring advanced disconnect recovery, WinView maintains and password-protects disconnected remote sessions. This allows remote users to reconnect from any workstation and, with the proper password, pick up right where they left off.

WinView Simplifies Your Administrative Tasks.   Network administration is made easy with WinView for Networks. Its extensive set of integrated network utilities, including resource management, event logging, user messaging, and print spooling, enables administrators to manage all the resources and security privileges of the WinView application server from any workstation on the network or from any remote PC. And when your remote users need help, WinView provides remote assistance, administration and DOS session shadowing capabilities. WinView also eases the support burden of software installation and distribution, since software is installed once at the server and automatically distributed to your users.

WinView Gives New Life to Your Existing Hardware and New Network PC Hardware.   WinView enables you to deploy Windows to your existing environment, saving you the time and expense of upgrading your current hardware base. WinView's unique distributed Windows architecture (under license from Microsoft) allows your existing PCs, even 286 machines not capable of running Windows, to access sophisticated Windows applications. That's because the processing power required to run the small WinView client program is minimal (640KB RAM on a 286 system) and is much less than that needed to actually run Windows 3.1 natively. As a result, you can instantly deploy off-the-shelf Windows applications to your current systems. WinView also works with the most common hardware configurations (modems and displays), so you don't have to reconfigure your systems.

A.5.2.1 HOW WINVIEW FOR NETWORKS WORKS.   With traditional networks, applications are loaded and executed on client workstations. But with WinView for Networks, Windows and DOS applications are loaded and executed on the application server, while just the results are displayed on user PCs. This is made possible through the exclusive

WinView Independent Console Architecture (ICA). That means, for the first time, high speed is achieved over low-bandwidth connections and wireless networks. In its most basic application, WinView for Networks addresses the communication and application needs of remote/mobile LAN users. In more advanced configurations, it facilitates the deployment and management of Windows applications and provides the flexibility and functionality to meet a whole host of other PC-LAN problems, such as giving underpowered hardware the power to run Windows applications and eliminating bottlenecks over wide area networks. (See Fig. A.5.)

A.5.2.2 PRODUCT HIGHLIGHTS

- Provides fast remote Windows by minimizing data traffic across telephone lines, LANs, and WANs
- Fully integrates into NetWare environments and supports industry-standard Windows and DOS applications
- 32-bit preemptive operating system runs in protected mode, so misbehaving applications cannot affect system integrity or other users' work
- A single 486 application server can support up to 10 concurrent Windows users (actual performance is application dependent)
- Incorporates multilevel system security, including classifications, login and password management, security auditing, and reboot protection
- Maintains and password-protects disconnected remote sessions, allowing users to reconnect from any workstation, with the proper password
- Includes extensive network utilities, including DOS session shadowing, resource monitoring, event logging, user messaging, and resource management
- Gives underpowered PCs (286 and higher) full access to Windows applications, maximizing your hardware investment
- Eliminates WAN bottlenecks and significantly increases performance by reducing the amount of data traveling over the wire

A.5.3 Xcellenet's RemoteWare

RemoteWare provides application-specific functions for generalized remote access. However, it must be implemented on a RAS such as Shiva's NetRover. In fact, Shiva and Xcellenet have a cooperative marketing relationship to bundle each other's products and support it as an integrated package.

Figure A.5
(a) Citrix extending
Windows applications
to remote locations;
(b) WinFrame—
Remote Windows
application. (Source:
Citrix.)

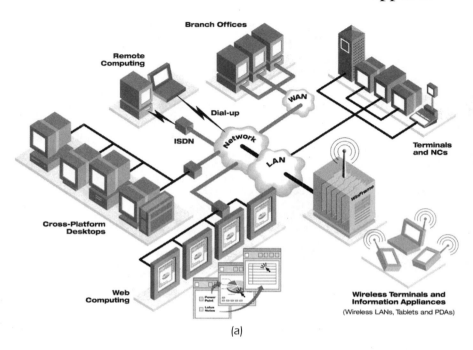

(a)

(b)

A.6 Remote Control Client Software Products

There are a number of remote control client software products available in the market. Table A.5 describes the main features of some of these products.

A.7 Miscellaneous RAS Software Products

In this category, we have listed several software products that can be used for data replication, file synchronization, or for horizontal remote access applications such as sales force automation.

A.7.1 Novell's Replication Software

Novell's Replication Services (NRS) gives NetWare administrators a powerful means to disseminate corporate data or roll out Web sites to geographically dispersed servers, automatically and transparently. One server can replicate to 50, each of which can replicate to 50 more and so on, allowing the file updates to be done easily in a hierarchical organization.

NRS is implemented as a NetWare Load Module (NLM). File and folder level security is provided. Although it does not replicate open files, NRS can force users to log off in order to close those files.

A.7.2 Puma's IntelliSync Software

The IntelliSync for Windows CE software from Puma Technology enhances the power and convenience of your Windows CE handheld device. IntelliSync for Windows CE enables you to synchronize your handheld PC directly with your favorite PC-based personal information management, contact management, and group scheduling applications all in one easy step.

As a mobile user, you frequently need to synchronize the information stored on your handheld PC with one or more PC applications. You may

TABLE A.5

Popular Remote Control Software Products

Product	Vendor Contact Information	OS Supported	Functions Supported	Networks Protocols Supported	Network Options	Internet Support	Security
pcAnyWhere	Symantec 408-253-9600 www.symantec.com	Windows 3.1, 95, and NT	File transfer, help desk, remote database access	TCP/IP, IPX, NETBIOS	PSTN, ISDN, cable, Infrared	Yes	Access control passwords, data encryption, callback, screen and keyboard locking
ReachOut	STAC Electronics, San Diego 619-794-4300 www.stac.com	DOS, Windows 3.1 and 95		TCP/IP, IPX, NETBIOS	PSTN, ISDN, cable	Yes	Extensive access control, audit, callback, screen blanking, entrust support
CarbonCopy	Microcom 800-822-8224 www.microcom.com	DOS, Windows 3.1 and 95	File transfer, help desk, remote drives, remote database access	IPX, NETBIOS	Direct connect, PSTN (modem link), ISDN, wireless	No	Access control, passwords, callback, screen and keyboard locking

Product	Company	Platforms	Functions	Protocols	Connections	Remote control	Security
LapLink for Windows	Traveling Software 206-683-8088 www.travsoft.com	Windows 3.1 and 95	File transfer, help desk, remote database access	TCP/IP, IPX, NETBIOS	PSTN, ISDN, Infrared	Yes	Access control passwords, session logs, data encryption, callback, screen and keyboard locking
Remotely Possible/32	Avalan Technology 508-429-6482 www.avalon.com	Windows 3.1, 95, NT	File transfer, help desk, remote database access	TCP/IP, IPX	PSTN, ISDN, Infrared	Yes	Access control passwords, session logs, data encryption, callback, screen and keyboard locking
Timbuktu for Windows	Farallon Communications 510-814-5000	Windows 3.1, 95, NT, and Macintosh	File transfer, remote database access	TCP/IP, IPX, NETBIOS	PSTN, ISDN, LAN	Yes	Access control password

SOURCE: *Information Week*, September 23, 1996, and *Network Computing*, October 1996 issue.
Note: Only a subset of commercially available products have been described.

make a series of changes or additions to the handheld PC information while you are out of the office, but also make changes or additions directly to your corresponding PC information when you are in the office. Intelligent synchronization of handheld PC data with the data in your PC applications in an essential housekeeping function that ensures that both will contain the most up-to-date information. This synchronization must be simple, accurate, and direct.

Automatically, IntelliSync for Windows CE links your PC contact and scheduling applications with your handheld PC when connected, guaranteeing that crucial scheduling and informational data is accurately reflected on both devices. IntelliSync's advanced architecture lets you decide whether to synchronize your handheld PC handheld PC data with a single PC application, or multiple PC applications. The synchronization is direct and comprehensive. The result is the necessary confidence that your vital information is current and accurate, enabling a smooth and productive workday.

Intellisync has the following key features:

Direct Automatic Synchronization.   Simply connect a handheld PC to your PC. IntelliSync for Windows CE synchronizes your handheld PC directly with your PC's personal information management, contact management, or group scheduling applications automatically. There are no extra steps and no intermediate data conversions are required. The synchronization with your PC applications is direct and clean. (See Fig. A.6.)

Synchronize Multiple Applications at One Time.   Many PC users have one PC application for personal information or contact management, a second for scheduling appointments, and sometimes a third for keeping track of *to do* items and notes. With IntelliSync for Windows CE you can synchronize your handheld PC applications with multiple PC applications at one time. For example, bring the handheld PC contacts into conformity with Lotus Organizer 2.1, and the calendar Microsoft Schedule+ for Windows 95 all in a single step.

Conflict Resolution.   What happens when information is changed both on your handheld PC and on your host-based system? Altering the contents of multiple files can lead to confusion when you attempt to synchronize data. While other synchronization software may take an easy way around the issue, often yielding less-than-desirable results, IntelliSync for Windows CE always resolves conflicts smartly and accurately. You can control conflict resolution options, for example, by marking a particular application (either on the handheld PC or the PC) as the source, or reference, for

Figure A.6
(a) Puma startup
screen shot; (b) Puma
synchronization
options. (Source:
Puma.)

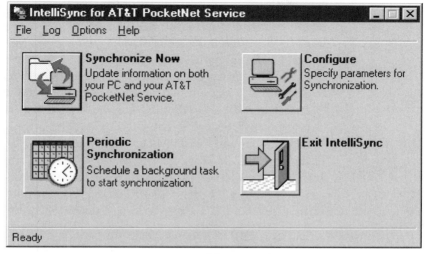

(a)

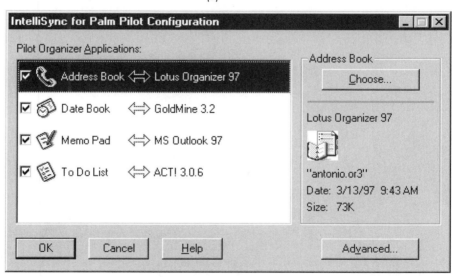

(b)

automatic conflict resolution. Or, you can elect to be notified when a conflict has occurred, then compare the fields in question, and make the decision yourself. IntelliSync for Windows CE prompts you to confirm any changes before the action is executed.

Customization. You can customize IntelliSync for Windows CE to work exactly the way you want. Once you specify which handheld PC applications are to be synchronized with which PC applications, you may not need anything else. However, should you want to, you can specify and modify the individual field mappings between your handheld PC and your PC applications. You don't have to worry about mapping invalid field combinations, since IntelliSync has incorporated special typing rules that block the mapping of invalid combinations. In addition, you can specify the date range to include timed items, and turn off synchronization for complete *to do* items.

IntelliSync works with Microsoft Schedule+ for Windows 95, Microsoft Exchange Address Book, Lotus Organizer 2.1 and 97, Sidekick for Windows and Windows 95, NetManage ECCO 3.03, now up-to-date for Windows and Windows 95, DayTimer Organizer 2.0, and so forth. Furthermore, it runs under both Windows 95 and Windows NT.

A.7.3 Sales Force Automation Software

Review the name of various sales force automation packages mentioned in Table 2.3 of Chap. 2.

A.7.4 Sybase's SQL-Anywhere Replication Server Software

Sybase Replication Server is one of the leading data replication solutions in the industry. It's been deployed at more than 7000 sites around the globe by organizations in every major industry, including banking, finance, insurance, manufacturing, telecommunications, healthcare, government, and others.

The latest version of Replication Server provides users with significant performance gains. Specifically, improvements to Replication Server's subscription materialization capability enable it to perform the initial load on new sites more quickly in order to get those new sites up and running much faster (as much as ten times, according to the vendor) than in previous versions.

Replication Server Manager 11.0 manages and monitors heterogeneous environments from a single console. Sybase Replication Server Manager version 11.0 provides system administrators with a GUI console, single point of management for their geographically distributed replication environment. Replication Server Manager 11.0 offers optimized control over the distributed enterprise, including functions for events monitoring and alarms on the status of network components; monitoring of replication latency events with enhanced queues and definitions; diagnostic functions for improved problem definition and suggested resolutions; and new troubleshooting features for setting performance thresholds and optimizing network usage.

Sybase Replication Server solution has interfaces with two other leading databases—IBM's DB2 and Oracle's version 7.

A.7.5 Symantec's AutoXfer Synchronization— A Feature in pcAnywhere

Symantec is shipping this file transfer and synchronization capability in its Norton pcAnywhere version 8.0. While this feature of pcAnywhere is similar to Traveling Software's LapLink for simple tasks, LapLink's Xchange agent technology is meant for more sophisticated requirements.

A.7.6 Traveling Software's LapLink's Xchange Concept

Traveling Software Inc.'s LapLink for Windows 95 is a popular file transfer and synchronization software used by mobile users to exchange data between notebooks, desktops, and other handheld devices. The latest version 7.5 features agent technology called *Xchange*, so that it updates entire folders automatically.

The Xchange concept is similar to an off-hours batch-processing job. For example, a remote salesperson using Xchange Agent can set up LapLink to upload sales reports or call sheets from a mobile PC to an office PC at a set time or at set time intervals. The job runs unattended; the notebook dials up the office, transfers files, and disconnects when the transfer has been completed.

Traveling Software achieved this function through integration with Microsoft Windows 95's System Agent scheduling technology. Xchange Agent's wizard also lets users set up synchronization tasks so that they need click only once to perform a synchronization.

LapLink uses adaptive compression technology—the slower the line speed, the higher the data compression LapLink automatically uses.

A.8 Firewall Software Products

We have described the role of firewall software in remote access solutions in Chap. 14. Here we list a few popular products; a more complete list can be obtained from trade magazines. In particular, we would mention an article in the March 21, 1997, issue of *Data Communications* magazine. The reader should refer to App. B, which lists various firewall vendors.

- Altavista Firewall 97 from Altavista Software Inc.
- Ascend Communications's SecretAccess Firewall
- Borderware Firewall from Secure Computing Corp.
- Check Point Firewall-1 from Check Point Software Technologies Inc.
- Cyberguard Firewall from Cyberguard Corporation
- EagleNT firewall from Raptor Systems Inc.
- Firewall/Plus from Network Software and Technology Inc.
- Guardian version 2.0 software from NetGuard
- IBM's Internet Connection Secured Network Gateway
- NetGuard's Ltd.'s Guardian 1.3
- Secure Computing Corp's BorderWare Firewall
- SideWinder Security Server
- SunSoft's Solistice Firewall

A.9 Value-Added Network Services

There are several network services providers which have started offering value-added remote access network services. Many of these companies are Telcos, ISPs, or traditional network outsourcing vendors. The services range from remote access connection to use Internet as a virtual private network to complete outsourcing of remote access services, including network management and support of the mobile users. MCI, GTE, and major ISPs have been mentioned in the trade press. You should contact your network supplier and ISP to inquire whether they offer these services.

B

Remote Access Vendors

Vendor Name and Address	Telephone Number and/or Internet URL	Products and Services
Advanced Computer Communications (ACC) Santa Barbara, CA	805-444-7854 www.acc.com	RAS vendor (Tigris high-end RAS Switch, Congo Voice Router as personal ISDN)
ADSL Forum	www.adsl.com	A forum for ADSL hardware vendors and communications service providers
AirSoft Inc. 1900 Embarcadero Road, Suite 204 Palo Alto, CA 94303	800-708-4243 415-354-8123	Powerburst remote access software implemented in Shiva remote access products
AllPen Software, Inc. 51 University Avenue, Suite J Los Gatos, CA 95030	408-399-8800	Pen-based applications for PDAs and pen computers
Altavista Software Inc. Littleton, MA	508-486-2308 www.altavista .software.digital.com	Altavista Firewall 97
ANS Communications Reston, VA	703-758-7700 www.ans.net	ANS Interlock Version 4 Firewall
ARDIS 300 Knightsbridge Parkway Lincolnshire, IL 60069	700-913-1215 700-913-1453	Wireless Packet Data Network services, wireless modems, mobile computing integration services
Ascend Communications Inc. Alameda, CA	510-769-6001 www.ascend.com	Full product line for RAS (MAX and Pipeline product families); especially strong for carrier-quality RAS switches and infrastructure products
Attachmate 3617 131st Ave. S.E. Bellvue, WA 98006	206-644-4010 800-426-6283	Remote control and node software
AT&T Wireless Group 1700 South Patterson Blvd. Dayton, OH 45479	206-990-4481	Full Service Cellular and CDPD Network Supplier
Avalan Technology P.O. Box 6888 7 October Hill Road Holliston, MA 01746	508-429-6482 800-441-2281	Remotely Possible/32 remote control software

Bay Networks Inc. Santa Clara, CA	408-998-2400 www.baynetworks.com www.xylogics.com	Full product line for RAS; Annex product line
Bell Atlantic Mobile 180 Washington Valley Road Bedminster, NJ 07921	908-306-7520	Circuit Switched Cellular and CDPD Services
Bell Mobility Canada 20 Carlson Court Etobicoke, Ontario, Canada M9W 6V4	416-674-2220	Cellular, paging, and packet data network services
Cable Industry Forum National Cable Television Association—NCTA 1724 Massachusettes Ave., N.W. Washington, DC 20036	202-775-3669 (Barbara York) http:/cable-online.com/ ncta.htm	Remote Access using Cable Modems
Check Point Software Technologies Inc. Redwood City, CA	415-562-0400 www.checkpoint.com	Check Point Firewall-1 Version 2.1
Cisco Systems Inc. San Jose, CA	408-526-4800	A leader in internetworking and remote access; major supplier to ISPs
Citrix Systems, Inc., 6400 NW 6th Way Fort Lauderdale, FL 33309	Phone: 954-267-3000 Fax: 954-267-9319 www.citrix.com	Software for remote invocation of Windows applications
CDPD Forum Liz Durham, Marketing Manager 401 N. Michigan Ave Chicago IL 60611-4267	1-800-335-CDPD	Vendor Consortium for CDPD network service providers
Cubix 2800 Lockheed Way Carson City, NV 89706	800-829-0550	Remote control and remote node Server
Cybergguard Corp. Fort Lauderdale, FL	954-977-5478 www.cyberguardcorp .com	Cyberguard Firewall Version 3.0
Cycon Technologies Fairfax, VA	703-383-0247 www.cycon.com	Cycon Labryinth Firewall
Cylink 910 Hermosa Court Sunnyvale, CA 94086	408-735-5800	Security Products; also AirLink spread spectrum microwave radios @ 2 Mbps in point-to-point configurations
Develcon Electronics Ltd. Saskatoon, Saskatchewan, Canada	306-933-3300 www.develcon.com	RAS products, including personal ISDN

Vendor Name and Address	Telephone Number and/or Internet URL	Products and Services
Digi International Inc. Minitonka, MN	212-912-3444 www.dgii.com	Serial I/O cards used in RAS servers and RAS products
Emulex Dept NC9604 14711 NE 29th Place Bellevue, WA	800-EMULEX1 714-662-5600 www.emulex.com	ConnectPlus RAS (low-end)
ERICSSON GE Mobile Communications Inc., a division of Telefon AB LM Ericsson Wireless Computing 15 E. Midland Ave. Paramus, NJ 07652	1-800-223-6336 201-890-3600 201-265-6600 201-265-9115 (Fax)	Complete suite of mobile computing infrastructure products (Mobitex packet, PCS, GSM and cellular), including modems, etc.
Ex-Machina, Inc. 11 East 26th Street, 16th Floor New York, NY 10010-1402	800-238-4738	Software between PCs and Servers to Pagers (e.g., NOTIFY, Broadcast of software to notebooks)
Farallon Communications 2470 Mariner Square Loop Alameda, CA 94501	510-814-5000	Timbuktoo remote control and node software
Flowpoint Corp. Los Gatos, CA	408-252-6470 www.flowpoint.com	Flowpoint 100 personal ISDN Router
Frontier Technologies Corporation 10201 North Port Washington Road Mequon, WI 53092	Sales: 800-929-3054 Fax: 414-241-7084 www.frontiertech.com FTP: ftp.frontiertech.com	Firewall software
Gandalf Technologies Napean, Ontario, Canada	613-274-6500	High-end XpressWay RLAN product, also Xpressconnect 5242I (personal ISDN)
Global internet Software Group Inc. San Mateo, CA	415-513-0635 www.globalinternet.com	Centri Firewall
Global Technology Associates Inc. (GTA) Orlando, FL	407-380-0220	Gnat Version 1.1 firewall
Golmine Software 17383 Sunset Blvd., Suite 301 Pacific Palisades, CA 90272	310-454-6800 800-654-3526	Goldsync remote control and remote node software

GTE MobileNet 245 Perimeter Center Parkway Atlanta, GA 30346	770-391-8000	Cellular and CDPD network services, modem pool access, wireless voicemail, paging
IBM Corporation 700 Park Office Drive, Building 662 Research Triangle Park, NC 27709	800-IBM-CALL5	Internetworking and RAS product line Firewall software for remote access
InfoNet El Segundo, CA	310-335-4700	DialXpress Remote Access products
Iridium Inc. 1401 H Street N.W., Suite 800 Washington, DC 20005	202-326-5600	Overall coordinating organization for Iridium satellite-based wireless network
ISDNet Inc. Sunnyvale, CA	408-522-5090 support@isdnet.com	NetRouter 1080
Livingston Technology UK US office in Pleasonton, CA	510-426-0770 sasek@livingston.com UK 44-0-1-322-407-892	Radius Security software leader FireWall IRX, Office Router PortMaster 3 remote access products
LOTUS Corporation Cambridge, MA (Offices in most large cities)	800-205-9333	cc:Mail mobile, paging-based messaging and Notes groupware software
Metricom 980 University Avenue Los Gatos, CA 95030	408-399-8200 (Phone) 408-354-1024 (Fax)	Network Service Provider
MicroCom 500 River Ridge Dr. Norwood, MA 02062	617-551-1000 800-822-8224	CarbonCopy remote control and LANExpress
Microsoft Redmond, WA	206-882-8080 www.microsoft.com	NT 3.51 RAS
MMTF Task Force mmtf-request@ www.epilogue.com	505-271-9933	Mobile Management Task Force for SNMP-compatible MIB standards
MobileWare Corporation 2425 North Central Expressway, #1001 Dallas, TX 75080-2748	214-952-1200 (Phone) 214-690-6185 (Fax)	MobileWare Software for accessing LAN applications from remote users

Vendor Name and Address	Telephone Number and/or Internet URL	Products and Services
Motorola Informations Systems Group (ISG) Huntsville, AL	205-439-8500 www.mot.com/ mims/isg	Extensive line of modems and ISDN adapters; also Vanguard 311 personal ISDN
Multitech	800-328-9717 www.multitech.com	Remote access servers
NCR Wireless Communications & Network Division 1700 S. Patterson Blvd. Dayton, OH 45479	800-CALL-NCR 513-445-5000 (Phone) 513-445-4184 (Fax)	WaveLAN wireless LAN products
NEC Technologies Inc. Internet Business Unit San Jose, CA	408-433-1549 www.privatenet.nec .com	Privatenet Secure Firewall
Netguard Ltd. Migdal, Ha-Emek, Israel	972-66-44-99-44 www.guardian-security.com	Guardian Firewall
Nomadic Software	415-335-4310	SmartSynch 2.0 software (synchronizes files between remote systems)
Northern Telecom P.O. Box 833858 Richardson, TX 75083-3858	800-4-NORTEL	Remote access servers Entrust security software—remote access in-building wireless, digital cellular (CDPD), PCS and wireless access network infrastructure
Notable Technologies Inc. 1221 Broadway, 20th Floor Oakland, CA 94612	800-732-9900	AirNote messaging and financial information services on pagers
Novell Inc. 122 East 1700 South Provo, UT 84606-6194 (Office in every major city)	800-258-5408	Novell NetWare NOS, NetWare Connect (RNA solution), NetWare Mobile, and Novell Replication Services
Penril Data bility Networks Gaithersburg, MD	301-921-8600 800-473-6745 www.penril.com	Access Beyond 4400

Persoft Inc. 465 Science Drive Madison, WI 53711	800-368-5383 608-273-6000 (Phone) 608-273-8227 (Fax)	Intersect NCR WaveLAN OEM bridges up to 3 miles, omni/bidirectional antenna, wireless LAN
Palm Computing Inc. California	415-949-9560	PIM software for Zoomer devices, Palm Connect— desktop PC connection utility
PairGain Technologies 14402 Franklin Avenue Tustin, CA 92780	800-370-9670 www.pairgain.com	DSL Modems
Proxim, Inc. 295 North Bernardo Ave. Mountainview, CA 94043	800-229-1630 415-960-1630 (Phone) 415-964-5181 (Fax)	Wireless LAN Products (902 to 928 MHz and 2.400 to 2.483 GHz)
PenRight 47358 Fremont Blvd. Fremont, CA 94538	510-249-6900	Pen-based software development kit
Puma Technology 2550 North First Street, Suite 500 San Jose, CA 95131	408-321-7650 408-433-2212 (Fax) www.pumatech.com	*Intellisync* synchronization software between desktops and handheld devices
QUALCOMM 10555 Sorrento Valley Rd. San Diego, CA 92121	619-587-1121 email: info@ qualcomm.com ftp.qualcomm.com	OmniTRACS satellite data radio, CDMA cellular, PCS
Raptor Systems Inc. Waltham, MA	617-487-7700 www.raptor.com	EagleNT Firewall
RNS (formerly Rockwell Network Systems) Meret Communications Inc. 7402 Hollister Avenue Santa Barbara, CA 93117-2590	805-968-4362 805-968-6478 800-262-8023 www.rns.com	DataShuttle ISDN Adapters NetHopper family of dial-up router products
RACOTEK 7301 Ohms Lane, Suite 200 Minneapolis, MN 55429	612-832-9800 612-832-9383	Mobile software for application and network integration, supports most wireless networks; experience in distribution and manufacturing applications
RAD Data Communications Tel-Aviv, Israel	972-3-645-9410 www.rad.com	Web Ranger as personal ISDN
RNS Santa Barbara, CA	805-268-4262 www.rns.com	NetHopper NH-BRI-611

Vendor Name and Address	Telephone Number and/or Internet URL	Products and Services
RAM Mobile Data 10 Woodbridge Ctr. Drive, Suite 950 Woodbridge, NJ 07095	800-736-9666 908-602-5500 908-602-1262 www.ram.com	Wireless WAN provider in U.S., UK, and Australia; paging, messaging services, mobile computing systems integrator
River Run Software Group 8 Greenwich Office Park Greenwich, CT 06831	203-861-0090	Application development tools for mobile applications
SBE Inc. San Ramon, CA	510-355-2000 www.sbei.com	netXpand SOHO RAS solution
Shiva Corporation Northwest Park 63 Third Avenue Burlington, MA 01803	617-270-8320	Remote Network Access products—software and hardware, including remote Web server
SkyTel Division of Mobile Telecommunications Technologies—Mtel 1350 I Street NW, Suite 1100 Washington, DC 20005	1-800-SKY-USER 800-456-3333 202-408-7444	SkyPager, SkyWord, SkyTalk, SkyStream, SkyFax regional, national, and international paging services, including two-way paging service
Stac Inc., San Diego, CA	800-522-7822 www.stac.com	Reachout 6.0 remote control software (32 bit, FTP client, also supportable by popular Web browsers)
Stampede Technologies 65 Rhodes Centre Dr. Dayton, OH 45458	800-763-3423 www.stampede.com	Remote Office Communications Server
Sun Microsystems Palo Alto, CA	415-960-1300 www.sun.com/ security	Sunscreen EFS Firewall
Symantec 10201 Torre Ave. Cupertino, CA 95104	800-441-7234 408-253-9600 www.symantec.com	pcAnywhere remote control
Technology Development Systems 2300 N. Barrington Road, Suite 603 Hoffman Estates, IL 60195	708-781-1800	WorldLink 2.2—database query, file synchronization

Travelling Software 18702 North Creek Parkway Bethwell, WA 90801	800-342-8080 206-483-8088 206-487-1284 (Fax) www.travsoft.com	Laplink wireless software— popular for small offices, and mobile laptops; user file synchronization
Telebit Cororation Chelmsford, MA	508-441-2181 www.telebit.com	MICABlazer
Trusted Information Systems Inc. Rockville, MD	301-527-9500 www.tis.com	Gauntlet Internet Firewall
3-COM Corp. Santa Clara, CA	408-764-5000 www.3COM.com	Full line of remote access products, including personal ISDN for SOHO
Xcellenet 5 Concourse Parkway, Suite 850 Atlanta, GA 30328	Tel: 770-804-8100 Fax: 770-804-8102 www.xcellenet.com	RemoteWare application- level remote access solution
Xircom 2500 Corporate Centre Drive Thousands Oaks, CA 91320	Sales: 800-438-4526 805-376-9300 (Phone) 805-376-9311 (Fax) www.xircom.soft	Full line of PC Card (PCMCIA), RAS hardware, and wireless LAN products
Xyplex Littleton, MA	508-952-4700 800-338-5316 www.xyplex.com	Full line of RAS products plus ISDN RouterRunner
Zyxel Anaheim, CA	714-693-0808 www.zyxel.com	RAS products, including Prestige 28641 personal ISDN

C

Remote Access Information Resources

Information Resource	How to Get Information	Type of Information Available	Comments
ADSL Forum	www.adsl.com	Tutorial, status of pilots and ADSL hardware and service availability	Excellent resource for ADSL but watch for industry bias
Cable Modem Industry Forum	National Cable Television Association (NCTA) 1724 Massachusettes Ave., N.W. Washington, DC 20036 202-775-3669 (Barbara York) http:/cable-online.com/ncta.htm	Cable modem information	
Cable Labs for testing compatibility of cable modems on the Internet	Mike Schwartz VP Communications 303-661-9100 http://www.cablelabs.com	Cable modem information	
Computer Security Institute (CSI)	http://www.gocsi.com 415-905-2626	Security-related information	
FIRST— Forum of Incident Response and Security Teams	http://www.first.log	Security-related information	
Frame-relay Forum	http://frame-relay.indiana.edu	Information on frame-relay	
ISDN home page (Dan Kegel's)	http://alumni.caltech.edu/ ~dank/isdn		Excellent resource for ISDN
Internet Newsgroups	comp.dcom.isdn comp.dcom.modems comp.dcom.telecom.tech comp.dcom.framerelay comp.security.misc alt.security		
Motorola Cable Modems	847-632-5954 (SuzanneBoggs) http://www.mot.com/ multimedia		
NCSA— National Computer Security Association	http://www.ncsa.com 717-258-1816	Security-related information	

RAS vendors	Web info in Appendix B	Product information and white papers	Mostly, marketing information; look for good white papers
Security resources	Computer Emergency Response Team 412-268-7090 e-mail: cert-advisory-request@ cert.org	Security information	
Shiva's homepages	www.shiva.com	Good white papers on remote access issues	

APPENDIX **D**

Remote Access
Related Standards

Standard	Description
ANSI X9.26	Secure sign-on standard
DES	Security standard
ITU-T -ex CCITT modem standards	See Chap. 6
MMAP	Mobile Management Application Protocol—supports mobility via various air interfaces on existing wireline switched networks
MMTF MIB	Mobile Management Task Force MIB for SNMP
Modem standards	See Chap. 6 for V-series and other standards
Multilink PPP—RFC 1717	PPP support for multiple ISDN links
RADIUS-IETF standard (RFC 2058)	Remote Access Dial In User Services. Latest version is RADIUS 2.0
RSA	Security standard
SNMP	Simple Network Management Protocol Standard
TACACS (IETF standards)	Terminal Access Controller Access Control System standard (TACACS is based on Cisco development and includes authentication only)
TACACS+ (IETF standards)	Terminal Access Controller Access Control System standard (TACACS+ includes authentication, authorization, and accounting)
TCP/IP	De facto standard for transport layer between remote nodes

E

Rule-of-Thumb Costs for Various Remote Access Components*

---

* These costs are based on many assumptions and should be used only for a very preliminary cost estimate prepared during initial stages of a feasibility study. This estimate should be refined further after determining capacity requirements and contacting the vendors listed in appendix B. Costs will generally decrease by 15 to 20 percent annually with the preceding numbers representing 1997 figures.

Component	Rule-of-Thumb Cost (1997 circa)
Remote access	Client PSTN modem or ISDN adapter or ADSL adapter or cable modem costs
Solution component costs	■ Client software, if any, costs ■ Telecommunication usage charges from the client to ISP or RAS location ■ ISP monthly service charges, if using the Internet ■ RAS or MCSS (for wireless) costs ■ Internet PVN (Private Virtual Network) component costs ■ Client application package costs ■ LAN infrastructure upgrade costs ■ Systems integration costs ■ Application integration costs ■ Network support costs ■ Ongoing operational support costs
RAS cost (server hardware/software)	Full-function enterprise RAS products go from $10,000 (entry level: 1 ISDN-PRI and 4 modems for supporting up to 100 users) to $40,000 (fully configured with 60 modems and 8 PRI-ISDN cards for supporting up to 1000 users) RAS and client software are generally bundled with the hardware. Reader should obtain budgetary quotes from vendors to update these rule-of-thumb costs
Personal (SOHO) RAS products (ISDN)	For small office/home office, assume $600 to $1200 per site with ISDN connection capability for hardware only; add telecommunications charges and host RAS hardware upgrade costs
RAS ISDN components	BRI cards $200 to $400 (the lower number is for the desktop, the higher range is for PC Card for notebooks) PRI cards $3000 to $4500
E-mail-related costs	See Table 2.1 in Chap. 2 Also see ARDIS, RAM sections in Chap. 8 of *Mobile Computing—A Systems Integrator's Handbook*
Sales force automation packaged software costs	$200 to $700 per seat for packaged software Ask an application development vendor to provide estimates for integration of sales force automation package to operational database applications
Custom application development costs	Ask an application development vendor or in-house development staff to provide estimates for custom application development

Remote access end-user devices	(Source: Use Mobile office magazine for current street prices) Full-function notebooks—$3000 to $4000 for Pentium II 233 MHz class, active matrix, 32 MB, 2.1 gigabyte hard disk, and CD-ROM Ruggedized notebooks—add $1000 to above. Note ruggedized notebooks do not have the same processing and storage capacity as ordinary notebooks Handheld computers—$2500 to $4500 per unit (ask Telxon and Norand) PDAs—$400 to $800 per unit Cellular PC Card modems—$300 for 28,800 bps per unit ARDIS or Mobitex PC Card modems—$300 to $400 per unit CDPD PC Card modem—$500 to $600 per unit (expected to fall rapidly to meet competition)
Wireless LAN-related costs	Typical Ethernet LAN wiring—$75 to $100 per node (100 ft average) Wireless PC Card LAN adapter (Ethernet)—$400 to $500 Wireless LAN Access Point (Ethernet)—$700 to $1000 Wireless PC Card LAN adapter (Token Ring)—$500 to $600 Wireless LAN Access Point (Token Ring)—$800 to $1200
Wireless or switched network operational (monthly for usage) costs	Network operational costs may vary from $150 to $300 per month per user, depending on the application—e-mail to OLTP. Network usage costs may be as high as 20 to 50% of the total bundled operational costs (hardware lease, software, maintenance, support, etc.) per user Calculate telecommunication charges by assuming 15 cents per minute, one hour per day as average per mobile sales force type mobile user. Add $3000
MCSS costs (for wireless networks)	Low-end (100 users)—$25,000; medium-capacity (100 to 500 users)—$50,000; high-capacity (1000 to 2000 users)—$100,000 to $125,000
ISDN Costs	See Table 10.2
Annual maintenance costs	7 to 12% of purchase costs for hardware; 15 to 25% of one-time license for software

Component	Rule-of-Thumb Cost (1997 circa)
Installation-related costs	Allow $100 to $200 per user for initial telephone-based installation-related help if users are not familiar with computers (Shiva charges this for ISDN handholding help for installation)
	$150 to $200 for handholding of trained systems administrators for RAS installation
	$500 plus for handholding of access server switches
	$1000 per day for on-site help
Installation-related costs (in-car installation for wireless connections)	In-car installation—$200 to $500 per remote workstation
Systems integration costs	Allow $200 per user for low-end configurations to $100 per user for large configurations
Application integration costs	Very difficult to estimate, since they depend on type and number of platforms, number of applications, communications interfaces supported
	Ask a seasoned consultant or in-house specialists to provide "ball-park" estimates based on experience or a systems integration firm to quote on your specific requirements
Network and technical support costs	Refer to studies by research firms such as Gartner, IDC, and Infonetics
	Our own research suggests that we should allow $1500 to $3000 per user on an annual basis

Glossary

100Base-T An IEEE 802.3 extension for providing Ethernet transmission at 100 Mbps on twisted-pair and powered signal-regenerative hubs.

100Base VG-AnyLAN An IEEE 802.12 committee extension for providing Ethernet transmission at 100 Mbps on twisted-pair with quartet (MLT-5) signaling, with a new MAC-layer protocol that does not support collision detection.

10Base-T Ethernet IEEE 802.3 standard for transmission at 10 Mbps specifically for twisted-pair wiring and connectors and signal-regenerative powered hubs.

10Base2 Ethernet standard for baseband networks with transmission rates of 10 Mbps, using coaxial cable segment lengths of 2×100 meters (200 meters).

10Base5 Ethernet standard for baseband networks with transmission rates of 10 Mbps, using coaxial cable segment lengths of 5×100 meters (500 meters).

Access equipment An access device with built-in basic routing protocol support, specifically designed to allow remote LAN access to corporate backbone networks. Not designed to replace backbone routers or to build backbone networks.

Adapter A PC board, usually installed inside a computer, that provides network communication capabilities to and from that computer system. The term *adapter* is often used interchangeably with *NIC* (Network Interface Card).

ADPCM Adaptive Differential Pulse Code Modulation.

ADSL *See* Asymmetric Digital Subscriber Line

Agent Generally, a software that performs tasks, processes queries, and returns replies on behalf of a client application. In the network management context, it gathers information about a network device, and executes commands in response to a management console's requests. In network management systems, agents reside in all managed devices and report the values of specified variables to management stations. In SNMP, the agent's capabilities are determined by the management information base.

AIM *See* Ascend Inverse Multiplexing protocol

AMI Alternate Mark Inversion. A line encoding scheme for transmitting data bits over T1 transmission systems.

Analog Signal A transmission in which information is represented as physical magnitudes of electrical signals.

ANI Automatic Number Identification.

ANSI (American National Standards Institute) The coordinating body for voluntary standards groups within the United States that is a member of the International Organization for Standards.

API Application Programming Interface.

AppleTalk Apple Computer's proprietary local area network for linking Apple computers and peripherals.

Application Functional system made up of software, hardware, or combination of both that performs some useful task. Database managers, spreadsheets, word processors, videoconferencing systems, LANs, and fax machines are examples of applications.

ARQ (Automatic Repeat Request) Communication method where the receiver detects errors and request retransmissions.

Ascend Inverse Multiplexing (AIM) protocol An in-band protocol used to manage interconnection of two remotely located inverse multiplexers. AIM is a feature-rich, widely used inverse multiplexing protocol developed and supported by Ascend Communications.

ASCII American Standard Code for Information Interchange.

Asymmetric Digital Subscriber Line (ADSL) Access line with up to 6.14 Kbps downstream bandwidth and up to 640 Kbps upstream bandwidth; can use either carrierless amplitude/phase (CAP) modulations or discrete multitone (DMT) modulation.

Asynchronous A process where overlapping communications operations can occur independently and do not have to wait for previous operation to be finished.

Asynchronous Transfer Mode (ATM) A high bandwidth, controlled-delay, fixed-size packet switching and transmission system. Uses fixed-size packets, also known as *cells;* ATM is often referred to as *cell relay.* ATM will provide the basis for future broadband ISDN standards.

ATMD (Asynchronous Time Division Multiplexing) A method of sending information in which normal time division multiplexing (TDM) is used, except that slots are allocated as needed rather than to specific transmitters.

ATMP Ascend Tunnel Management Protocol. A protocol that allows transparent tunnels to be dynamically created, as needed, between Ascend remote access servers, located in ISP or carrier POPs, across public carrier networks, such as the Internet or public frame-relay services. ATMP allows corporations to build virtual private networks (VPNs) using the Internet or carrier frame-relay networks instead of expensive dedicated leased lines.

Attenuation Loss of communication signal energy.

B8ZS Binary Eight Zero Suppression. An encoding scheme for transmitting data bits over T1 transmission systems.

Backbone The part of the communications network architected specifically and primarily to carry the bulk of traffic. Provides connectivity between subnetworks in an enterprise-wide network.

Backbone router Routers designed to be used to construct backbone networks using leased lines. Typically do not have any built-in digital dial-up WAN interfaces.

Backup The process of creating a copy of computer data on an external storage medium, such as floppy disk, tape, or external hard drive. If the external storage medium is remotely located, some form of data communications channel must be established between sites.

BACP Bandwidth Allocation Control Protocol. An IETF proposed standard to provide Dynamic Bandwidth Allocation.

Band A portion of the radio frequency.

Bandwidth (1) The range (band) of frequencies that is transmitted on a channel. The difference between the highest and lowest frequencies is expressed in hertz (Hz) or millions of hertz (MHz). (2) The range of frequencies on the electromagnetic spectrum allocated for wireless transmission. (3) The wire speed of the transmission channel.

Base station The low power transmitter/receiver and signal equipment located in each cell in a cellular service area.

Basic Rate Interface An ISDN subscriber line, consisting of two 64-Kbps B channels, or *bearer* channels, and one 16-Kbps D channel, used for both data and signaling purposes.

Baud (1) A unit of signaling speed represented by code elements (often bits) per second. (2) A French language term that represents the transfer of one bit.

BERT (Bit Error Rate Test) A test to determine the percentage of received bits in error compared to the total number of bits received. Usually expressed as a number to the power of 10.

Bit Contraction of the term *Binary digIT.* The smallest unit of information a computer can process, representing one of two states (usually indicated by *1* and *0*).

Block (of frequencies) A group of radio frequencies within a band set aside for a particular purpose. Cellular telephony uses four blocks of frequencies within the 800-MHz portion of the UHF band. Nonwireline and wireline carriers are assigned separate blocks of frequencies. *See also* Band

Block A The block of 800-MHz cellular radio frequencies assigned to the nonwireline or Block A carrier.

Block B The block of 800-MHz cellular radio frequencies assigned to the wireline or Block B carrier.

BOC (Bell Operating Companies) The local telephone companies that existed prior to deregulation, under which AT&T was ordered by the courts to divest itself in each of the seven U.S. regions.

BONDING Bandwidth On Demand Interoperability Group. A consortium of over 40 data communications equipment vendors and service providers who joined together to create a standardized inverse multiplexing protocol, so that inverse multiplexers from different vendors could interoperate. Also refers to the resultant specification, sometimes known as the *BONDING specification.*

BPR (Business Process Reengineering) Business Process Reengineering is the discipline of first analyzing and then redesigning current business processes and their components in terms of their effectiveness, efficiency, and added value contribution to the objectives of the business.

BPS Bits per second. Also, Kbps stands for kilo bits per second, and Mbps stands for million bits per second.

BRI, ISDN BRI *See* Basic Rate Interface

Callback An access security system where a calling party is identified by the central (called) site, after which the call is terminated by the central site and redialed as an outbound call to the original calling site. The calling number used by the central site is maintained in a security database.

Call setup time The time required to establish a switched call between DTE and devices.

CAP *See* Carrierless Amplitude Phase

Carrier (1) A company that provides telephone (or another communications) service. Also, an unmodulated radio signal. (2) A signal suitable for modulation by another signal containing information to be transmitted.

Carrierless Amplitude Phase One of two line coding techniques used in ADSL implementations [*see* Discrete Multi-Tone (DMT)].

CCITT (Consultative Committee for International Telephone and Telegraph) An international organization that makes recommendations for networking standards like X.25, X.400, and facsimile data compression standards. Now called the International Telecommunications Union Telecommunication Standardization Sector; this is abbreviated as ITU, ITU-T, or ITU-TSS.

CDMA Code Division Multiple Access.

CDPD (Cellular Digital Packet Data) Uses idle moments on voice channels to send pure data over the channel without affecting quality of voice transmissions.

Cell The basic geographic unit of a cellular system and the basis for the generic industry term *cellular.* A geographical area is divided into small *cells,* each of which is equipped with a low-powered radio transceiver. The cells can vary in size depending on terrain and capacity demands. By controlling the transmission power and the radio frequencies assigned from one cell to another, a computer at the Mobile telephone switching office monitors the movement and transfers (or hands off) the phone call to another cell and another radio frequency as needed. The region in which RF transmission from one fixed transmission site can be received at acceptable levels of signal strength.

Cell splitting Dividing one cell into two or more cells to provide additional capacity within the original cell's region of coverage.

Cellular (1) Using cellular phone technology. (2) A reference to the wireless switched circuit network consisting of overlapping coverage cells that provides analog voice and CDPD.

Central site A location that acts as a data collection point for remote and branch offices as well as mobile users.

Channel An individual communication path that carries signals at a specific frequency. The term also is used to describe the specific path between large computers (e.g., IBM mainframes) and attached peripherals.

Channel bandwidth The frequency range of an RF channel; for example, in a CDPD, it is 30 KHz.

Channel hop The process of changing the RF channel supporting a channel stream to a different RF channel on the same cell.

Channel hopping A radio frequency transmission method whereby transmissions *hop* from one channel to another. The channels are visited in a predefined order specified by a hopping sequence. Typically this uses the ISM band from 2.4000 to 2.4835 GHz with 85 one-megahertz channels or hops, but at least 50 different frequencies must be used by FCC regulation. Also, CDPD uses frequency hopping on analog cellular systems to take advantage of unoccupied voice channels.

CHAP Challenge handshake authentication protocol. Protocol that describes how to authenticate incoming data calls using a password from the calling end. Password is encrypted over the access line.

CIR *See* Committed Information Rate

Circuit switching An open-pipe technique that establishes a temporary dedicated connection between two points for the duration of the call. A switching system in which a dedicated physical circuit path must exist between sender and receiver for the duration of the call. Used heavily in the phone company network, circuit switching often is contrasted with *contention* and *token passing* as a channel-access method, and with *message switching* and *packet switching* as a switching technique.

Class of Service (COS) An indication of how an upper-layer protocol wants a lower-layer protocol to treat messages. In SNA subarea routing, COS definitions are used by subarea nodes to determine the optimal route to establish a given session.

CLID Calling Line ID.

CO (Central Office) The telephone-switching station nearest the customer's location. A local telephone company office to which all local loops in a given area connect and in which circuit switching of subscriber lines occurs.

CODEC Coding/Decoding Device. In the videoconferencing world, a video codec converts analog video signals from a video camera to digital signals for transmission over digital circuits, and then converts the digital signal back to analog signals for display. In the audio world, an audio codec converts analog signals to digital signals for transmission over digital circuits, and then converts the digital signals back to analog signals for reproduction.

Code Division Multiple Access (CDMA) (1) A division of the transmission spectrum into codes, effectively scrambling conversations. Several transmissions can occur simultaneously within the same bandwidth, with the mutual inference reduced by the degree of orthogonality of the unique codes used in each transmission. (2) Wireless transmission technology that employs a range of radio-frequency wavelengths to transport multiple channels of communication signals. *See also* Spread-spectrum technology

Committed Information Rate (CIR) A frame-relay parameter that defines the minimum throughput that should be expected on a given virtual circuit.

Compression The process of reducing information size or transmission without affecting information content.

Control-lead dialing The initiation of a dialed call over the network using signals on leads within the interface cable between an application and the network access equipment. Thus, an application instructs the network access equipment to dial a call by toggling one or more leads within the cable between the application and the network access equipment.

Crosstalk A technical term indicating that stray signals from other wavelengths, channels, communication pathways, or twisted-pair wiring have polluted the signal. It is particularly prevalent in twisted-pair networks or when telephone and network communications share copper-base wiring bundles. A symptom of interference caused by two cell sites causing competing signals to be received by the mobile subscriber. This can also be generated by two mobiles causing competing signals that are received by the cellular base station. Crosstalk sounds like two conversations and often a distortion of one or the other or both.

CSMA/CA Carrier Sense Multiple Access with Collision Avoidance.

CSMA/CD (Carrier Sense Multiple Access with Collision Detection) A communications protocol in which nodes contend for a shared communications channel and all nodes have equal access to the network. Simultaneous transmissions from two or more nodes result in random restart of those transmissions. Used in Ethernet protocol.

CSU (Channel Service Unit) A device used to connect a digital phone line coming in from the phone company to network access equipment located on the customer premises. A CSU may also be built into the network access equipment. It is often paired with a digital service unit.

CTI Computer Telephone Integration.

CTIA (Cellular Telecommunications Industry Association) The organization created in 1981 to promote the cellular industry, address the common concerns of cellular carriers, and serve as a forum for the exchange of nonproprietary information.

Customer premises equipment (CPE)   Terminal equipment located on the customer premises that connects to the telephone network.

D₄   A T1 framing format.

Data compression   A reduction in the size of data by exploiting redundancy. Many modems incorporate MNP5 or V.42bis protocols to compress data before it is sent over the phone line.

dB (decibels)   A value expressed in decibels is determined as 10 times the logarithm of the value taken to base 10.

DCE   (1) Data Communications Equipment. (2) In software architecture, it implies distributed computing environment.

DDS 56   Dataphone Digital Service, 56 Kbps. Private line digital service at 56 Kbps. Offered on an inter-LATA basis by AT&T, and on an intra-LATA basis by the RBOCs.

Dead spot   A location in a radio/cellular system where, for one reason or another, signals do not penetrate.

Decompression   The restoration of redundant data that was removed through compression.

DECT (Digital European Cordless Telephone)   The specs for future European cellular, as yet not fully defined.

Dedicated channel   An RF channel that is allocated solely for the use of a particular user or service. For example, in CDPD, a channel may be dedicated to data.

De facto standard   A standard by usage rather than official decree; a default standard.

De jure standard   Literally, "from the law." A standard by official decree.

DES (Data Encryption Standard)   An encryption/decryption algorithm defined in FIPS Publication 46. The standard cryptographic algorithm developed by the National Institute of Standards and Technology.

Digital dial-up bandwidth   Communications channels created by signaling to the network from the caller's site the intended destination of the connection. These channels may be terminated whenever the caller or called party chooses. The user pays for the bandwidth only when it is used. Digital dial-up bandwidth operates in a fashion similar to the dialed voice telephone network, but the resultant connections are digital and of specified bandwidths.

Digital modem   A system component that allows communication over digital access facilities with a remotely located system connected to the public network over analog facilities. Converts the incoming digital data stream containing PCM-encoded modem waveform into actual data contained in the waveform at the data rate transmitted by the far-end modem. It performs the inverse function for the outgoing data stream.

Digital Subscriber Line (DSL)   Digital service provided over a local loop (i.e., ISDN BRI).

Disaster recovery   The use of alternate network circuits to reestablish communications channels in the event that the primary channels are disconnected or malfunctioning.

Discrete Multi-Tone (DMT)   One of two coding techniques used in ADSL implementations (*see* Carrierless Amplitude Phase).

Downlink   The process of receiving information from a source computer.

Driver   A software program that controls a physical computer device such as an NIC, printer, disk drive, or RAM disk.

Drop and insert   A process of adding data (inserting) to a T1 data stream, or terminating data (dropping) from a T1 data stream to other devices connected to the drop and insert equipment.

Ds_0   A 64-Kbps unit of transmission bandwidth. A worldwide standard speed for digitizing one voice conversation, and more recently, for data transmission. Twenty-four Ds_0's (24×64 Kbps) equal one Ds_1.

Ds_1   A 1.544-Mbps unit of transmission bandwidth in North America; a 2.048 Mbps unit of transmission elsewhere. A telephony term describing a 1.544 or 2.048 Mbps digital signal carried on a T1 facility.

DSU (Data Service Unit)   A device used in digital transmission for connecting data terminal equipment (DTE), such as a router, to a digital transmission circuit (DTC) or service.

DTE (Data Terminal Equipment)   A computer terminal that connects to a host computer. It may also be a software session on a workstation or personal computer attached to a host computer.

DTMF   Dual Tone Multifrequency.

Dual 56   Two switched 56 calls made between videoconferencing equipment to allow data transfer at 112 Kbps. The videoconferencing equipment performs a two-channel inverse-multiplexing procedure to ensure channel alignment.

Dual-mode New cellular phones that work with both digital and analog switching equipment. Digital cellular offers the benefits of more channels, clearer-sounding calls, and ensured privacy.

Duplex (1) The method in which communication occurs, either two-way as in full-duplex, or unidirectional as in half-duplex. (2) Cellular phones, using separate frequencies for transmission and reception, allow for duplex communications by allowing both parties to talk and listen at once. Push-to-talk systems are not duplex.

Dynamic bandwidth allocation The process of determining current traffic loads over a channel and automatically increasing or decreasing the bandwidth of the channel to optimize overall utilization efficiency.

E1 A digital transmission link with a capacity of 2.048 Mbps, used outside North America. For telephony or remote access applications, often channelized into 30 Ds_0s, each capable of carrying a single voice conversation or data stream. It uses two pairs of twisted-pair wires.

E1-R2 A channelized E1 line (30 channels of 64 Kbps each) configured to use a signaling method known as R2 signaling.

Encryption The processing of data under a secret key in such a way that the original data can be determined only by a recipient in possession of a secret key.

Erlang 1 hour, 300 seconds, and 36 CCs. If a channel is occupied (used) constantly for 1 hour, that circuit has carried 1 Erlang of traffic. Also known as a carried load.

ESF Extended Super Frame. A T1 framing format.

ESMR Extended Specialized Mobile Radio.

ETC (Enhanced Throughput Cellular) AT&T Paradyne protocol for data transmission over analog cellular connections consisting of enhancements to V.42 and V.32bis for compression, error detection, and error correction.

Ethernet A 10 Mbps local area network technology.

Fading The combination of out-of-phase multiple signals that results in a weaker or self-canceling data signal.

FCC Federal Communications Commission.

FDDI (Fiber Data Distributed Interchange) FDDI provides 125 Mbps signal rate with 4 bits encoded into 5-bit format for a 100-Mbit/s transmission rate. It functions on single- or dual-ring and star network with a maximum circumference of 250 km.

FDMA (Frequency Division Multiple Access)   The analog communications technique that uses a common channel for communication among multiple users allocating unique time slots to different users.

FHSS (Frequency Hopping Spread Spectrum)   IEEE 802.11 Wireless LAN Standards Committee approval for Lannair Ltd.'s concept for at least 2 Mbps transmission rate with dynamic data rate switching.

Firewall   A device, mechanism, bridge, router, or gateway that prevents unauthorized access by hackers, crackers, vandals, and employees from private network services and data.

Fractional T1   Service offering rates between 64 Kbps (DS_0 rate) and 1.536 Mbps (DS_1 rate), in specified intervals of 64 Kbps.

Frame   A segment of a digital signal that has a repetitive characteristic in that corresponding elements of successive frames represent the same things. In a time-division multiplex system, a frame is a sequence of time slots, each containing a sample from one of the channels served by the multiplex system. The frame is repeated at the sampling rate, and each channel occupies the same sequence position in successive frames.

Frame-relay   A form of packet switch, but using smaller packets and less error checking than traditional forms of packet switching (such as X.25). Now a new international standard for efficiency handling high-speed, bursty data over wide area networks.

Frequency hopping   A radio frequency transmission method under spread-spectrum technology used in wireless communication. Typically this uses the ISM band from 2.4000 to 2.4835 GHz with 85 one-megahertz channels or *hops*. Also, CDPD uses frequency hopping on analog cellular systems to unoccupied voice channels. Transmissions hop from one channel to the other, staying only $^1/_{10}$ of a second on any given channel. The channels are visited in a predefined order specified by a hopping sequence.

FT1   *See* Fractional T1

Geosynchronous orbit   Orbit taken by satellites where the satellite's orbit velocity matches the rotation of the earth, causing the satellite to remain stationary relative to a position on the earth's surface. Geosynchronous orbit demands a position about 23,000 miles above the earth's surface over the equator.

GIS (Geographic Information System)   Generally refers to a database of geographical data. In some circles, it refers to Graphics database.

GloBanD   The name given for a set of European network services which offer digital dial-up bandwidth on demand in 64 Kbps increments, accessed from the customer premise over PRI lines. These services are offered under different names in each participating country.

GLS (Global Locationing System)   A triangulation system used to locate a vehicle and convey that information to a central management facility.

GPS (Global Positioning System)   A satellite-based triangulation system used to ascertain current location.

GSM (Global System for Mobile Communications)   The pan-European digital cellular system standard.

Handoff   The transfer of responsibility for a call from one cell site to the next. The process by which the MTSO, sensing by signal strength that cellular mobile is reaching the outer range of one cell, transfer or *hands off* the call to an adjacent cell with a stronger signal. *See also* Cell

H channel   A transmission channel, defined in the CCITT ISDN standards, made up of multiple B channels. Currently defined H channels include H_0 (384 Kbps), H_{10} (1.472 Mbps), H_{11} (1.536 Mbps), and H_{12} (1.920 Mbps).

High Data Rate Digital Subscriber Line (HDSL)   An improved way to provide T1 or E1 bandwidth digital connections.

Host   Any computer, although typically a mainframe, midsized computer, minicomputer, or LAN server, servicing users and their processing at the centrally based processor but distributing the results to terminal-based or client connections.

H.221   A CCITT standard describing a method of inverse multiplexing for videoconferencing terminals, to be used with Px64 videoconferencing.

H.261   A CCITT standard describing a protocol for digitally encoding and decoding video images to allow videoconferencing terminals from different manufacturers to interoperate.

H.320   A set of CCITT standards describing methods to allow video-conferencing terminals from different manufacturers to interoperate.

H_0, ISDN H_0   An H channel made up of 6 B channels to create a 384 Kbps ISDN channel.

H_{10}, ISDN H_{10}   An H channel made up of 23 B channels to create a 1.472 Mbps ISDN channel.

H₁₁, ISDN H₁₁   An H channel made up of 24 B channels to create a 1.536 Mbps ISDN channel.

H₁₂, ISDN H₁₂   An H channel made up of 30 B channels to create a 1.920 Mbps ISDN channel.

Hybrid private/public networking   The creation of a network using both private leased lines and public switched facilities (digital dial-up bandwidth). The goals of combing both networking technologies are increased performance and flexibility at reduced cost.

Hz (hertz)   Signal frequency use for voice, data, TV, and other forms of electronic communications represented by the number of cycles per second.

IDSL   *See* ISDN Digital Subscriber Line

IEC (Inter Exchange Carrier)   Common carrier providing communications channels between local telephone companies (LECs, or Local Exchange Carriers). Also known as long distance carriers, such as AT&T, MCI, Sprint, WilTel, and so forth.

IEEE (Institute for Electrical and Electronic Engineers)   A membership-based organization based in New York City that creates and publishes technical specifications and scientific publications.

IEEE 802   An Institute of Electrical Engineering standard for interconnecting of local area networking computer equipment. The IEEE 802 standard describes the physical and link layers of the OSI reference model.

IEEE 802.1   A specification for media-layer physical linkages and bridging.

IEEE 802.3   An Ethernet specification derived from the original Xerox Ethernet specifications. It describes the CSMA/CD protocol on a bus topology using baseband transmissions.

IEEE 802.4   Broadband and baseband bus using token passing as the access method and physical interface specifications.

IEEE 802.5   A Token-Ring specification derived from the original IBM Token-Ring LAN specifications. It describes the token protocol on a star/ring topology using baseband transmissions.

IEEE 802.6   A token bus specification for metropolitan area networks with star/ring topology using baseband transmissions.

IEEE 802.11   A physical- and MAC-layer specification for wireless network transmission based on direct and frequency hopping, SST, and infrared at transmission speeds from 1 to 4 Mbps. This specification includes the basic rate set for fixed bandwidths supported by all wireless stations (for compatibility) and an extended rate set for optimal speeds.

IEEE 802.12   A specification for wireless network transmission based on SST.

IETF   Internet Engineering Task Force. An international committee of vendors, users, and others drive the standards for the Internet.

IMTS (Improved Mobile Telephone Service)   Cellular telephone predecessor that uses a single central transmitter and receiver to service a region. A two-way mobile phone system that generally uses a single high-power transmitter to serve a given area and is automatically interconnected with a land-line telephone system.

In-band signaling   Transmission within a frequency range normally used for information transmission. Contrasted with out-of-band signaling, which uses frequencies outside the normal range of information-transfer frequencies.

Infrared   Electromagnetic waves whose frequency range is above that of microwave but below the visible spectrum. LAN systems based on this technology represent an emerging technology.

Infrastructure   The physical and local components of a network. Typically, this includes wiring, wiring connections, attachment devices, network nodes and stations, interconnectivity devices (such as hubs, routers, gateways, and switches), operating environment software, and software applications.

Intranet   A private Internet for the exclusive use of a selected group, such as a corporation or government body.

Inverse multiplexing   The creation of a single higher-speed data channel by combining and synchronizing two or more lower-speed data channels.

IP   Internet Protocol.

IPX   Internet Packet Exchange (Novell NetWare LAN protocol).

IrDA (Infrared Data Association)   A group of wireless infrared product vendors that promotes serial infrared linkages and interoperability between vendor products.

IS 54   Interim Standard developed by CTIA for introduction of TDMA in conjunction with FDMA.

IS-41   TIA cellular standard for seamless roaming with intersystem handoff, call delivery, validation, and authentication.

IS-54   TIA cellular standard defining the air interface to TDMA and digital handsets to base station communications.

IS-95   TIA cellular standard defining the air interface to CDMA and digital handsets to base station communications.

ISDN   Integrated Services Digital Network. A concept embracing a set of standardized worldwide digital data and telecommunications networks. The standards are based on recommendations created by the CCITT and are implemented on a national basis. Currently available ISDN offerings include: Basic Rate Interface and Primary Rate Interface access lines, and numerous new network-based services. Future offerings include broadband ISDN (B-ISDN), a set of very high speed networks.

ISDN Digital Subscriber Line   Technological innovation from Ascend to provide dedicated 128 Kbps data service on a local loop using existing ISDN BRI terminal adapters and routers.

ISDN Multirate   A network-based ISDN service which allows users' network access equipment to dial network channels of bandwidth in increments of 64 Kbps, up to 1546 Kbps. Access to ISDN Multirate service is obtained over ISDN PRI lines.

ISM   Instrumentation, Scientific, and Medical band.

ISO   International Standards Organization.

IXC   Inter eXchange Carrier. *See* IEC

Ka-band   A high-bandwidth satellite wireless communication frequency using the 30-GHz spectrum.

KHz (kilohertz)   A measure of audio and radio frequency (a thousand cycles per second). The human ear can hear frequencies up to about 20 KHz. There are 1000 KHz in 1 MHz.

LAN   *See* Local Area Network

LAN internetworking   The reach of local area networks (LANs) to other networks, so users can get access to other applications. Bridges and routers are the devices which typically accomplish the task of joining LANs.

Laser   Light amplification by stimulated emission of radiation.

Layer 2 Tunneling Protocol   Proposed tunneling protocol standard being defined by the IETF.

LDC   Long Distance Carrier. *See* IEC

Leased line   A dedicated common carrier circuit providing point-to-point or multipoint network connection, reserved for the permanent and private use of a customer. Also called a *private line.*

LEC   Local Exchange Carrier. Local telephone company, providing connections between local points or to long-distance carriers for extended connections. Examples are Pacific Bell in California, Illinois Bell in Illinois, and GTE in Hawaii.

LEO   Low-Earth Orbit Satellite.

Line of sight   The connection between communication devices characteristic of certain transmission systems (such as laser, microwave, and infrared systems) in which no obstructions on a direct path between transmitter and receiver may exist.

Local Area Network   A data communications network spanning a limited geographical area, usually within a single facility or campus. It provides communications between computers and peripherals.

Local loop   The line from a telephone subscriber's premises to the telephone company CO.

Local loopback   A loopback performed between an application and network access equipment. The signal is sent from the application to the network access equipment and back to the application without being sent out over the network.

Location directory   The repository of information specifying the current forwarding address of a collection of mobile hosts to be accessed by the redirectors.

Loopback   A diagnostic test or test state in which the transmitted signal is returned to the sending device after passing through a communications link or network.

LSU   LAN Service Unit. *See* Multiband LSU

L2TP   *See* Layer 2 Tunneling Protocol

MAN (Metropolitan Area Network)   A network that spans buildings, or city blocks, or a college, or corporate campus. Optical fiber repeaters, bridges, routers, packet switches, and PBX services usually supply the network links.

MAX   Media Access Exchange. The Ascend MAX is a system-level network access unit, with a cage and backplane into which Multiband or Pipeline cards can be inserted to configure it for various application requirements. It supports up to 32 host ports or direct Ethernet connections, and up to 8 Mbps to the network. It supports multiple applications, including remote LAN access, leased-line backup, and individual videoconferencing units, as well as connecting videoconference MCUs to the digital dial-up network.

Mbps (megabytes per second) Speed of data transmission.

MCP/1 Mobitex compression protocol.

MCSS (Mobile Communications Server Switch) A hardware/software configuration that provides communications connection and message switching functionality. It sits between the wireless network and information servers.

MCU Multipoint Control Unit.

MDBS (Mobile Data Base Station, a CDPD term) The hardware used by a cellular provider to convert the data streams into a valid signal and route cellular switched data calls through the wired phone network or to the cellular destination. This station manages and accesses the radio interface from the network side. It relays and retransmits packets sent from the mobile data intermediate system.

MES (Mobile End System) The portable wireless computing device that can roam from site to cell while communicating with the MDBS via CDPD. An end system that accesses the CDPD network through the airlink interface.

MHX (Mobitex Main Hierarchical Exchange) Part of Mobitex network hierarchy. Each MHX exchanges information with other MHXs.

MHz (megahertz) Measures frequency in million cycles per second.

MIB (Management Information Base) An SNMP term. A directory listing the logical names of all information resources residing in a network and pertinent to the network's management. A key element of SNMP management systems.

Microwave Electromagnetic waves in 1 to 30 GHz range.

MIN Mobile Identification Number.

MMTF (Mobile Management Task Force) A vendor-organized body that has undertaken to create SNMP-based MIB for mobile network management.

MNP (Microcom Network Protocols) A set of modem-to-modem protocols that provide error correction and compression.

MNP5 Microcom Network Protocols with simple data compression. Dynamically arranges for commonly occurring characters to be transmitted with fewer bits than rare characters. It takes into account changing character frequencies as data flows. Also encodes long runs of the same character. Typically compresses text by 35 percent.

MNP10   Microcom Network Protocols for cellular or wireless transmission applying compression, error detection and error correction, data rate fallback, and readjustment.

MOX (Mobitex Area Exchange)   A node in the internal Mobitex network.

MPAK   Mobitex packet that is routed through Mobitex network. A 512 octet of user data.

MPP   Multichannel Point-to-Point protocol. A protocol similar to PPP (Point-to-Point Protocol), but operable over multiple network channels in an inverse multiplexed scenario.

MPT/1   Mobitex Transport Protocol.

MSA (Metropolitan Statistical Area)   The 30 U.S. urban areas (markets) as defined by the FCC, using SMSA (Standard Metropolitan Statistical Area) data. All are licensed for two cellular operators, and almost all have both operators on the area. MSAs comprise 76 percent of the U.S. population, but only 22 percent of its land surface area.

MTSO (Mobile Telephone Switching Office)   The cellular system's switching computer, located between a cell site and a conventional telephone switching office.

Multiband LSU   LAN Service Unit. It allows LAN bridges and routers to be interconnected, creating WANs, using a combination of dedicated leased circuits and digital dial-up circuits. By creating such hybrid networks, users can match bandwidth to real-time traffic loads, thereby saving money and maximizing performance.

Multiplexing   The process of combining a number of individual channels into a common frequency band or into a common bit stream for transmission. The converse equipment or process for separating a multiplexed stream into individual channels is called *demultiplexing.*

NAMPS (Narrowband Advanced Mobile Phone System)   Using a radio frequency transmission on a single, preset frequency.

Narrowband   PCS frequency in the 900 to 931 MHz range for two-way paging.

NDIS (Network Device Independent Specification)   A Microsoft network interface specification for operating system and protocol-independent device drivers. An effort to create a standard for bridging different types of network adapter cards and multiple protocol stacks.

This network-level protocol is supported by IBM LAN Manager and new Microsoft networking products, such as MS Windows for Workgroups and NTAS.

NFAS Non-Facilities-Associated Signaling. Allows a D channel on one ISDN PRI to control channels located on other PRIs.

NIC (Network Interface Card) The network access unit that contains the hardware, software, and specialized PROM information necessary for a station to communicate across the network. Usually referenced as *network interface controller.*

NOS (Network Operating System) A platform for networking services that combines operating system software with network access.

Octet Eight data bits.

ODI (Open Data-link Interface) A protocol that supports media- and protocol-independent communication by providing a standard interface allowing network layer protocols to share hardware without conflict. Presently used in PC software, mostly.

OLTP On-Line Transaction Processing.

Out-of-band signaling Signaling that is separated from the channel carrying the information and sent over an independent (out-of-band) channel.

PAP Password Authentication Protocol. Protocol that describes how to authenticate incoming data calls using a password from the calling end. The password is not encrypted over the access line.

PBX *See* Private Branch Exchange

PCCA (Portable Computer and Communications Association) A nonprofit association of vendors to develop and promote software and hardware for mobile computing applications.

PC Card Standard Latest PCMCIA specification PCMCIA 5.0. Adds support for low-voltage 3.3-volt operation, DMA, multifunction capability, and CardBUS, which provides higher performance, bus mastering, and 32-bit data path.

PCMCIA (Personal Computer Memory Card International Association) A standard for a computer plug-in, credit card—sized card that provides about 90 percent compatibility across various platforms, BIOS, and application software.

PCS (Personal Communications Services)/PCN (Personal Communications Network) A term used to describe emerging wireless/

portable network technology where subscribers carry their own personal communication numbers with them, and the system locates them wherever they are.

Permanent virtual circuit   A point-to-point circuit through the network where the call setup and call clearing do not occur or are not visible to the end user (i.e., X.25 or frame-relay).

POP   Point-of-Presence.

POPS   One unit of population. The POPS concept is used to measure relative market sizes.

POTS   Plain Old Telephone Service.

PPP   Point-to-Point Protocol. A protocol that allows a single node to access a LAN backbone network constructed of leased lines and routers. Often used for dial-up remote LAN access.

PRI, ISDN PRI   Primary Rate Interface. An ISDN subscriber line, consisting of twenty-three 64-Kbps B channels in North America (thirty 64-Kbps channels elsewhere), and a 1-Kbps D channel, used for signaling purposes.

Private Branch Exchange (PBX)   A private switching system, usually serving an organization such as a business or a government agency, and usually located on the customer's premises.

PSDN   Public Switched Digital Network. A term used to describe the set of digital dial-up services offered by carriers (IECs and LECs).

PSTN (Public Switched Telephone Network)   The telecommunications network traditionally encompassing local and long-distance landline carriers and now also including cellular carriers. Refers to the telephone network.

PVC   *See* Permanent virtual circuit

PVDN (Private virtual dial-up network)   Same as PVN but emphasizes dial-up nature of access into the Internet.

PVN (Private virtual network)   A network that uses a combination of the Internet services and private lines into the corporate network infrastructure to give the appearance of a private network.

Remote loopback   A loopback performed between an application and remotely located access equipment or application. The signal is sent from the application over the network to the remote access equipment or application, from where it is looped back to the originating equipment.

RF   Radio Frequency.

RJ-11   Standard four-wire connectors for phone lines.

RJ-22   Standard four-wire connectors for phone lines with secondary phone functions (such as call forward, voice mail, or dual lines).

RJ-45   Standard eight-wire connectors for networks. Also used as phone lines in some cases.

RNA (Remote Network Access)   Term used to define hardware and software solutions used for remote access of information resources using dial-up and ISDN solutions. This is terminology based on current usage; may include wireless networks in the future.

Roaming   The ability to access a network anywhere and move freely while maintaining an active link through a wireless connection to a network. Roaming usually requires a handoff when a node (user) moves from one cell to another.

Router   A device that interconnects networks that are either local area or wide area.

RS-232   A set of EIA standards specifying various electrical and mechanical characteristics for interface between DTE and DCE data communications devices. The standard applies to both synchronous and asynchronous binary data transmission at rates below 64 Kbps.

RS-366   An EIA standard for providing dialing commands to network access equipment. Uses RS-232 electrical specifications, but different connector pinouts and signal functions.

RS-422   An EIA standard describing electrical characteristics for balanced-voltage digital interface circuits. Typically used for high-speed synchronous data connections between DTE and DCE data communications devices.

RS-423   An EIA standard describing electrical characteristics for unbalanced-voltage digital interface circuits. Typically used for high-speed synchronous data connections between DTE and DCE data communications devices.

RS-449   An EIA standard for a 37-pin data communications connector, usually used with RS-422 or RS-423 electrical specifications.

Rubber bandwidth   A term used to describe a communications channel whose bandwidth can be increased or decreased without terminating and reestablishing the channel. Typically used with inverse multiplexing.

SDSL *See* Symmetric Digital Subscriber Line

SMDR Station Management Detail Recording. The ability of network access equipment to output call statistics and performance information for tabulation and analysis.

SMDS Switched Multimegabit Data Service. A packet-based network service allowing the creation of high-speed data networks (up to 45 Mbps). Now in the testing and initiation implementation phases.

SMP Symmetric Multiple Processing. Processor hardware architecture that allows multiple processors to share processing workload, using common memory.

SMR Specialized Mobile Radio.

SNMP Simple Network Management Protocol. A protocol governing network management and the monitoring of network devices and their functions. Originally developed in the TCP/IP environment.

Spread-Spectrum Technology Technology used in wireless LANs or metropolitan area networks where a wide spectrum of frequencies is used simultaneously by several users. The wide spectrum makes it difficult for unwanted parties to intercept information.

SPX System Packet Exchange, a protocol used in Novell's NetWare network operating system.

SS#7 Signaling System #7, channel for network control.

SST Spread-Spectrum Technology, used in wireless LANs.

Switched 56 (SW56) A dial-up network-based service providing a data channel operating at a rate of 56 Kbps. Also a type of network access line, used to provide access to switched 56 network services.

Switched 64 (SW64) A dial-up network-based service providing a data channel operating at a rate of 64 Kbps.

Switched 384 (SW384) A dial-up network-based service providing a data channel operating at a rate of 384 Kbps.

Switched 1536 (SW1536) A dial-up network-based service providing a data channel operating at a rate of 1536 Kbps.

Switched Digital Services Applications Forum (SDSAF) A consortium of equipment vendors, service vendors, and users, with the goal of advancing the state of switched digital services.

Symmetric Digital Subscriber Line Provides one half of HDSL (768-Kbps bandwidth) on a local loop.

T1 (1) Bell technology referring to a 1.544-Mbps communications circuit provided by long-distance carriers for voice or data transmission through the telephone hierarchy. Since the required framing bits do not carry data, actual T1 throughput is 1.536 Mbps. T1 lines may be divided into twenty-four 64-Kbps channels. This circuit is common in North America. Elsewhere, the T1 is superseded by the ITU-TTS designation DS-1. (2) A 2.054-Mbps communications circuit provided by long-distance carriers in Europe for voice or data transmission.

T3 An AT&T standard for dial-up or leased-line circuits with a signaling speed of 44.736 Mbps. Superseded in Europe by the ITU (ITU-TTS) DS-3 designation.

Tariff Documents filed by a regulated telephone company with a state public utility commission or the Federal Communications Commission. Document details services, equipment, and pricing publicly offered by the telephone company.

TCP/IP Transaction Control Protocol/Internet Protocol. A set of protocols developed by the Department of Defense to link dissimilar computers across networks.

TDMA Time Division Multiple Access.

Telecommuter A work-at-home computer user who connects to the corporate LAN backbone using remote access technologies (i.e., using a modem over analog lines, ISDN TA over ISDN lines, or CSU/DSU over switched 56 lines).

TELNET Terminal-to-remote host protocol developed for ARPAnet. It is the TCP/IP protocol governing the exchange of character-oriented terminal data.

TIA Telecommunications Industry Association.

Tunnel The path followed by a datagram when it is encapsulated.

UTP Unshielded Twisted Pair. Wiring with one or more pairs of twisted insulated copper conductors bound in a single plastic sheath.

V.25bis An automatic calling and answering command set for use between DTE and DCE which includes both in-band and out-of-band signaling.

V.32 An international standard for synchronous and asynchronous transfer of data of up to 9600 bps over dial-up telephone lines.

V.32bis An international standard for synchronous and asynchronous transfer of data of up to 14,400 bps over dial-up telephone lines.

V.34 An international standard for synchronous and asynchronous transfer of data of up to 28,800 bps over dial-up telephone lines.

V.35 Commonly used to describe electrical characteristics and connector characteristics for a high-speed synchronous interface between DTE and DCE. Originally, V.35 described a 48-Kbps group band modem interface with electrical characteristics defined in an appendix. Although V.35 is no longer published by the CCITT, its legacy lives on in the data communications world.

V.42 An international error correction protocol that uses Link Access Procedure Modem (LAP-M) as the primary protocol, and MNP2-4 as back-up protocols.

V.42bis An international data compression protocol that can compress data by as much as 4 to 1.

VDSL Very High Data Rate Digital Subscriber Line. Using fiber-optic cable to provide 52 Mbps downstream and 2.3 Mbps upstream.

VPN Virtual Private Network. Using a public network, such as the Internet, for private or corporate purposes by creating protected paths called tunnels.

VSAT Very Small Aperture Terminal.

VT-100 An ASCII character data terminal, consisting of screen and keyboard. Manufactured by Digital Equipment Corporation (DEC), the VT-100 has become an industry standard data terminal. VT-100 emulation software allows a standard PC to act as a VT-100 terminal.

WAN Wide Area Network. A data network extending a LAN outside a building or beyond a campus, over IXC or LEC lines to link to other LANs at remote sites. Typically created by using bridges or routers to connect geographically separated LANs.

X.21 A set of CCITT specifications for an interface between DTE and DCE for synchronous operation on public data networks. Includes connector, electrical, and dialing specifications.

INDEX

B

C